Raz Jelinek
Biomimetics

I0038069

Also of interest

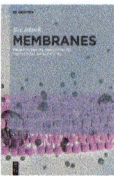

Membranes.
From Biological Functions to Therapeutic Applications
Jelinek, 2018
ISBN 978-3-11-045368-3, e-ISBN 978-3-11-045369-0

Materials for Medical Application
Heimann (Ed.), 2020
ISBN 978-3-11-061919-5, e-ISBN 978-3-11-061924-9

Microencapsulation
Tylkowski , Giamberini, Fernandez Prieto (Ed.), 2020
ISBN 978-3-11-064176-9, e-ISBN 978-3-11-064207-0

Nickel-Titanium Materials.
Biomedical Applications
Oshida, Tominaga, 2020
ISBN 978-3-11-066603-8, e-ISBN 978-3-11-066611-3

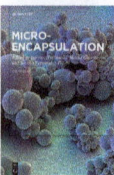

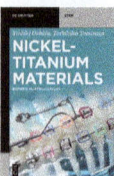

Raz Jelinek

Biomimetics

—

A Molecular Perspective

2nd Edition

DE GRUYTER

Author
Prof. Raz Jelinek
Ben-Gurion University
Dept. of Chemistry
1 Ben Gurion Avenue
84105 Beer Sheva
Israel

ISBN 978-3-11-070944-5
e-ISBN (PDF) 978-3-11-070949-0
e-ISBN (EPUB) 978-3-11-070994-0

Library of Congress Control Number: 2021938200

Bibliographic information published by the Deutsche Nationalbibliothek
The Deutsche Nationalbibliothek lists this publication in the Deutsche Nationalbibliografie;
detailed bibliographic data are available on the Internet at http://dnb.dnb.de.

© 2021 Walter de Gruyter GmbH, Berlin/Boston
Cover image: Gettyimages/Kaan Sezer
Typesetting: Integra Software Services Pvt. Ltd.
Printing and binding: CPI books GmbH, Leck

www.degruyter.com

Contents

1 Introduction

How one is to define "Biomimetics"? The term was likely coined by Janine Benyus in her 1997 book "Biomimicry: Innovation Inspired by Nature." Biomimicry (from *bios*, meaning life, and *mimesis*, meaning to imitate) has been generally defined as "the examination of nature, its models, systems, processes, and elements to emulate or take inspiration from in order to solve human problems." Solving human problems has certainly been a major driving force in this burgeoning field, especially with regards to biomedical applications and therapeutic avenues to treat disease, that have borrowed concepts, molecules, and physiological processes from nature. However, biomimetics as a distinct scientific concept has attracted the imagination of researchers well beyond the goals for producing better drugs, artificial organs, or new bio-inspired products, but rather as a foundation and inspiration for new innovative ways to design and understand Matter on a deeper level.

When considering the possibilities of biomimetics to advance human knowledge and technology, one faces the enormous diversity of natural phenomena, processes, and materials – all essentially coalesce to the eons it has taken for the creation of the enormous variety of biological structures and processes around us. It is thus quite expected in light of the "impatience" of our species that progress in the field of "biomimetics" would occur much faster compared to the millions of years evolution has so far dictated the design of new and improved biological features through trial and error. Indeed, for a relatively young scientific discipline, the achievements of biomimetic sciences have been remarkable.

This book aims to outline the broad current contours of *biomimetics* as a distinct field of research, its practical applications, and potential. It is important to emphasize that I focus here on biomimetic *molecular* systems, rather than the broad aspects of "macro-scale" systems such as bio-inspired water pumps, wind turbines, "green buildings", and many others wonderful and useful concepts and systems which are beyond the scope of this book. Also, this is not a book about overall historical inspiration from the living world (not included are Icarus and Daedalus mimicking the bird's flight). Rather, this book focuses on molecules, molecular assemblies, and molecular processes and concepts which elegantly exploit biology towards solving non-biological challenges, or implement non-biological solutions for biological and biomedical problems. Nevertheless, while I have tried to include as many aspects as possible of this extremely broad and diverse field, for practical reasons of space and scope varied topics have been excluded. I generally did not discuss pertinent and prominent disciplines that are related to biomimetics, for example the huge arsenal of pharmaceutical compounds designed to mimic biological molecules or to interfere with biological processes.

This book is not intended just for "experts" in chemical and biological sciences, but rather to a broader scientific readership. Still, knowledge of biological and

https://doi.org/10.1515/9783110709490-001

chemical concepts is a prerequisite for grasping many of the concepts and examples presented. Importantly, discussions of the various topics covered here closely follow experiments reported in the scientific literature which represent the specific concepts and ideas highlighted in the different chapters. Indeed, *examples* from pertinent studies are the primary didactic tool employed in this book. My hope, however, is that even non-experts will find this book useful for attaining a better understanding of this burgeoning field. In this context, it should be emphasized that this book is aimed to be a "starting point" for further reading on the different subjects. More thorough discussions will naturally be found in more specialized texts and publications, accordingly some further references are outlined at the end of the book.

Thematically, the book is divided into several chapters, each concentrating on a specific general topic in which *biomimicry* plays major roles. Since many scientific phenomena defy compartmentalization, a certain overlap exists between chapters. For example, biomimetic *surfaces* are discussed both in Chapter 3 dedicated to Biomimetic Surfaces as well as in Chapter 4 (Tissue Engineering). Similarly, biomimetic *membranes* are analyzed in Chapter 7 (Artificial Membranes), Chapter 8 (Artificial Cells), and also Chapter 9 (Drug Delivery). Some subjects, such as "nanobiotechnology", are not presented in specific chapters but rather weave through several topics. Indeed, biomimetics and nanotechnology, both relatively young disciplines, have been closely linked. The increasing sophistication of manipulating matter in the nanoscale (e.g. molecular and atomic levels) has allowed bringing the synthetic and natural worlds ever closer, and this aspect is a prominent theme in this book.

While this book is designed to be as inclusive as possible, the enormous breadth and diversity of biomimetic systems and applications naturally imposes constraints in terms of selecting subjects to present and discuss in more detail. I hope that the reader will not find certain topics too detailed, while others absent. It should be also emphasized that even though this book is naturally aimed at celebrating the enormous progress science has made in harnessing biology for creating useful devices and materials, conversely, I also highlight the weaknesses, limitations, and barriers encountered. Like many scientific disciplines and research avenues, alongside the successes and hype, *biomimetics* has had its share of disappointments and dead-ends.

Overall, I hope that this book will reveal to the reader the beauty, impact, and enormous potential of a field which inherently spans the interface of many "traditional" scientific disciplines. I am certain that *biomimetics* in its various definitions, emphases, and lines of study will continue to expand and evolve in the very same way Nature itself has inspired human imagination and progress for millennia.

2 Bio-inspired and bio-hybrid materials

Nature, with its extraordinary diversity, provides seemingly endless possibilities for new materials. This statement encompasses biomimetic materials spanning the macroscale, through micro- and nanoscale dimensions, and all the way to single molecules. Exploiting natural concepts and materials can take different forms, from using actual biomolecules for human applications, such as "prospecting" the rain forest for substances having pharmaceutical potential, to the fabrication of new materials of which the structural features and hierarchical organization, for example, are inspired by Nature. This chapter presents and discusses examples of both systems, with the emphasis on new functionalities that embody the integration of biological concepts on the one hand and the world of materials science on the other. As such, this exciting multidisciplinary field at the interface of biology, nanotechnology, materials science and related disciplines continuously produces new materials, processes, and concepts used in a variety of applications.

2.1 Biomimetic functional materials

Among the fundamental challenges of biomimetics research is the development of artificial entities that not only mimic biological structures or organization, but perhaps more importantly – successfully perform biological *tasks* and *functions*. In some instances, seemingly simple molecular assemblies can mimic the functionality of biological entities participating in complex physiological processes. Such materials have been presented both in the scientific literature and in varied practical applications and everyday systems

A case in point is the construction of "synthetic platelets" for aiding blood clotting in situations of trauma and blood loss (Fig. 2.1). Platelets (Fig. 2.1a) constitute the primary blood clotting factors, circulating in the bloodstream and rapidly aggregating in sites of injury, reducing and eventually preventing blood leakage. While varied materials have been developed for enhancing blood clotting, limitations have been often encountered, in particular in situations of severe bleeding. Accordingly, introducing synthetic species that could effectively substitute natural platelets might have a significant therapeutic impact, for example in battlefield situations where immediate care is required.

The synthetic platelets designed by E. Lavik and colleagues at Yale University and shown schematically in Fig. 2.1b are aimed to attach to and aggregate onto initial, naturally-forming platelet plaques and thus expedite and enhance the clotting process. The particles comprise a biocompatible, biodegradable core, coated by biologically-inert polymer (polyethyleneglycol) arms terminated with the tripeptide motif arginine-glycine-aspartate (RGD) known to recognize and bind to recep-

https://doi.org/10.1515/9783110709490-002

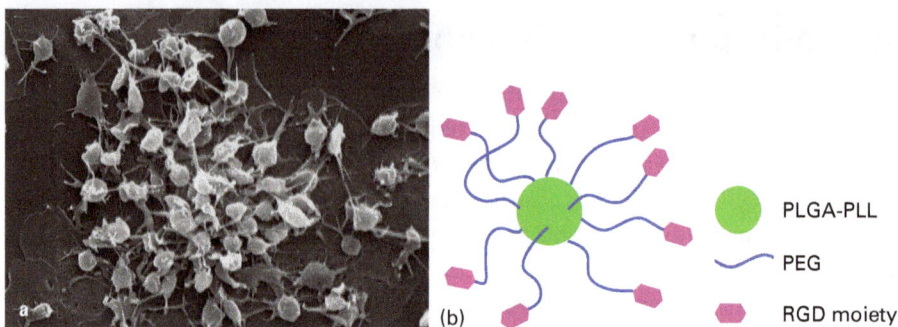

Fig. 2.1: Synthetic blood platelets. **(a)** Native platelets, diameters of 2–3 mm. Scanning electron microscopy image courtesy of Professor J.M. Gibbins, University of Reading; **(b)** schematic drawing of synthetic platelets comprising a biocompatible polymer cores (polylactic-co-glycolic acid (PLGA) and poly-L-lysine (PLL)), a polyethylene glycol (PEG) arm, and a cell binding peptide module – arginine-glycine-aspartate (RGD, single letter code).

tors displayed on the surface of activated platelets at a site of injury. These synthetic, biomimetic particles are thus designed to quickly produce a dense mesh which is *functionally* similar to a physiological clot – blocking further blood loss. Importantly, the synthetic platelets were also found to activate endogenous platelets (i.e. further promote the creation of a blood clot), a crucial requisite for possible clinical use.

Another interesting blood-related biomimetic application concerns the creation of *artificial oxygen carriers*. Such synthetic substances could have extremely important applications in blood transfusion, substituting defective red blood cells (and/ or dysfunctional hemoglobin – the protein that actually binds oxygen and is hosted in these cells). Synthetic oxygen carriers could be similarly used as the primary components of artificial blood. Diverse artificial oxygen carrier systems have been constructed over the years by chemical means. All such substances, however, have to adhere to several important conditions, besides exhibiting satisfactory O_2 transport capabilities. As materials that will ultimately operate inside the human body, they should exhibit minimal pathogenicity, high stability in physiological conditions, and undergo biodegradation after performing their tasks.

Scientists have utilized synthetic chemistry methods to produce biomimetic oxygen carriers based upon human hemoglobin (the oxygen-carrying protein). Fig. 2.2, for example, depicts a synthetic spherical assembly designed to encapsulate hemoglobin, enable the protein to bind oxygen, and efficiently transport it in blood. To achieve this task, J. Li and colleagues in the Beijing National Laboratory for Molecular Sciences have used a porous inorganic scaffold ($CaCO_3$) to enclose hemoglobin, coating the entire spherical assembly with polyethyleneglycol (PEG) for further protection from disintegration in the bloodstream. This configuration mimics the natural configuration of red blood cells transporting oxygen, providing a potentially useful artificial and biocompatible vehicle for oxygen transport.

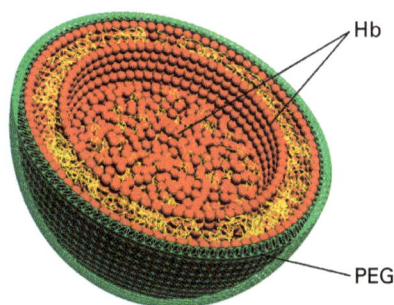

Fig. 2.2: Artificial oxygen carriers. Schematic drawing of a synthetic "hemoglobin sphere" for potential use as an artificial oxygen transporter in blood. Reprinted with permission from Duan L. et al., *ACS Nano* 2012 *6*, 6897–6904. Copyright (2012) American Chemical Society.

In the example above, researchers have used the actual functional (biological) molecule (hemoglobin, which transports oxygen in the bloodstream) in a completely new setting – the spherical porous calcium carbonate. Porous matrixes, in fact, have been used in many instances as a scaffold for biological molecules, together exhibiting unique functionalities. In particular, the pore structure and internal surface of the host matrix enable significant adsorption of varied guest molecules and creation of localized "nanoreactors" within the pores for carrying out chemical reactions. Furthermore, nano- and "meso"-porous materials (exhibiting wider pore structures) allow selective encapsulation of guest compounds through tuning of the pore sizes.

An intriguing demonstration of a biomimetic porous material application has been the development of "artificial enzymes" consisting of porous host materials (usually transparent silica-based matrixes) inside which amino acids have been chemically-attached to the inner walls. These biological or inorganic hybrids have often displayed unique catalytic, enzyme-like activities associated with the dense surface coverage of the functional molecular units, i.e. amino acids which mimic enzyme active sites. It is likely that the catalytic properties of these materials can be traced to the compact organization of the surface-displayed residues combined with the constrained physical environment inside the pores.

Similar applications of porous guest-host systems make use of the intrinsic *transparency* of the silica-based assemblies. Fig. 2.3, for example, depicts a chemical system designed in the laboratory of T. Itoh at the Toyota Central R&D Laboratories, Japan, in which chlorophyll molecules were inserted into the nanometer-size cavities of mesoporous silica, yielding an artificial *photosynthesis* system. This bio-hybrid system conformed to the basic parameters of photosynthesis: *transparency* to light, molecular geometries allowing charge separation, and energy transfer (other biomimetic photosynthesis systems are discussed in Chapter 11).

The ability to synthesize pores in transparent silica matrixes with dimensions of proteins and other biomolecules has opened many research possibilities for

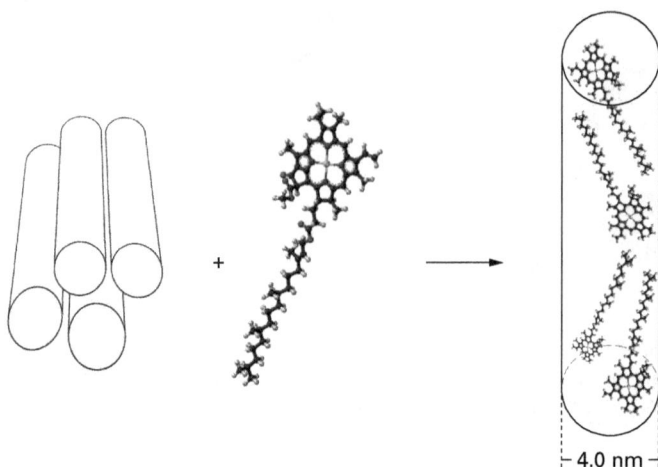

Fig. 2.3: Artificial photosynthesis in porous silica. Chlorophyll molecules are encapsulated within the porous transparent silica matrix. The positioning and orientation of the chlorophyll guest molecules facilitate light absorption, electron excitation and transfer – the essential constituents of a photosynthetic process. Reprinted with permission from Ruiz-Hitzky, E. et al., *Adv Mater* **2010** *22*, 323–336. Copyright (2010) John Wiley and Sons.

biomolecule-silica hybrids. Specifically, numerous studies have demonstrated that enzymes, proteins, DNA, and even whole cells could be encapsulated inside the pores without adversely affecting the biological functions. This remarkable observation combined with the high loading capacity of the porous matrixes enable utilization of these guest-host assemblies for diverse applications, such as long-term storage of proteins and enzymes, protein immobilization and separation, and analysis of molecular properties and interactions in constricted spaces.

One does not necessarily need to venture far to identify means for exploiting nature for creating useful biomaterials. *Wood* is, in fact, a wonderful example for an abundant and extremely diverse biomaterial that can be used for varied applications. Wood – specifically its most abundant molecular building block *cellulose* – has been used since early civilization for various non-biological applications, most notably perhaps – paper. Wood, however, has been also exploited for diverse uses nominally far from its natural "environment". For example, ceramic materials have been generated through chemical reactions (mainly liquid or gas infiltration processes) involving wood as a "synthesis template". Such templates can be produced through *pyrolysis* – a process in which the organic substance in wood is digested, leaving behind the rigid, porous template upon which new materials can be assembled. The resultant *biomorphic materials* exhibit a hierarchical *macroscopic* structure similar to original wood, albeit with totally different molecular compositions.

Cellulose has in fact inspired numerous biomimetic materials aiming to capture some of its properties, such as resilience, long-term stability, and rigid molecular framework, and complement them with additional capabilities. *Cellulose composites,* in particular, have opened new horizons for diverse materials-science applications. New materials comprising cellulose and other polysaccharides or proteins have been found to exhibit useful properties, forming fibers, gels, films, and foams. Some of these new substances combine the resilience to mechanical stress of cellulose (exemplified in the extraordinary strength of many wood species and wood-barks) with the formation of molecular barriers against pathogens and degrading molecules. Like many other composite materials, the enhanced properties emanate from tuning the molecular architectures of the molecular cellulose components and added species.

Fig. 2.4 presents examples of a novel *foam* structure, produced through manipulation of a cellulose derivative. The material comprises *nanocrystalline cellulose* (NCC) – essentially carefully grown cellulose single crystals. The foam material depicted in Fig. 2.4, developed in the laboratory of O. Shoseyov at the Hebrew University, Israel, was assembled from a *liquid crystalline* suspension of NCCs, enhancing the oriented morphology of the foam assembly and maintaining a high-porosity structure (Fig. 2.4b). The new foam assembly exhibits light weight and high compressibility, underscoring the potential of using molecular cellulose derivatives for constructing new materials with properties that are, in fact, quite different to the original macroscopic source – wood.

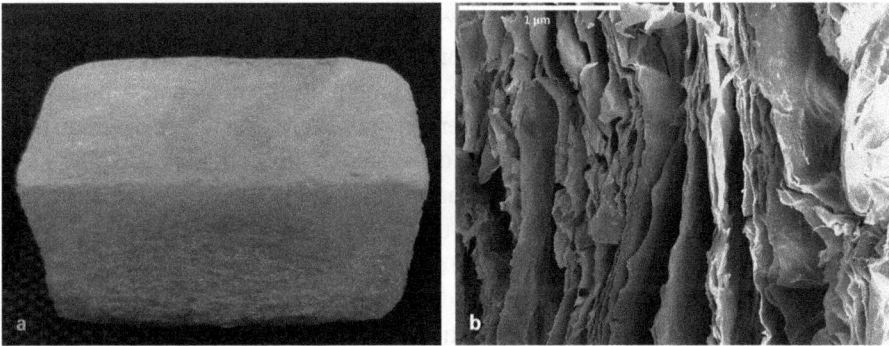

Fig. 2.4: Foam produced from nano-crystalline cellulose. **(a)** Picture of a foam specimen precipitated from a solution of NCC; **(b)** scanning electron microscopy (SEM) image depicting the internal foam structure; the image underscores the laminated porous structure bestowing the light weight and compressibility on the foam. Images courtesy of Professor O. Shoseyov, Hebrew University, Israel.

2.2 Protein-based functional materials

Peptides and proteins present a diverse universe from which one is able to select and construct a variety of new materials. Biomimetic protein-based functional materials constitute, in particular, a fine example of the advantages of "bottom-up" materials science. Such generic molecular approaches contrast with the traditional "top-down" engineering methods for which properties emanate from the overall global design. Indeed, multi-functionality in natural protein-based materials is derived from the assembly of simple elements (e.g. amino acids or small peptide motifs) into hierarchical structures that are characterized by multiple length scales (from the nanoscale to microscale and beyond). Peptide chemistry offers additional broad benefits for materials design – *molecular specificity* on the one hand and *chemical diversity* on the other. These aspects, discussed in more detail below, have already yielded numerous functional material designs.

The considerable advances in our ability to design new protein sequences and control protein properties through molecular manipulation has been the key factor responsible for the dramatic expansion of the field of protein-based materials science. Research over the past several decades has thoroughly enhanced our understanding of the molecular and energy landscapes of proteins and peptide structural elements. Through integrating computational chemistry and biology, specifically modeling methods, sophisticated protein expression systems, synthetic schemes, and post-expression manipulation techniques, scientists are currently able to achieve a remarkable degree of control over protein assemblies leading to the introduction of novel protein-based materials.

An important aspect of peptide- and protein-based materials concerns the manner by which weak *non-covalent* intermolecular interactions rather than the chemical bonds generate mechanically strong materials. Indeed, non-covalent *molecular self-assembly*, shaped by hydrogen bonding, electrostatic, hydrophobic, and van der Waals interactions, is probably the most ubiquitous phenomenon exploited for peptide biomaterial design. Many biomaterials whose core structure is based on non-covalent interactions, however, display short-lived stability, particularly in physiological environments – attributed to degradation induced by salts, natural surfactants, other amphiphilic molecules, or enzymes that lead to dissolution. Accordingly, many biomaterials constructed through peptide self-assembly rely on particularly resilient configurations such as *fibrils*, outlined in detail below.

Self-assembling peptides hold particular interest as building blocks for new biomaterials. Indeed, the spontaneous association of peptides into larger-scale architectures can be harnessed, producing interesting properties derived from the intrinsic features of the individual peptide units. Many of the systems reported so far in this context employ *amphiphilic peptides*. Similar to soap, such peptides exhibit both hydrophilic (water-attracting) and lipophilic ("lipid loving") properties, which overall give rise to formation of distinct architectures via self-assembly – such as fibers, porous gels, and even thin films having optical properties. Furthermore, self-assembled peptides can be induced to form more complex architectures, such as three-dimensional "mesh" structures which could function as membranes, or hosts for molecular cargoes (Fig. 2.5). Peptide-based materials exhibit other notable properties, including biocompatibility, and the possibility of coupling of diverse structural and functional biological units that might be exploited in biomimetic applications.

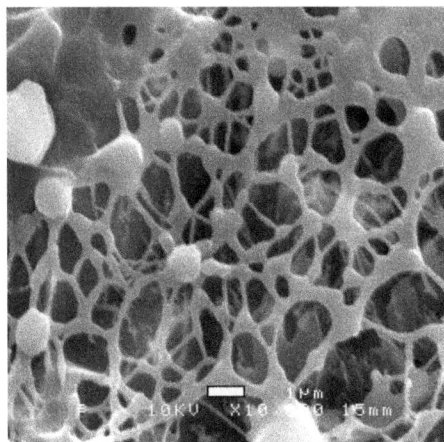

Fig. 2.5: Peptide-derived porous gel. Cryogenic scanning electron microscopy (cryo-SEM) image of a porous gel constructed through self-assembly of a Fmoc-Leu-Gly-OH dipeptide. Reprinted with permission from Johnson, E.K. et al., *Journal of the American Chemical Society* **2010** *132*, 5130–5136. Copyright (2010) American Chemical Society.

Designing peptide-based materials through tweaking the amino acid sequences of self-assembled peptides has been a thriving research field. Specifically, distinct structural motifs associated with specific amino acid sequences have been identified and employed as building blocks comprising higher-scale hierarchical materials. These *de novo* ("from the beginning" in Latin) approaches have had a dramatic impact on the development of new classes of protein-based materials (discussion of *de novo* techniques as a tool for *protein design* is presented in Chapter 6).

Varied peptide modules have been used as molecular construction elements in self-assembled biomaterials (Fig. 2.6). The *β-sheet* is an abundant motif found in numerous proteins and generally employed to stabilize fibrillar peptide assemblies through hydrogen bond formation. The *coiled-coil α-helix* motif is another fundamental protein structural unit which has been employed in protein-based materials design. This motif combines a hydrophobic interface, providing the means for protein self-assembly, and charged-residues localized in strategic locations within the sequence which enable electrostatic control of, for example, chemical triggering of oligomerization (for example through protonation or deprotonation in acidic or basic solutions, respectively). Other motifs utilized in artificial protein designs include the *triple helix* and *β-spiral*.

The concept of modifying the peptide sequence as a fundamental structural tool for designing new materials has led to the development of ever more sophisticated experimental schemes. "Symmetry-based" peptide design is a new technique aimed at producing higher-order hierarchical structures and patterns based upon short peptide building blocks (Fig. 2.7). In that study, M. Noble and colleagues at Oxford University have shown that complex long-range protein patterns can be created

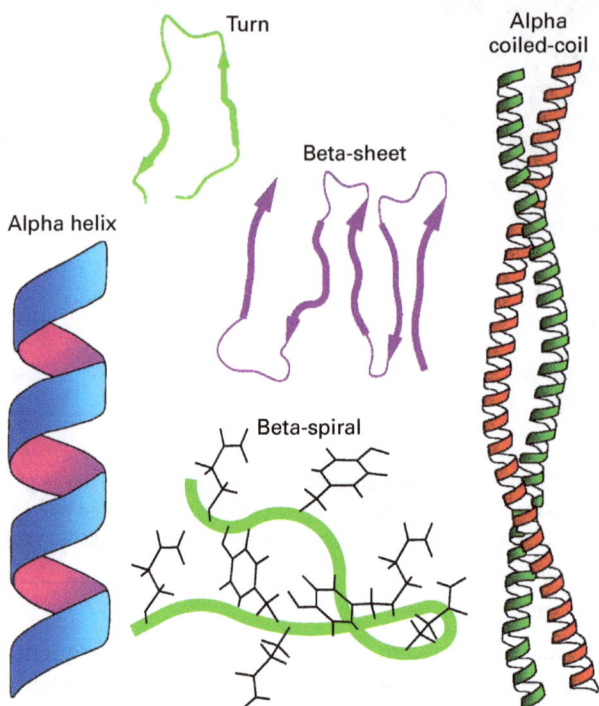

Fig. 2.6: Protein structural motifs used in biomaterials. Structural units within proteins and peptides constitute building blocks in *de novo* design of peptide-based biomaterials.

through self-assembly of peptide monomers that exhibit two important properties: they can be covalently connected to form *oligomers* through placement of peptide-fusion linkers at specific desired locations, and the monomers further share *axes of symmetry*. Consequently, similar to popular children's games, the modular monomer units do not self-assemble randomly, but rather self-assemble into arrays of hierarchical configurations, constrained by the axis of symmetry that they both share.

In a broader context, methodological developments along the lines of the monomer-symmetry approach might open new avenues for protein-based materials design in two crucial directions. First, a modular approach using *peptide* building blocks could be a platform for fabricating materials in predictable extended geometries. Second, similar to the proliferation of DNA-based nanostructures (see Chapter 10), the development of technologies for construction of *organized, controlled* three-dimensional protein-based assemblies could substantially expand the "structural universe" of synthetic protein-based assemblies.

Collagen is an excellent example of an abundant natural peptide which has inspired and been incorporated within varied material applications. Collagen (which essentially comprises a family of several peptides), abundant in mammalian cell

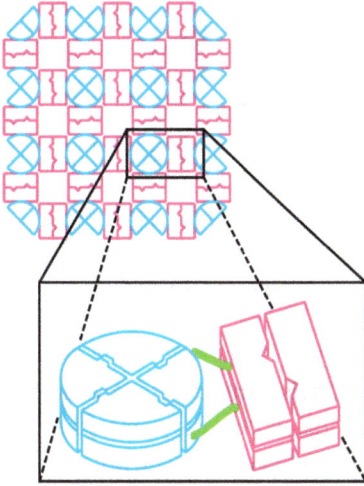

Fig. 2.7: Symmetry-based ordered peptide assembly. Hierarchical structure constructed from peptide oligomers comprising a *four-fold symmetry* domain (blue) and a *two-fold symmetry* region (purple). The individual oligomers organize according to their symmetry axes, while a higher-order "pattern" is produced through peptide-bond formation between adjacent oligomers. [Drawing based upon work of Sinclair et al., *Nature Nanotech.* **2011** *6*, 558–562.]

systems, constitutes an important structural component of the extracellular cell matrix, and also plays a prominent role in determining the structural and mechanical properties of soft tissues such as ligaments, tendons, cartilage, and other connecting tissues. The core feature of collagen that makes this peptide attractive for biomimetic materials is its propensity to form highly stable *fibers*. This property is derived from the prevalence of polyproline peptide motifs within the collagen sequence, which assemble into right-handed triple helixes. Collagen fibers themselves often further assemble into higher-order water-containing porous gel structures (i.e. *hydrogels*) which exhibit other interesting functionalities related to this morphology.

New synthetic chemistry techniques have been developed in recent years towards creation of modular, functional collagen-based nano- and microdimensionality materials for biomedical applications mostly focused upon regenerative medicine. These approaches aim at mimicking the hierarchical self-assembly of natural collagen. Specifically, the design process begins with modeling peptide chains that form *triple helixes*; continuing with self-assembly of the helixes into *nanofibers*; and finally fiber polymerization yielding *hydrogels* (Fig. 2.8). It should be noted, however, that while "on paper" this design process appears straight-forward, adhering to the structural and organizational targets has proved to be quite a difficult task, making the development of collagen-mimic materials a challenging and highly active scientific and technological field.

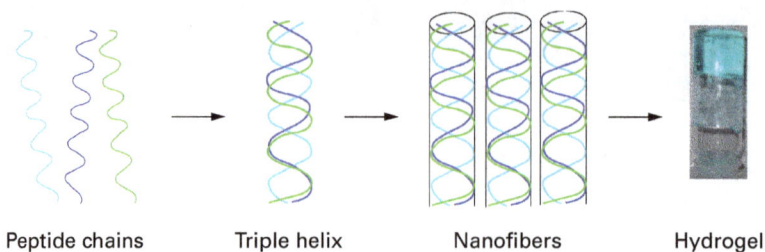

Peptide chains   Triple helix   Nanofibers   Hydrogel

Fig. 2.8: Collagen-mimic hierarchical biomaterials. Shown schematically are the stages for construction of collagen-mimic materials.

It should be emphasized that collagen-mimic materials do not necessarily conform to the exact amino acid sequence of the natural peptide. Rather, the primary precondition that needs to be fulfilled is the adoption of the *triple-helix* structure by the synthetic peptides. Furthermore, such peptides forming collagen-like triple helixes can be further modified to include varied functional units. In particular, specific modifications of the amino acid sequence of the synthetic peptide monomers enable tuning of the physical and chemical properties of the macroscopic materials.

A variety of elegant synthetic strategies have been introduced for constructing materials with the desired controlled self-assembly features of collagen. Fig. 2.9 depicts an example of a collagen-mimetic system in which much shorter peptides than collagen were shown to adopt a collagen-mimic triple-helix structure. The peptides, designed by J. Hartgerink and colleagues at Rice University, contained a critical peptide motif – proline-hydroxyproline-glycine (Pro-Hyp-Gly) repeating unit – complemented by inserting the polar residues lysine and aspartate in strategic positions to form a sticky-ended collagen-like triple helix. In sufficient concentrations, the short peptides formed triple helixes which could bundle into a homogeneous population of

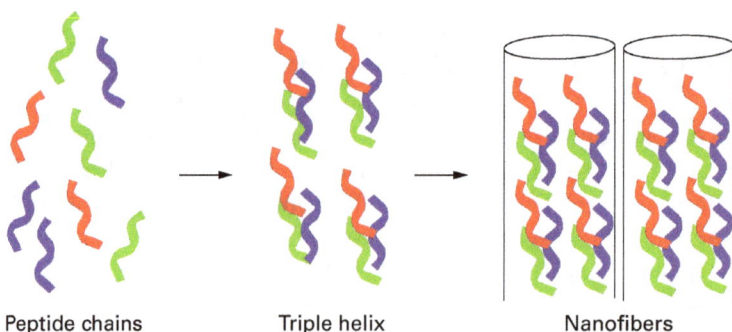

Peptide chains   Triple helix   Nanofibers

Fig. 2.9: Biomimetic collagen assembly from small-peptide agglomeration. Short peptides can self-assemble into the collagen-mimic triple-helix structure [drawing inspired by the work of O'Leary et al., *Nature Chemistry*, **2011**, *3*, 821–828].

nanofibers, which themselves subsequently condensed into high-quality hydrogels – nicely recapitulating the multi-hierarchical organization of natural collagen.

While collagen has been used in many instances as a *structural* component in new biomimetic materials, the *mechanical* properties of the peptide have important uses in regenerative medicine, as the collagen framework lends itself to association with molecules participating in varied cellular and biochemical processes (see the detailed discussion of this topic in Chapter 4). Collagen scaffolds have been shown, for example, to host cell adhesion units and biological signaling molecules designed to enhance and direct cell proliferation. Fig. 2.10 depicts a viable cell grown inside an artificial collagen-mimic peptide matrix.

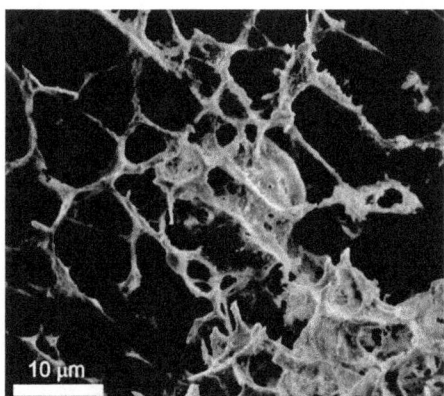

Fig. 2.10: Cell in a biomimetic collagen matrix. Cryo-SEM image of a HeLa cell encapsulated within a self-assembled peptide network mimicking collagen structure. Reprinted with permission from Przybyla, D.E. and Chmielewski, J., *Biochemistry* **2010** *49*, 4411–4419. Copyright (2010) American Chemical Society.

Risks of utilizing collagen from natural sources need to be considered as well, particularly in materials to be used for therapeutic purposes, such as implants or tissue scaffolds. Specifically, there might be undesired and potentially harmful substances attached to the collagen fibers, unwittingly imported into the body. Other deficiencies of collagen-based materials designed for human therapy involve eliciting of immunogenic response and possible degradation through the activity of collagen-digesting enzymes that are physiologically active in the extracellular space.

Silk is another notable example of a unique natural proteinaceous substance for which humanity has continuously found many uses, beginning thousands of years ago with fabric, and continuing in the present with intriguing biomedical and materials applications. Probably the most remarkable property of silk is its extraordinary tensile strength, which combined with its light weight can be exploited for fabrication of extremely strong nanofibers for various applications. Complementing the physical strength of silk, its biocompatibility and biodegradability as a naturally-occurring protein contribute to it being a very attractive component in advanced materials.

Molecularly, silk comprises mainly fibrous proteins which contain several repetitive modular units linked together. The amino acid motifs within the proteins confer specific structural and mechanical functions to silk, such as stiffness, tensile strength, flexibility, etc. Sequence motifs identified in silk proteins (particularly *silk fibroin*, the main protein constituent produced by the silkworm) include poly-alanine and alanine-glycine repeats which promote β-sheet structures that are central components in fibers due to inter-strand hydrogen bonding, and glycine-glycine-X or glycine-X-glycine (where X corresponds to another amino acid) which are associated with fiber elasticity and "shape memory" (Fig. 2.11).

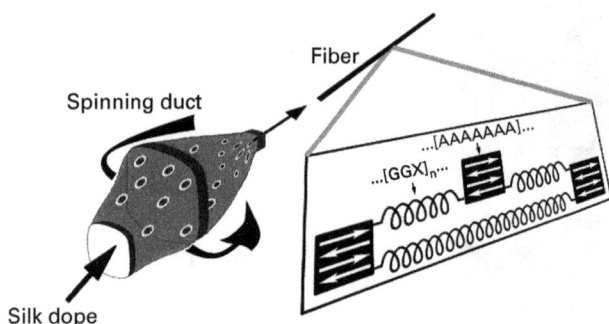

Fig. 2.11: Silk production and its modular structure. Schematic depiction of the silk spinning process in the spider duct, and the modular peptide structure of the resultant silk fiber. [Drawing inspired by Kushner, A.M. and Guan, Z., *Angew. Chem. Int. Ed.* **2011**, *50*, 9026–9057.]

The modular structure is a key feature underlying potential utilization of silk-based materials in diverse applications. Specifically, it is possible to design artificial biomimetic proteins in which the *number* and *relative position* of the structural motifs is tuned in order to modify the mechanical properties. Such engineered proteins could be produced in sufficient quantities using conventional recombinant DNA technology and expression vehicles, such as genetically-engineered bacteria. The artificial silk-like proteins could be assembled into a variety of materials – thin films, porous gel scaffolds for tissue engineering, artificial ligaments, transparent optical materials, and others.

The extraordinary stability of the β-sheet forming motifs, however, has a "flip side" – significantly reducing the *water solubility* of the protein aggregates and consequently imposing a major hurdle for practical utilization of silk fibroin or of synthetic fragments comprising the peptide sequence. Several approaches, however, have been devised to circumvent the solubility barrier, for example through the use of *ionic liquids* to solvate silk fibroin instead of conventional water-based solutions.

Ionic liquids (ILs) are a class of diverse organic salts with relatively low melting points (below 100 °C) attracting great interest as a promising "green" source for alternative solvents. In the context discussed here, ILs have been shown to effectively dissolve silk fibroin, enabling the creation of ordered silk "patterns" following solvent removal (Fig. 2.12). Such patterns could then be used for varied applications, such as directed cell growth, drug delivery, and others – all based upon the biocompatibility of the patterned silk.

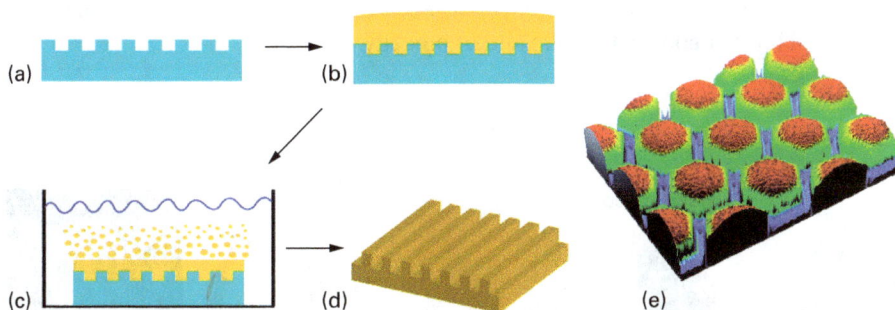

Fig. 2.12: Patterned silk films. Schematic illustration of patterned silk films produced through dissolution in ionic liquids. **(a)** PDMS stamp in the desired design; **(b)** an ionic-liquid solution containing dissolved silk fibroin is spin-coated onto the PDMS stamp; **(c)** the coated stamp is submerged in a methanol bath to extract the ionic liquid solvent, resulting in crystallization of the silk film; **(d)** the crystallized silk film is separated from the stamp; **(e)** white light interferometry scan of a patterned silk film. Reprinted with permission from Gupta, M.K. et al., *Langmuir* **2007** *23*, 1315–1319. Copyright (2007) American Chemical Society.

Engineering proteins to mimic silk functionalities is an active research field. Proteins comprising sequence elements of silk have been manufactured through conventional bacterial expression platforms, but also by using mammalian cells (for example expressing silk proteins in the milk of transgenic goats), enabling industrial-scale production. Protein design, particularly the length and organization of the peptide modules associated with silk formation and properties, enables fine-tuning the properties of both individual expressed proteins as well as the macroscopic fibers produced through physical association or chemical cross-linking of the protein units (e.g. fiber thickness, elasticity, stiffness, etc.). Difficulties, however, have been encountered in the pursuit of practically useful artificial silk-like proteins. The primary challenges have been the premature aggregation indicated above, and the realization that the superior properties of silk are not just related to the amino acid sequence, but also post-expression processing, particularly the quite complex *spinning* mechanism of the silk fibers.

Silk-based hollow *microcapsules* represent a non-natural construct comprised solely of a natural species – silk protein fibers as the external coating layer. Such shell-type aggregates offer distinct advantages as drug delivery vehicles. The micro-capsules could encapsulate and transport drug molecules, minimize early biodegra-dation (taking advantage of silk protein resilience), facilitating tunable permeability through controlling the silk fiber density to achieve consequent slow release of the capsule content. *Layer-by-layer silk microcapsules* (Fig. 2.13) are a specific example of silk-based biomimetic molecular containers. To manufacture the capsules, β-strand-containing silk-protein sheets are deposited consecutively (layer upon layer) around a silica core, followed by removal of the silica through chemical processes such as acidic digestion. The generic layer-by-layer technique enables control of both the microcapsule's core dimensions as well as the silk shell thickness (through the number of layers deposited) and resultant porosity.

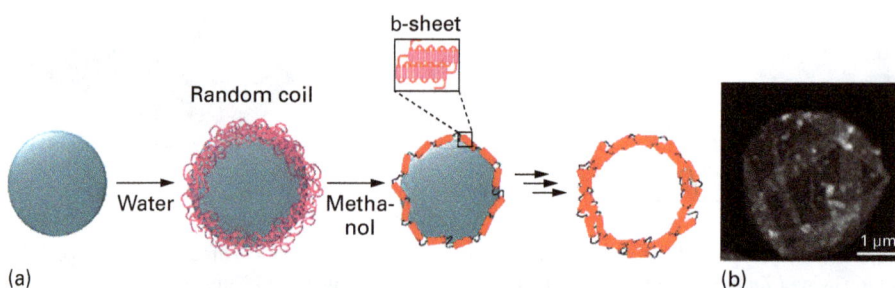

Fig. 2.13: Layer-by-layer silk microcapsules. **(a)** Microcapsule preparation scheme: Silk proteins in random coil configurations are absorbed onto the surface of silica particles, followed by the forma-tion of a β-sheet-rich structure through methanol treatment. This procedure is repeated several times creating layered silk protein coating around the silica core. The silica core is eventually dis-solved, leaving behind an empty silk microcapsule. **(b)** Confocal laser scanning microscopy image of a silk microcapsule. Reprinted with permission from Shchepelina O. et al., *Adv. Mater.* **2011** *23*, 4655–4660. Copyright (2011) John Wiley and Sons.

Other interesting applications of silk have been reported. Researchers have devel-oped, for example, a novel drug delivery platform based upon short fragments of silk fibers – a biomimetic "microneedle". These microneedles, prepared via a stamping technique similar to the one depicted in Fig. 2.12 above, can be "spiked" with drug molecules resulting in a vehicle for slow release of the molecules when the needles are pressed through the skin. This configuration exhibits practical advantages – bio-compatibility, minimal disruption and interference with biological processes at the site of application – and excellent stability and durability.

Amyloid peptides, a unique and expanding family of diverse peptides, represent a class of biomolecules with yet unresolved functional and pathological functions. This class of peptides, which include widely-studied molecules such as amyloid-beta, synuclein, prions, amylin, and others, has gained notoriety as pathological hallmarks for devastating diseases such as Alzheimer's disease, mad cow disease, and others.

From a fundamental biological standpoint, however, amyloid formation is increasingly perceived as an alternative structural pathway for (misfolded) proteins and peptides. In parallel with the broad research efforts aiming to understand amyloid fibril formation and its biological implications, investigators have attempted to harness amyloid peptide aggregation for applications in materials science. In particular, the spontaneous *self-assembly* of amyloid peptides can be exploited for a variety of applications, taking advantage of the diversity of protein functions, structural modules, and macroscopic architectures.

Biomimetic materials based upon amyloid fibril systems aim to utilize both the *molecular* β-sheet organization of the peptides comprising the fibrils, as well as the *macroscopic* fibril structures themselves. The prevalent β-sheet motifs in amyloid fibrils, which can extend to thousands of molecular units, are considered the primary factor for the exceptional stability and lack of degradability of the fibril aggregates. This property is also attractive from a materials design point of view – utilizing the β-sheet assemblies as structural components provides extra physical strength, stability, and resistance to (bio) degradation. Moreover, since it has been shown that even peptides comprising only four or five amino acids form stable amyloid fibrils, the use of simple, short sequences as building blocks for more complex, hierarchical structures is feasible.

There have been many examples in recent years for synthetic assemblies and new materials based upon amyloid fibrils (and other peptide-based *fibrillar* assemblies) as the core structural units, and scaffolds. Such systems take particular advantage of the intrinsic stability of the peptide fibers even in harsh physical conditions – traced to the β-sheet organization of the peptide backbone – which might make fibrils amenable for industrial processes carried out in demanding environments. Furthermore, the use of *peptides* as the fundamental building blocks has certain advantages – utilizing the powerful arsenal of biology and biotechnology for introducing binding sites and exploiting molecular recognition, as well as controlling fiber properties such as thickness, and particularly *length*.

Amyloid fibrils can particularly serve as *patterning templates* for constituents that do not form organized nanostructures on their own, such as metal nanoparticles, biochemical dyes, and others. Indeed, a "natural" application for amyloid fibrils, which essentially resemble elongated wires, is their use as a platform for fabrication of conductive nanowires. The peptide fibers themselves are not conductive, however they can be employed as templates for deposition of metals through mostly established chemical and biochemical procedures. One potential approach in this context involves chemical modification of the fiber surface to attach (either covalently or noncovalently) molecular units that bind metals, consequently producing a conductive metal-coated fiber (Fig. 2.14). This goal can be also achieved via conventional *biotechnological* means – introduction of metal-binding amino acids such as *cysteine* (Cys, C) into the peptide sequence through genetic engineering or peptide synthesis. Consequently, silver or gold colloids can be attached to the fibril surface through the thiol moieties of the cysteine, ultimately producing a conductive wire. Fig. 2.14 presents a fine example of the above scheme, in which S. Lindquist and colleagues at the University of Chicago fabricated a conductive metal-coated nanowire deposited on a prion protein fiber.

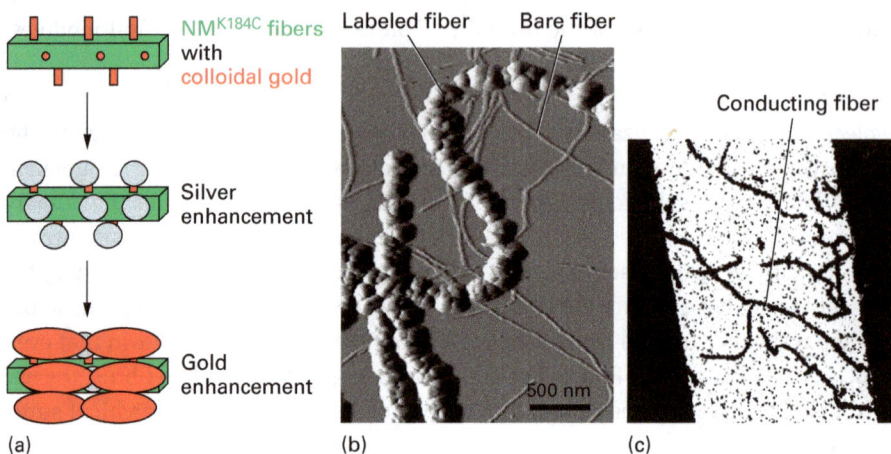

Fig. 2.14: Metal nanowires assembled on amyloid protein fibers. **(a)** Schematic drawing showing the metal deposition scheme: fibers were assembled from a yeast prion protein fragment in which a lysine residue was substituted with cysteine in position 184. Gold colloids derivatized with *maleimide* units were subsequently covalently attached to the cysteines at the fiber surface. The gold colloids acted as "seeds" for further deposition of metal layers, through a process denoted "metal enhancement". **(b)** Atomic force microscopy (AFM) image of bare protein fibers and metal-coated fibers, respectively. **(c)** Metal-coated prion fiber placed between two electrodes exhibiting electrical conductivity. Reprinted with permission from Scheibel, T. et al., *PNAS* **2003** *100*, 4527–4532. Copyright (2003) National Academy of Sciences, U.S.A.

Experimental parameters affecting the amyloid fiber-wire template scheme could make this concept a versatile and potentially useful platform for practical applications. The peptide fibers upon which the metal is deposited can be annealed (i.e. dissolved) without disintegrating the conductive nanowire. Moreover, the propensity of peptide fibers to align onto surface patterns might enable construction of complex metal nanowire designs. This observation suggests that actual circuit patterns might be constructed through integrating conventional lithography methods (or other techniques for surface patterning) with fiber-based nanowire assembly.

Amyloid fibril "templating" has been demonstrated in other chemical systems. *Insulin* protein fibers, for example, have been employed as substrates for deposition of electron *donor-acceptor* assemblies in organic solar cells. The efficiency of charge transport between an electron donor and electron acceptor is the fundamental factor affecting the overall performance and yield of an organic solar cell. In this context it has been shown that incorporation of the insulin fibers within the active layer of the solar cell (in which electrons are excited by light) gave rise to more effective charge transfer and electron transport. This result might be related to the fiber template concept: attachment of donor or acceptor molecules to the fibers improves their blending in the cell's active layer and consequent electron transport.

Peptide fibrils have been used as constituents for fabricating complex nanostructures. The two-residue peptide *diphenylalanine* can assemble into *hollow tubes* rather than filled fibers. This organization inspired construction of a coaxial metal nanotube, using the diphenylalanine assembly as a template (Fig. 2.15). Specifically, E. Gazit and colleagues at Tel Aviv University, Israel, demonstrated a clever approach for deposition of metal nanoparticles both *inside* and *outside* of the hollow tubes, yielding a conductive coaxial "nanocable" (e.g. conductor-insulator-conductor cable) – a feat that has been difficult to produce via more conventional engineering or chemistry techniques.

Amyloid fibrils lend themselves to other sophisticated applications. Self-assembled peptide fibrils have been used as scaffoldings for *light harvesting* through surface adsorption of functional units acting as molecular "antennae" (Fig. 2.16). Light harvesting materials which aim to reproduce the photosynthesis process generally require the presence of chemical units – pigments or chromophores – for absorbance of light energy, and this energy has to be further efficiently transmitted to perform important functions such as chemical reactions, physical work, etc. fiber-based light harvesting systems operate through covalent attachment of color pigments (e.g. photon absorbing moieties) to the fiber-forming peptides (which could be either natural fiber-forming peptides, fragments of larger amyloid peptides, or artificially designed sequences).

Through carefully controlling the conditions of the self-assembly process, the pigment-bound peptides can be incorporated into the native fiber while displaying the chromophore – essentially the light-absorbing antenna – upon the fiber surface. An *acceptor* unit, coupled to the same functionalized peptide or to a different peptide within the fiber, could be then used to demonstrate effective light harvesting through energy transfer. Indeed, controlling the distance and orientation of the donor and acceptor species is a fundamental challenge in the design of artificial light harvesting systems. Amyloid fibril aggregates provide, in fact, intrinsic control mechanisms for placement of the energy donors and acceptors, since both molecular units can be independently attached at different locations within the peptide sequence. Furthermore, the elongated structure of the fibers could dictate the relative orientation of the donor and acceptor, ultimately affecting the extent and efficiency of energy transfer.

Amyloid proteins have been also used as platforms for scalable two-dimensional functional films. Thin films comprising peptide fibers can be constructed through deposition of the fibers upon solid substrates and removal of the aqueous solvent. Interestingly, M. Welland and colleagues at Cambridge University discovered that addition of gelatinous co-solute molecules to the peptide fibers suspension resulted

A

| Diphenylalanine | Silver ions | Linker peptide | Gold nanoparticles | Gold ions |

| Formation of peptide nanotube | Reduction of silver inside the peptide nanotube | Binding of linker peptides to peptide nanotube surface | Attachment of gold nanoparticles | Reduction of gold over the peptide nanotube |

B

100 nm

100 nm

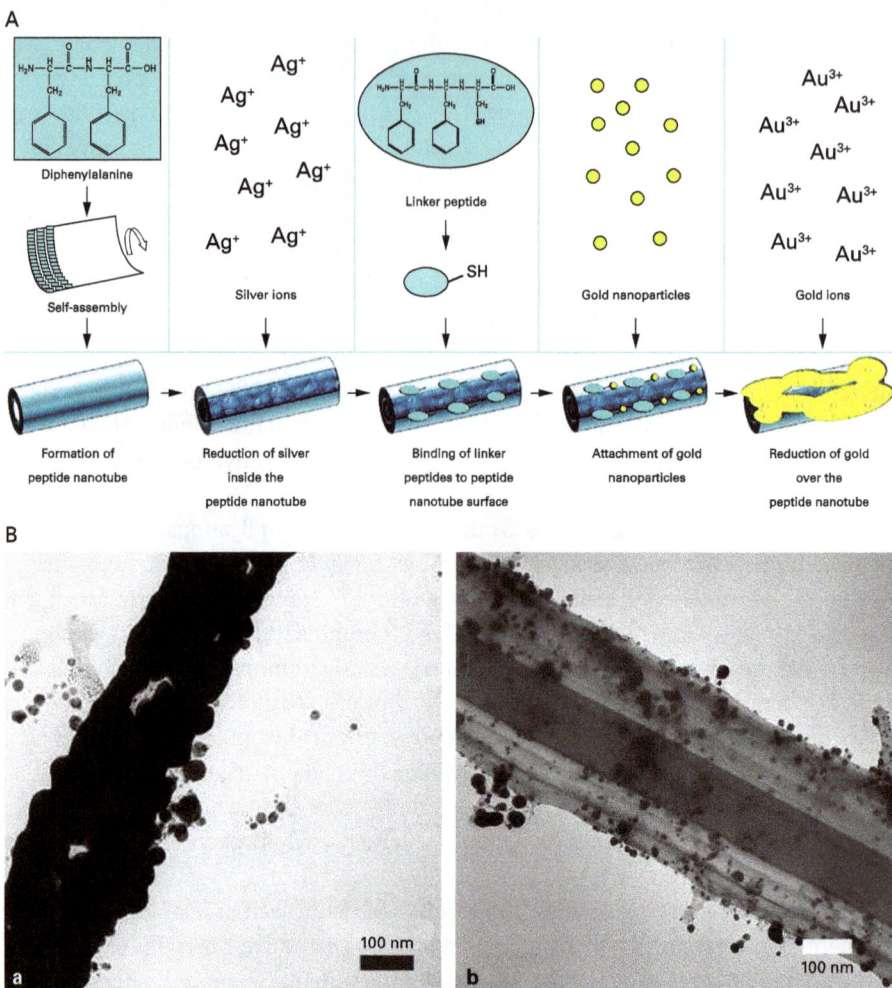

Fig. 2.15: Peptide-templated coaxial metal cable. **A.** Procedure for synthesis of a coaxial metal cable using a hollow diphenylalanine nanotube as a template. From left: self-assembly of the dipeptide forming a nanotube; Ag^+ ions are reduced inside the tube forming a metallic inner wire; linker peptides containing the diphenylalanine unit and *thiol* groups are bound to the nanotube surface (via hydrophobic association of the dipeptide component); Au nanoparticles are attached to the nanotube surface through binding with the SH units; further Au deposition on the nanotube surface, with the attached Au NPs serving as nucleation sites. Reprinted with permission from Carny, O. et al., *Nano Lett.* **2006** 6, 1594–1597. Copyright (2006) American Chemical Society. **B.** Transmission electron microscopy (TEM) images of diphenylalanine self-assembled tubes: coated with gold metal layer **(a)**; and showing the metallic silver core and partial attachment of Au NPs to the external surface **(b)**. TEM images courtesy of Prof. E. Gazit, Tel Aviv University.

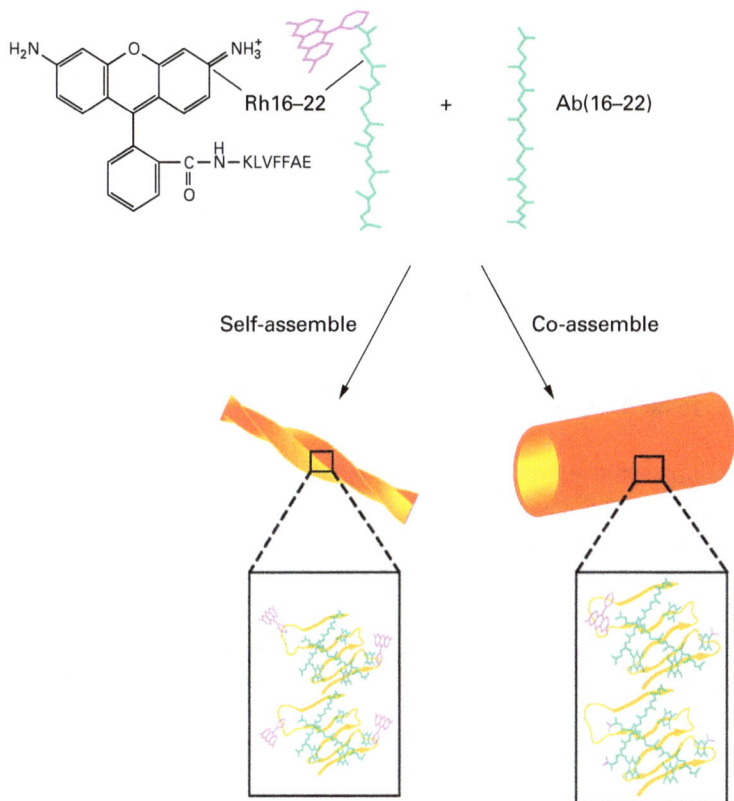

Fig. 2.16: "Amyloid antenna" for light harvesting. Native 7-residue fragment of the amyloid-beta peptide was co-assembled with the peptide chemically-functionalized – with rhodamine (Rh), a light harvesting pigment. The two peptides formed mixed assemblies in different morphologies (screw-type fibers or nanotubes), depending upon the experimental conditions. In both configurations the Rh pigment was displayed on the fiber or tube surface, enabling light absorption and energy transfer.

in formation of liquid-crystalline films, i.e. films in which the fibers were oriented with respect to a common axis (Fig. 2.17). This configuration is tied to the intrinsic spatial anisotropy of the fibers which dictates a macroscopic ordering in sufficiently viscous solutions. Such organized films could be used in turn as templates for guest molecule pointing to applications in which molecular orientation is important, such as optical devices.

While the examples above underscore the use of amyloid fibers as metal templates, peptide assemblies have been also employed for "indirect" patterning using the fibers as "masks" for generation of surface structures. Fig. 2.18 illustrates this concept, depicting protein fibers serving as masking agents similar to ones used in conventional lithography techniques. Essentially, the fibers are deposited on a metal-

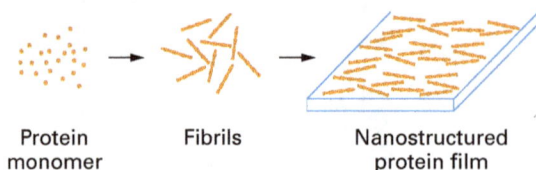

| Protein monomer | Fibrils | Nanostructured protein film |

Fig. 2.17: Liquid crystalline fibril film. Amyloid fibrils assembled from peptide monomers can orient upon slow drying of the fibrils dissolved in a viscous solvent, producing a liquid crystalline organization.

coated surface, blocking etching of the metal layer directly underneath. The etching process ultimately yields a nanowire pattern which traces the position of the surface-deposited fibers.

While the examples above underscore the use of amyloid fibers as metal templates, peptide assemblies have been also employed for "indirect" patterning using the fibers as "masks" for generation of surface structures. Fig. 2.18 illustrates this concept, depicting protein fibers serving as masking agents similar to ones used in conventional lithography techniques. Essentially, the fibers are deposited on a metal-coated surface, blocking etching of the metal layer directly underneath. The etching process ultimately yields a nanowire pattern which traces the position of the surface-deposited fibers.

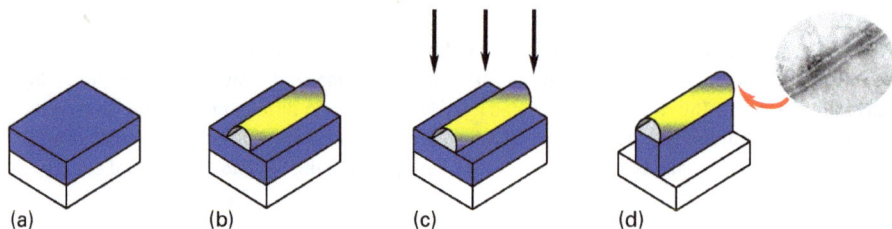

(a)   (b)   (c)   (d)

Fig. 2.18: Peptide fibers as masking agents in lithography. **(a)** Metal film (blue) is deposited on the substrate; **(b)** protein fiber is placed on the metal film; **(c)** an etching process removes the metal layer except for the area directly underneath the fiber; **(d)** the area masked by the protein fiber remains metal-coated, producing a conductive nanowire.

Certain practical issues need to be considered in the field of peptide fiber-based functional materials. A notable detriment is the inherent *low solubility* of fibrillar assemblies, ascribed to the hydrophobic properties of the peptides (which constitute, of course, the primary driving force for fibrillation). Since most chemical processes take place in solutions, the lack of solubility curtails the capability to process and control fiber properties through chemical means. Varied synthetic schemes, however, have been developed to overcome this barrier, mostly focused on chemical modulation of the peptide building blocks to increase their solubility. Studies have examined the effects of pH modulation; protonation (at low pH solutions) of amino acids within the fiber-forming peptide sequences disrupted the hydrogen bonding network which significantly contributes to fiber stability. In consequence, the propensity of the peptides to assemble into fibers was reduced and solubility was enhanced. Reversing the process was achieved in such systems through *increasing* the pH, thereby promoting fiber formation.

A common characteristic of almost all the fiber-based peptide materials discussed above is the *rigidity* of the macroscopic framework. The vast protein world, however, presents opportunities for tuning other mechanical parameters, such as *elasticity* and *flexibility*. Such properties can be achieved through harnessing specific protein families, very much like collagen and fibroin, discussed above, have been used or inspired numerous materials designs. *Elastin* and *titin*, for example, are among the main human proteins providing elasticity and storage of mechanical energy in organs and connective tissues. The mechanical properties of elastin and related proteins are believed to arise from unique "β-spiral" conformations associated with specific amino acid repeats – for example valine-proline-glycine-valine-glycine. According to some proposals, the macroscopic elasticity corresponds to both the intrinsic molecular flexibility of the peptide spiral (a sort of "molecular spring"), as well as hydration of the peptide environment enabling expansion or contraction of the spiral.

Elastin-mimic peptides with variations of the β-spiral peptide repeats have been synthesized. Such peptide modules function as "molecular springs"; when cross-linked or physically incorporated within other molecular entities they endow the resultant composite materials with remarkable elasticity and flexibility. A case in point is a *polymer* comprising β-spiral elastin repeats (Fig. 2.19). This type of configuration engenders distinct elasticity profiles, and underscores the important notion that organization of specific *molecular* units can significantly impact mechanical, cooperative properties of biomimetic materials.

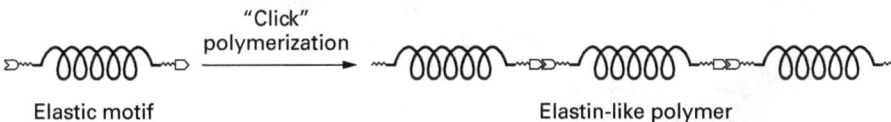

Fig. 2.19: Elastin-mimic polymers. Peptide "β-spiral" repeats linked in a polymeric chain produce spring-like properties.

While secondary structure elements, such as the β-spiral motifs, give elastin and elastin-mimic molecules their mechanical elasticity, the *tertiary* structure of proteins such as titin is the underlying factor responsible for their exceptional mechanical resilience and has been reproduced in different materials. Titin, a major protein constituent in muscles, has a modular structure with hundreds of repeating folded protein subunits (Fig. 2.20). Upon application of a pulling action, the individual subunits gradually unfold; subsequent contraction induces refolding of the connected titin modules. This remarkable feature provides extraordinary elasticity and contraction capacities to the protein. The concept of mechanical responsiveness linked to a modular arrangement of folding and unfolding subunits has been mimicked in synthetic polymer-based materials. The core requirements in such designs are the presence of chemically-linked units that can be reversibly folded and unfolded, and triggering of these molecular transformations through *mechanical* (rather than chemical) stimuli.

Among the more exciting facets of biomimetic protein-based materials design has been the use of functional proteins as templates for structures or systems that are totally different to their physiological origins. *Chaperonins* are a case in point. These biomolecules have critical biological roles as they assist other proteins to fold

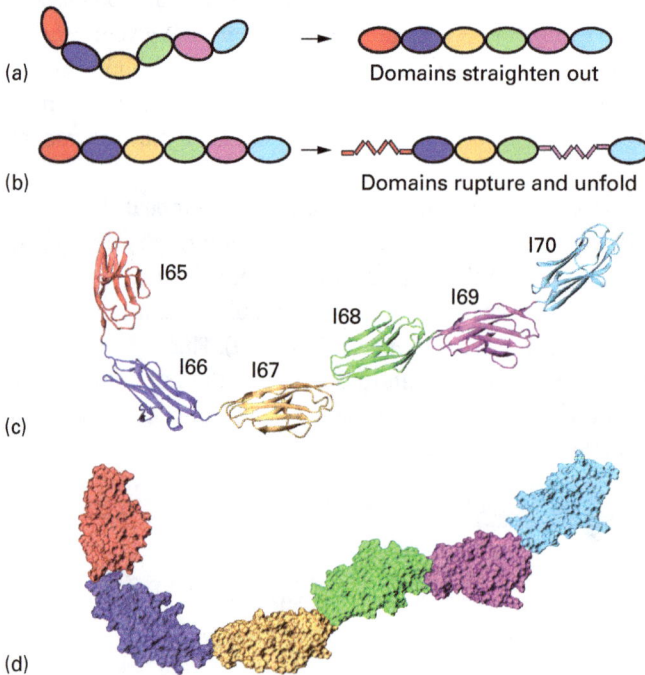

(a)   Domains straighten out

(b)   Domains rupture and unfold

(c)

(d)

Fig. 2.20: Titin elasticity associated with its tertiary structure. **(a)** Elasticity originating from "straightening" of the tertiary domain organization of titin; **(b)** "rupturing" of individual domains induced by an external mechanical force, elasticity maintained through refolding of the ruptured units; **(c)–(d)** structures of domains 165–170 of titin. Reprinted from Lee, E.H. et al., Tertiary and secondary structure elasticity of a six-Ig titin chain, *Biophysical Journal* **2010** *98*, 1085–1095. Copyright (2010), with permission from Elsevier.

correctly within the cellular cytosol. Intriguingly, some chaperonins form periodic "doughnut-shape" lattices when isolated and crystallized, thus forming "templates" for ordering deposited nanostructures. Indeed, chaperonins appear well-suited for such roles since they usually contain large structural cavities or pores that can host other molecular species. J. Trent and colleagues at NASA Ames Research Center, for example, applied protein engineering techniques for creating periodic chaperonin superlattices in which the pores of the ring-shaped chaperonins were used to immobilize metal or semiconductor nanoparticles (NPs) (Fig. 2.21). The researchers showed that through manipulating the protein sequence they could display numerous *histidine* amino acids for targeting and binding metal ions and their subsequent reduction. The researchers were also able to modulate the pores' dimensions, thus hosting NPs with different sizes or shapes.

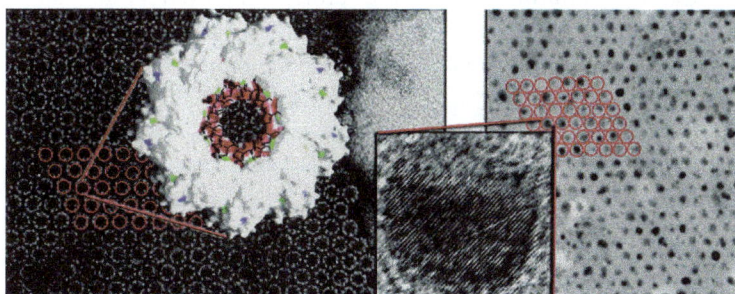

Fig. 2.21: Metal nanoparticles grown in a *chaperonin template*. Ordered array of donut-shaped chaperonins used as a template for growing metal NPs. The high-resolution transmission electron microscopy (HR-TEM) image in the middle demonstrates the formation of a crystalline metallic Ni-Pd NP inside the chaperonin pore. Reprinted with permission from McMillan, R.A. et al., *JACS* **2005** *127*, 2800–2801. Copyright (2005) American Chemical Society.

The expanding field of peptide-based biomimetic materials presents many opportunities as well as technical and conceptual challenges. On the one hand, we can successfully exploit the significant diversity and "firepower" of design principles based upon both covalent bonds and *weak* non-bonding molecular forces guiding self-assembly of individual peptides into high-order structures. However, it should still be emphasized that while our synthetic expertise in manipulating *atoms and molecules* is quite advanced, we are still in the early stages of developing the tool-box for controlling and modulating weak molecular forces in peptide assemblies. Nevertheless, it would probably be safe to predict that our understanding of the fundamental parameters affecting peptide self-assembly would make it possible to generate increasingly diverse and sophisticated peptide-based functional materials.

2.3 Bioelectronics

The use of biomolecules in micro- and nanoelectronics is a recent intriguing develop-ment, part of the broad progress in nanobiotechnology applications. On a biologi-cal level, electron transfer occurs naturally in physiological systems, for example in neural networks. Some research avenues indeed explore the use of biological mol-ecules as actual components in electronic circuits and devices. Moreover, biological molecules and assemblies can be employed as *scaffolding* upon which electronic components can be constructed (discussed for example above in the context of fiber-forming proteins).

Filamentous proteins are promising candidates for nanowire fabrication since they often form nanoscale fibrous configurations with dimensionalities (diameter, length) that conform to potential utilization in nanoelectronic circuitry. **Actin** is an interesting example of a (non-fibrillar) filamentous protein employed for fabrication of conductive nanowires. The core feature of actin which can be exploited for effi-cient metallization and wire formation is its *polymerization* properties. As depicted in Fig. 2.22, an actin chain is produced from polymerization of the actin monomers through an adenosine triphosphate (ATP)-dependent process.

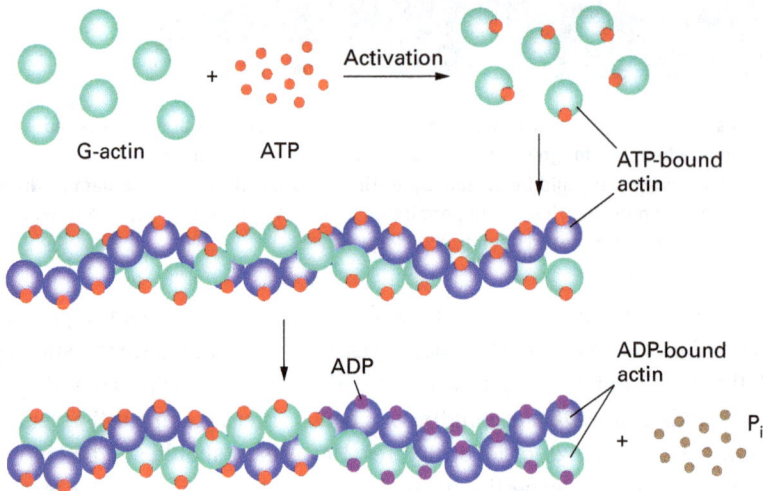

Fig. 2.22: Actin polymerization process. G-actin monomers are polymerized via an ATP-induced process. Following energy uptake during polymerization, the phosphate monomers are released.

The polymerization process for actin biosynthesis can be used as a basis for fabricat-ing metallic filaments. As schematically shown in Fig. 2.23, the "modular" organiza-tion of actin enables using *modified* actin monomers displaying gold colloids (or other metals). The resultant engineered actin filament subsequently serves as a "nucleation platform" on which the surface-displayed metal colloids promote further deposition or reduction of metal ions from solution. This generic approach can be used, in principle, for functionalization of actin monomers (and subsequently the polymerized form) with other types of guest molecules, not just metals.

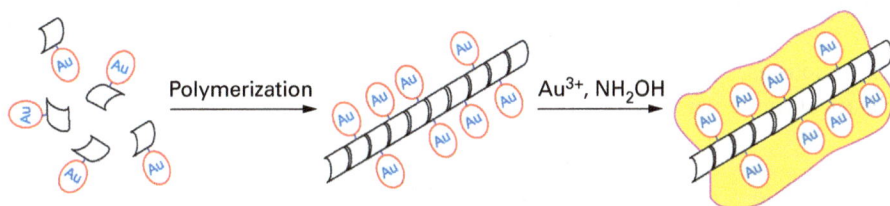

Fig. 2.23: Actin as a scaffold for metal deposition. Actin monomers functionalized with Au colloids are used in the polymerization process, producing gold-displaying actin filament. Subsequent incubation in a solution of gold salt and a reducing agent results in "gold enhancement", i.e. metal deposition upon the actin-displayed nucleation centers.

Actin, in fact, exhibits another important biological property that can be exploited in biomimetic designs – its spatial mobility (or *motility*). Actin motility is extremely important in biology as it is the primary phenomenon enabling cell migration, cell shape changes, response to mechanical stress, and other physiological processes involving mechanical transformations. Molecularly, actin filaments migrate on surfaces comprised of another protein species – myosin – using energy input by ATP molecules. Similar to the actin-metallization process described above, ATP-induced actin-myosin transport can also take place when the actin filaments are *metallized* – allowing in principle to assemble a biomimetic "nano-construction site", in which the metal actin-templated nanowires can be transported upon myosin surfaces from one location to another.

DNA molecules have been also used as templates for nanoelectronic circuitry. Beside the still-unresolved controversy as to whether DNA strands *by themselves* allow electron transport, the use of DNA as a template for conductive nanowires has been one of the earliest manifestations of the promising interface between biology, nanotechnology and electronics. Such applications take particular advantage of two important properties of DNA. First and foremost is the complementary or modular nature of the molecule (which enables "Lego-like" fabrication of the wire template employing the nucleotides as the building blocks). The second useful characteristic of DNA assemblies is the *negative charge* of the molecule, which provides the means for adsorption of metal ions through simple electrostatic attraction, with subsequent reduction into conductive metal wiring which traces the DNA scaffold.

The generic methodology for creating DNA-templated metal nanowires, schematically shown in Fig. 2.24, combines electrostatic attraction of charged metal-containing species, followed by reduction of the metal resulting in metal deposition on the DNA. The first step of DNA metallization involves attachment of metal ions or metal complexes onto the DNA framework, substituting positive counter-ions such as sodium or magnesium that neutralize the phosphate groups on the DNA backbone. The ion substitution process, also referred to as "activation", is followed by metal reduction facilitated by reducing agents co-added to the DNA solution. The result-ant "metal seeds" attached to the DNA strand subsequently serve as crystal nuclea-tion sites which direct further binding and reduction of metals, ultimately yielding continuous assemblies of metals which can function as heat and, more importantly, electrical conductors.

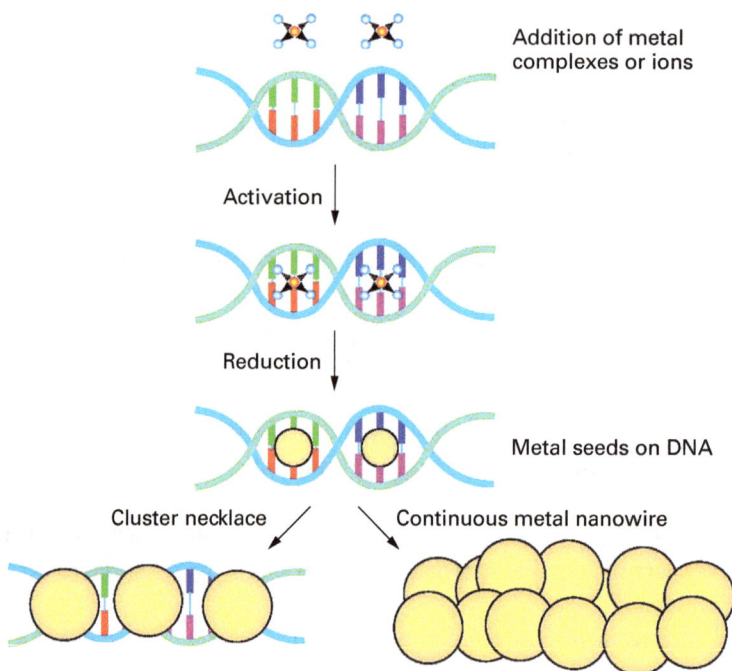

Fig. 2.24: Methodology for creating DNA-templated metal wires. Positively-charged metal ions or complexes are embedded within the DNA strand through electrostatic attraction; addition of a reduc-ing agent to the solution generates metal colloids attached to the DNA framework, which subse-quently constitute seeding or nucleation sites for further metal deposition. The final outcome of the process is a continuous (conductive) metal nanowire.

A classic experiment by E. Braun and colleagues at the Technion, Israel, was among the earliest demonstrations of DNA-based "nanoelectronics" (Fig. 2.25). The researchers first covalently attached two short single-stranded DNA to two adjacent gold electrodes. These strands served as "anchors" for connecting a single-strand DNA filament between the two electrodes, which bound to the two anchors through base-pair complementa-rity in its two ends. The DNA filament was subsequently incubated with silver ions to replace the native sodium counter-ions on its backbone, followed by reduction to silver metal seeds through addition of hydroquinone, a commonly used reducing agent. The Ag seeds promoted further accumulation and reduction of metal ions, resulting in formation of a continuous conductive silver wire.

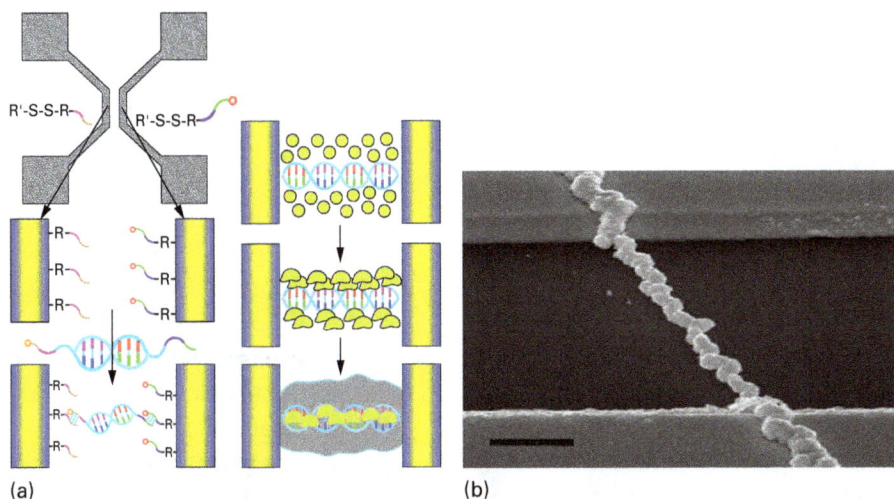

Fig. 2.25: Conductive metal nanowire assembled from DNA. **(a)** Schematic depiction of experiments in which a complementary DNA strand was used as a template for metal deposition, resulting in a conductive wire, see details in text. **(b)** SEM image of a DNA-templated metal wire assembled between the electrodes. Image courtesy of Prof. E. Braun, Technion, Israel.

Other biological molecules have been used in bioelectronic applications. *Alginate*, for example, a polysaccharide extracted from brown algae and widely used in biomedical and drug design applications, has been examined as the "host matrix" for metal ions in lithium-ion batteries. The effect of the alginate scaffold was to significantly enhance battery performance – improving recharging capacity and increasingly stability significantly as compared to the conventional "non-biological" counterparts. The effect of alginate in these battery systems is not entirely clear, however it is believed that the porous polysaccharide framework provides stable scaffolding, and enables efficient diffusion of the metal (lithium) ions.

2.4 Microorganism-synthesized biomimetic materials

Historically, chemistry has provided the most common and robust routes for creating new materials, developing new synthetic processes, or fashioning new functionalities to existing materials. Since the early days of scientific and technology research, however, there has been interest in using *biological systems* to accomplish the same goals through scalable, cost-effective, and easy-to-implement processes. Beside the abundance and diversity of biological systems, their structures, and reaction pathways, there are distinct *practical* advantages for using biology for materials design. Specifically, the synthesis conditions are generally milder in terms of temperature, pressure, solution conditions, and more environmentally-friendly compared to chemical methods which might involve toxic organic and inorganic reagents. Also, biological ingredients are in many cases abundant, and thus inexpensive and relatively easy to obtain.

A quite fascinating subset of nature-inspired materials science has been the use of *entire organisms* as templates, platforms, or synthetic vehicles for production of materials and device components. The majority of such efforts have focused on simple microorganisms, such as viruses and bacteria, although more complex organisms have been utilized as well (please refer to Chapter 5 for a more thorough discussion of *biomineralization* processes in living creatures). Probably the most common cases have been the use of microorganisms as *physical templates* for assembling inorganic structures and functional materials.

Viruses constitute a vast and diverse order of microorganisms, which unlike their (well-deserved) bad publicity have continued to inspire biomimetic R&D – from new therapeutic avenues all the way to biomimetic functional materials. The two primary features considered attractive in viral-based materials development are the relatively small and easily-manipulated viral genome, which allows modulating the virus properties and protein composition, and the diversity of nanometer-size viral structures and morphologies that make viruses convenient building blocks in materials design. *Higher-order* organization of viral populations is another noteworthy parameter, presented below in specific examples, which has contributed to construction of hierarchical biomimetic assemblies.

Rod-shaped viruses are a unique subclass which exemplifies many of the features outlined above. The *M13 filamentous bacteriophage*, described in more detail in Chapter 5 in a context of biomineralization applications, is a prominent representative of this diverse class of viruses. Several features need to be particularly emphasized. First is the intrinsic *anisotropy* of the viral structure (e.g. the elongated rod morphology) which can be exploited for creation of liquid-crystalline alignment of viral particle populations in certain conditions of the aqueous solution – concentration, ionic strength, and pH. Second, manipulation of the M13 genome allows expressing varied functional moieties and recognition elements that can be displayed on the viral surface. These binding sites essentially enable attachment of various non-biological functional substances to the viral particle – inorganic crystallites, metal nanoparticles, fluorescent "quantum dots", and others (Fig. 2.26). The attachment of foreign objects and molecules, combined with the macroscopic alignment of the virus rods, make filamentous bacteriophages a useful platform for creating higher-order three dimensional structures. Another noteworthy point is the fact that M13 is easily produced in large quantities by bacterial cells, providing a natural mechanism for large-scale manufacturing of viral particles for practical applications.

3-D QD alignment   3-D nanotube alignment

Viruses as ...   Building blocks

2-D electrodes   3-D devices

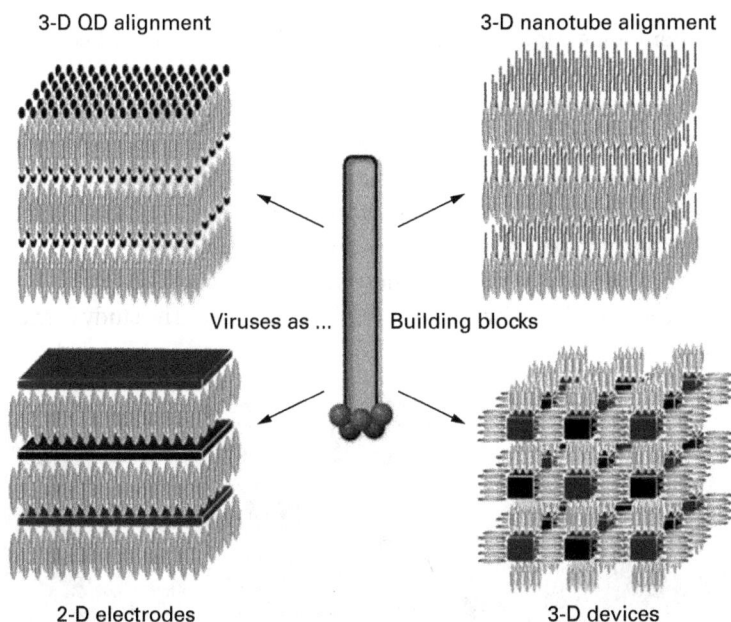

Fig. 2.26: Engineered filamentous bacteriophages as templates in materials science. The rod-shaped M13 filamentous bacteriophage (middle) can be used as a template for creating a variety of materials, taking advantage of both binding of distinct molecular species through engineering the coat proteins (spheres at the tip of the bacteriophage rod), as well as the macroscopically-ordered organization of the bacteriophage particles. Reprinted from Flynn, C.E. et al., Viruses as vehicles for growth, organization and assembly of materials, *Acta Materialia* **2003** *51*, 5867–5880, Copyright (2003), with permission from Elsevier.

The abovementioned factors – genetic engineering of phage proteins and macroscopic alignment of phage particles – can be exploited for other innovative applications. S.-W. Lee and colleagues at the University of California, Berkeley, have recently created a piezoelectric device combining the molecular properties of the individual phage particles with their liquid-crystalline organization. Specifically, the researchers reasoned that the electrical dipole moment of the filamentous phage can effectively produce electric current upon mechanical motion (i.e. *piezoelectricity*). Indeed, alignment of multiple phage particles in a liquid-crystalline film created a simple piezoelectric device that both produced significant electrical currents as well as exhibited *reversibility* of phage orientation after removal of the mechanical stimuli. Further control of the piezoelectric properties of the biomimetic device, i.e. the current generated by the phage assembly, was achieved through genetically engineering a display of charged amino acids upon the phage surface. This manipulation modified the electrical dipole and consequent current generated by the mechanical motion.

Bacteria are sometime referred to as the "workhorse" of biotechnology, as these microorganisms are widely used for expression and large-scale production of numerous proteins and other biological molecules. Bacterial cells, in fact, have been also used as interesting vehicles for production and processing of minerals, inorganic materials, and nanoparticles. In such applications, the biological machinery of the bacterium is harnessed to manufacture non-biological assemblies through clever manipulations. Furthermore, beneficial environmental applications have been demonstrated from "hijacking" bacterial biochemical processes for non-native applications towards processing and removal of toxic substances.

The diversity of bacterially-induced NP synthesis routes is exemplified in Fig. 2.27 depicting different Au NPs produced by filamentous cyanobacteria. The study carried

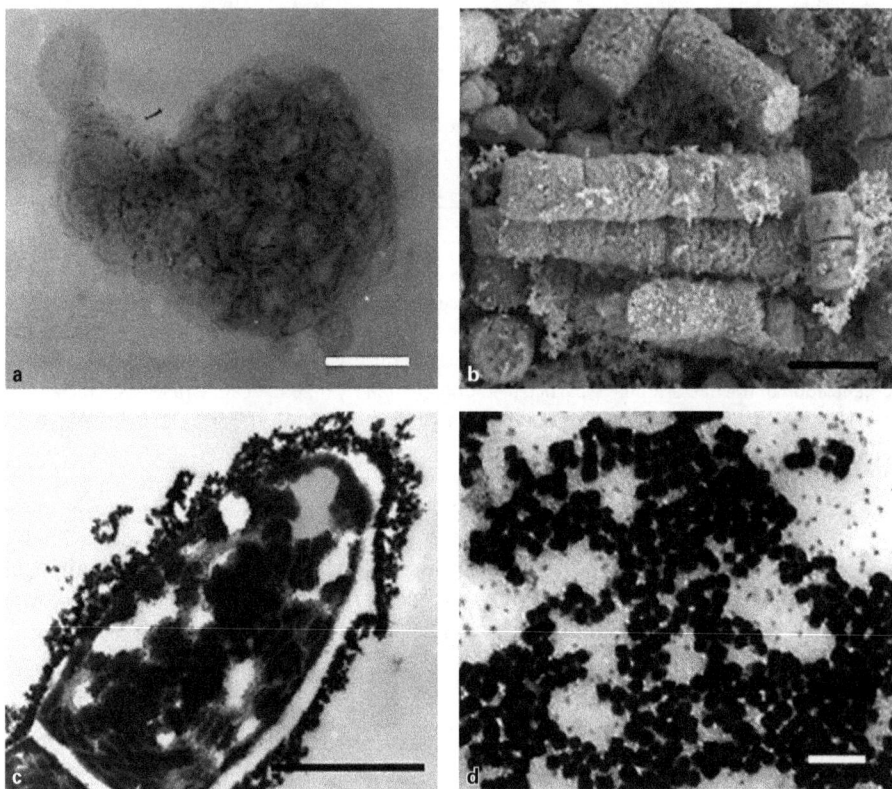

Fig. 2.27: Biogenic synthesis of gold NPs by bacteria. Electron microscopy images of cyanobacteria cells grown in the presence of $Au(S_2O_3)_2^{3-}$, highlighting distinct NP structures, sizes, and morphologies: **(a)** Thin nanoscale gold particles on membrane vesicles formed upon the external cell surface at 25 °C; **(b)** SEM micrograph of a cyanobacteria cell encrusted with gold and gold-sulfide NPs at 100 °C; **(c)** TEM micrographs of a thin section of cyanobacteria cells with NPs of gold and gold-sulfide deposited on the cell wall and inside the cell at 100 °C; **(d)** cubic Au NPs precipitated in solution at 100 °C with smaller particles of gold sulfide (< 10 nm) in the background. Scale bars in A–D are 0.5, 2, and 1 μm and 100 nm, respectively. Reprinted with permission from Lengke, M.F. et al., *Langmuir* **2006** *22*, 2780–2787. Copyright (2006) American Chemical Society.

out by M. Lengke and colleagues at the University of Western Ontario, Canada, demonstrated that incubation of the bacteria with several gold complexes yielded a remarkable range of crystalline gold aggregates having different sizes and morphologies, distributed both inside the cells, encrusted on the cell wall, as well as dissolved in solution. Such studies attest to the capability of bacteria to initiate and control complex chemical reactions and crystallization pathways that might be harnessed for assembly of metal NPs in desired shapes and morphologies.

Several studies have shown that metal nanoparticles (NPs, mostly gold and silver) could be produced by different bacterial species. Most of these processes rely upon intrinsic mechanisms in which specific proteins expressed by the bacterium are responsible for producing metal NPs through metal salt reduction. Several reports have described bacterial over-expression of proteins associated with reduction pathways in the presence of high concentrations of metal salts, resulting in production of large quantities of stabilized metal NPs. Furthermore, bacteria could supply hydrophobic substances, including hydrophobic peptides and lipopeptide conjugates, which function as stabilizers of metal NPs. Such additives have generally crucial roles for solubilizing and preventing aggregation of synthesized NPs. *Bacillus subtilis*, for example, has been shown to cap intracellular-produced Au NPs with surfactin, a known amphiphilic lipopeptide.

Biogenic generation of *silver NPs* has attracted particular interest, primarily due to the expanding use of these NPs as potent antimicrobial agents. While production of Ag NPs through chemical means is well-developed, microorganisms have been also recruited to produce such aggregates; the advantages of these approaches lie in their being environmentally-friendly, utilize widely available sources (when using abundant microorganisms such as algae), and the recognition that these are high-yield, mostly inexpensive processes. Specific examples include *fungi*, shown to produce Ag NPs from abundant precursors such as silver nitrate through different (and not completely understood) reduction pathways. In many cases fungi-mediated processes yielded a relative narrow distribution of Ag NP diameters. Particularly important, the fungi-synthesized NPs were readily available and exhibited pronounced toxic effects against bacteria. Interestingly, bacterial cells themselves were also used as vehicles for production of toxic Ag NPs. *Shewanella oneidensis*, for example, is a known metal-producing bacterium found to synthesize toxic-level concentrations of Ag NPs. Indeed, the mechanism or mechanisms enabling this bacterial species to tolerate the toxicity of self-synthesized NPs is still quite a mystery.

Magnetic nanoparticles (MNPs), primarily comprising iron oxide (Fe_3O_4), have attracted interest as useful constituents in the emerging biomedical field of "theranostics" (therapy + diagnostics), besides the more traditional applications of data recording and storage, industrial catalysis, and others. Similar to metal NPs discussed above, chemical synthesis of MNPs can be costly and environmentally-harmful, providing an incentive for development of alternative microorganism-based approaches for production of these molecular aggregates. *Microbiological* routes for magnetite-based MNP production include both *extracellular* and *intracellular* processes, relying on the discovery of diverse bacterial families that extract iron ions from their environments for use in their metabolic processes.

Extracellular iron reduction pathways, primarily utilizing redox-active enzymes secreted by bacteria to the extracellular space, have been demonstrated for bacteria such as *Shewanella* and *Geobacter*. These bacterial families use respiratory processes which could yield large quantities of MNPs through reduction of poorly soluble Fe(III) salts. Remarkably, *intracellular* production of MNPs has been observed by magneto-tactic bacteria such as *Magnetospirillum*, which use specialized organelles denoted 'magnetosomes' to capture and internalize magnetite from the cell environment, and produce magnetic oxide aggregates inside the bacterial cell. Magnetosomes might have in fact an important biological function – bestowing sensitivity to external magnetic fields and consequent navigation of the bacteria.

Bacterial-assisted production of MNPs can be further tuned through addition of metal dopants, such as cobalt, nickel, zinc, and others. These metal ions are occasionally used in MNP synthesis to modulate the magnetic and physical properties of the particles. Interestingly, *in vivo* metal doping of MNPs assembled in bacterial magnetosomes has been reported, impacting particle sizes and magnetic properties. High concentration loading of the dopants in those systems could be often achieved simply through addition of the metal ions into the bacterial growth medium. In addition to metal NPs and MNPs, bacterial cells and their metabolic pathways have been exploited for manufacturing other inorganic substances, including oxides and chalcogenides (ionic compounds containing sulfide, selenide, and telluride ions). Metal chalcogenide NPs such as CdS have attracted great interest in recent years due to their unique semiconducting and optical properties. Biogenic synthesis of chalcogenide NPs has been achieved by varied bacterial species, mostly pathogenic, employing intrinsic metabolic routes to reduce sulfur.

Bacteria and bacteria-associated processes have been employed for seemingly tangential applications. *Electrically-conductive* bacterially-produced molecules are a notable example. Conceptually similar to conductive polymers that are broadly used in microelectronics and optics, conductive biomolecules that could have practical applications represent an active field of research at the interface between biology and electronics. Naturally-produced conductive bacterial "nanowires" were identified several years ago in strains for which electronic transport is a fundamental metabolic pathway. Bacteria such as *Geobacter sulfurreducens* and *Shewanella oneidensis* utilize fibrous proteins called *pilins* (e.g. pili-forming proteins) to produce elongated filaments that enable electron transfer (Fig. 2.28) (*Geobacter sulfurreducens*, in fact, utilizes the conductive properties of the pilin filaments to transfer electrons to iron oxide in soil – a process akin to "breathing").

In some instances, *networks* of pilin filaments have been assembled and their charge transfer properties investigated. An interesting approach for controlling pilin assembly has been the use of bacterial *biofilms* grown between electrodes. Biofilms

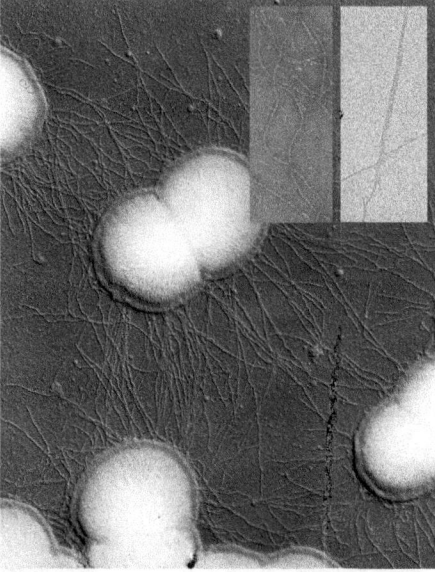

Fig. 2.28: Bacterially-produced conductive protein wires. Thin *pilin* filaments extended from *N. gonorrhoeae* bacterial cells. Reprinted from Craig, L. et al., Type IV Pilus Structure by Cryo-Electron Microscopy and Crystallography: Implications for Pilus Assembly and Functions, *Molecular Cell* **2006** *23*, 651–662. Copyright (2006), with permission from Elsevier.

comprise a highly rigid and degradation-resistant matrix formed by proliferating bacteria within which bacterial cells are eventually immobilized. It is notable that scientists could detect conductivity in such systems both when the biofilm assemblies comprised actual living cells (which produced pilin filaments), but also when films were made of just networks of dried pilin filaments without live cells present.

In a broad sense, while conductive nanowires are believed to be synthesized by bacteria to carry out specific, essential biological functions, such wires might open the door to diverse non-biological applications, such as environmental cleanup (through catalyzing redox reactions designed to eliminate toxic materials), bioelectronics, and biosensing. Perhaps the most intriguing feature of bacterially-produced filaments such as pilin nanowires is the fact that conductivity is achieved even though these assemblies are not metallic but rather consist of only biological molecules.

Bacterially-produced *S-layer proteins* are also interesting candidates for the generation of electrically-conductive networks. Bacterial surface layer (S-layer) proteins form ordered networks within the bacterial cell wall, and constitute an essential element in protection and communication of the cell with its environment. The remarkable periodic structure of the S-layer, dramatically visualized in Fig. 2.29, is a particularly desirable feature in the context of materials design in light of the patterned, symmetrical surface structure which allows external deposition of substances. Furthermore, experimental schemes enabling *laboratory* (i.e. *in vitro)* growth of S-layers could assist promoting their use as actual platforms in materials synthesis.

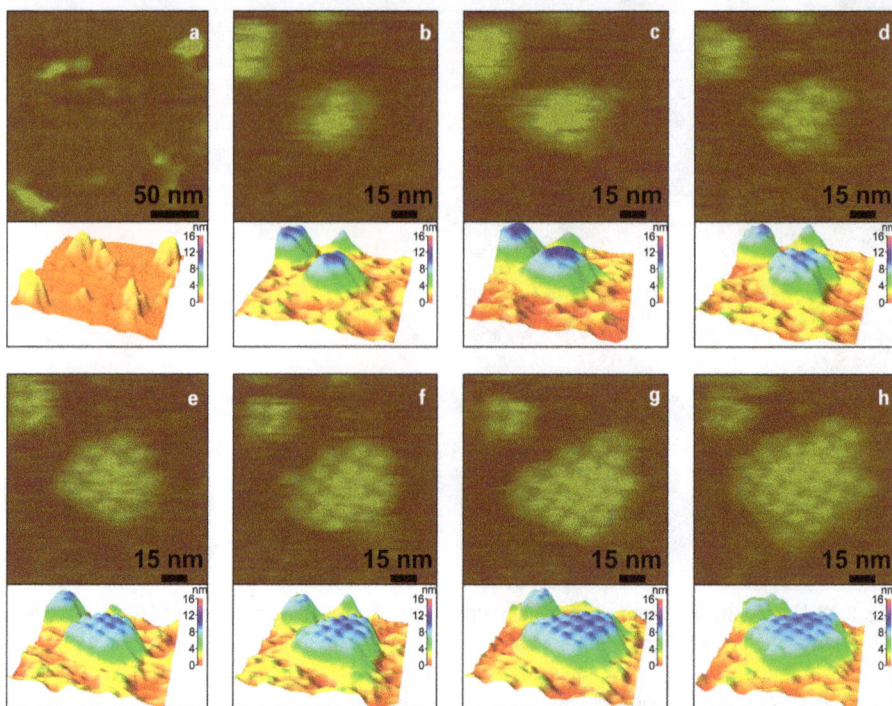

Fig. 2.29: Crystalline S-layer growth. *In-situ* atomic force microscopy (AFM) images and surface plots showing the sequential assembly of an S-layer protein array on a surface. Incubation times were **(a)** 0, **(b)** 62, **(c)** 65.8, **(d)** 70.3, **(e)** 88.5, **(f)** 95.5, **(g)** 102.6, and **(h)** 116.8 min. Images courtesy of Dr. J. De Yoreo, Lawrence Berkeley Laboratory, Berkeley, U.S.A.

Building upon the periodic structure of the S-layer, it can function as a *template* for placement of metal NPs (or other inorganic species such as semiconductor "quantum dots", silica microspheres, and others). The NPs can be either covalently attached to the protein surface, or immobilized through non-covalent forces; in either case the NP "superlattice" might exhibit unique optical and spectroscopic properties associated with the periodicity. An important aspect of this biomimetic field is that S-layer proteins can be isolated and recrystallized, thereby forming periodic superlattices in artificial, non-biological environments. Fig. 2.30 presents an example of an Au NP superlattice assembled by D. Pum and U. Sleyter at the Universität für Bodenkultur Vienna, Austria, using a purified, recrystallized S-protein layer on a silicon wafer substrate.

As this chapter has shown, the scope of applications and practical targets for biomimetic functional materials is broad – from biotechnology, to nanotechnology, to "hard-core" materials science. While the potential certainly exists for bio-inspired and biomaterials-induced functional materials, actual practical applications with near-term commercial horizons are still rare, reflecting the hurdles that still need to be overcome. Applications such as catalysis, tissue engineering, and sensing appear to be the more readily implemented in a commercial context due to the more direct relationship between the biological entities employed and the envisaged technical usage. Other ambitious biomimetic systems, such as bioelectronic devices, magnetic storage, and photonics, might take more time for introduction of viable technological alternatives, partly due to the rather "entrenched" nature of existing technologies.

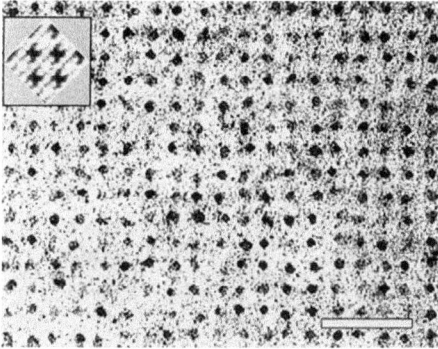

Fig. 2.30: Au NP superlattice formed on recrystallized S-layer protein. Electron microscopy image of a periodic Au NP structure formed on an S-protein template having square lattice symmetry. The mono-disperse NPs are embedded within the S-layer pores. The inset shows an image of the *native* S-layer at the same scale, confirming that the S-layer structure was retained after NP adsorption. Scale bar is 50 nm. Reprinted from Pum, D. and Sleytr, U.B., The application of bacterial S-layers in molecular nanotechnology, *Trends in Biotechnology* **1999** *17*, 8–12. Copyright (1999), with permission from Elsevier.

2.5 Biomimetic robotics

Robots, developed and employed in varied applications for decades, have become conduits for delegating diverse human tasks to machines Biomimetic robotics constitute an exciting frontier in multidisciplinary research combining human physiology, engineering, psychology, and other fields. In many respects, the conceptual and technical challenges in robotics stem from the complex practical requirements – mechanical flexibility together with robustness, long-term stability, fast response times, comfortable and easy implementation at the human interface. Furthermore, the emerging field of "microrobotics" calls for construction of tiny robots that could perform tasks in miniaturized environments, for example inside the human body.

Varied strategies and technologies have been devised to address the above challenges, drawing on expertise from engineering, material sciences, product design, and all the way to psychology and human perceptions. Nature and biological systems have contributed to this growing field, particularly in the emerging "soft robotics", i.e. the use of "soft" biomimetic materials to accomplish functions that are difficult to achieve using rigid components. "Soft robots" may be also energy-efficient, lightweight, and biocompatible (essential in applications requiring interface with living systems or the human body).

Among the more active research areas in soft robotics have been the construction of bio-hybrid systems that mimic mechanical traits of certain organisms. The thrust of such research efforts has been to accomplish desired mechanical properties through combining synthetic constructs with actual living cells. Fig. 2.31 presents an example of this concept. In that study, S.R. Shin and colleagues at Harvard Medical School aimed to produce an artificial "swimmer" that mimics the motion of a stingray. Stingrays exhibit a flexible cartilaginous structure and achieve directed motion through contraction and release of muscle tissues within the cartilage skeleton (Fig. 2.31a).

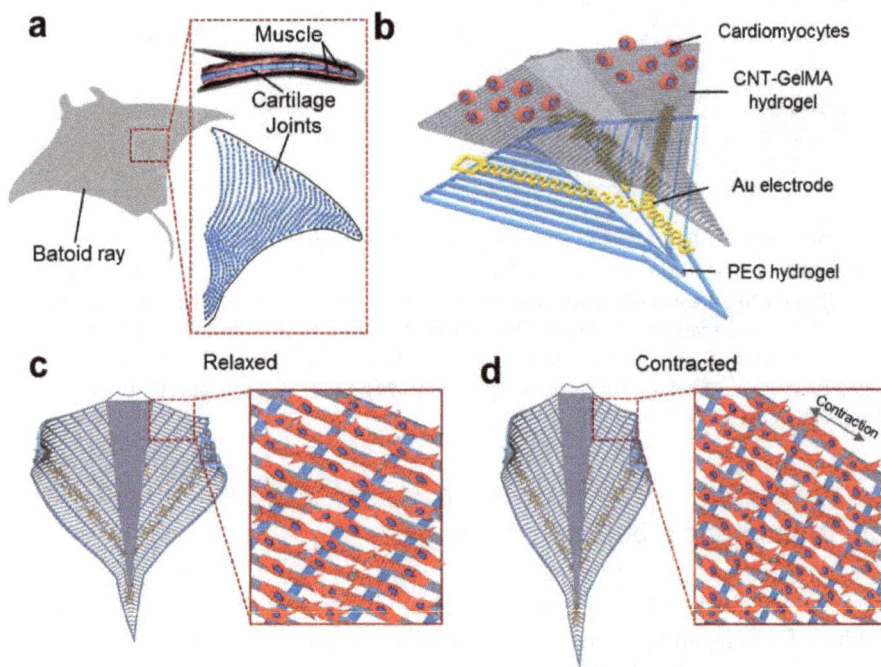

Fig. 2.31: Biomimetic swimmer based on stingray muscle action. (a) Schematic drawing of the stingray muscle organization, and the biomimetic synthetic construct **(b)**. **(c–d)** Motion of the bio-inspired stingray. Passage of electrical current in the Au electrodes causes contraction of the cardiomyocytes; the soft hydrogel facilitates the macroscopic contraction/release of the entire robot and consequent forward motion. Reprinted with permission from Shin et al., Electrically Driven Microengineered Bioinspired Soft Robots, *Advanced Materials*. 2018, *30*, 1704189, Copyright (2018) John Wiley and Sons.

The bio-inspired synthetic stingray is schematically depicted in Fig. 2.31b. Specifically, the artificial stingray comprised four distinct components, exhibiting complementary functions that together facilitated coordinated swimming motion. The first layer was a polyethylene-glycol (PEG) hydrogel, serving as a mechanically stable and biocompatible framework mimicking the cartilage skeleton of the stingray. Further embedded within the hydrogel was a flexible gold microelectrode network designed to efficiently propagate electrical signals triggering the biomimetic "muscle" contraction. Another gel composite layer, comprising carbon nanotubes (CNTs) and a porous gelatinous matrix (gelMA hydrogel), served as a biocompatible matrix for hosting mouse cardiomyocytes – heart muscle cells responsible for heart contraction and relaxation cycles through coordinated shape transformations. Importantly, as shown in Fig. 2.31, the hydrogel micropattern dictated alignment of the cardiomyocytes. Accordingly, the cooperative contraction of the cells, induced both by externally applied electrical pulses as well as through spontaneous "beats", results in macroscopic shape transformation of the stingray and its concomitant forward motion in water.

Varied studies have attempted to borrow mechanical traits and functionalities of organisms for therapeutic applications. Fig. 2.32 illustrates an innovative drug delivery strategy based upon a microdevice mimicking the action of the hookworm (Fig. 2.32A), a parasite latching onto the gastrointestinal (GI) tract using its sharp protruding "teeth" which penetrate through the protective mucus layer. This insertion mechanism aids the hookworm to withstand the considerable mechanical forces of the intestine. The biomimetic star-shaped microdevice depicted in Fig. 2.32B-D, developed by F.M. Selaru and colleagues at Johns Hopkins University, was termed "theragripper" and designed to "grip" the GI tract via multiple sharp bendable microtips.

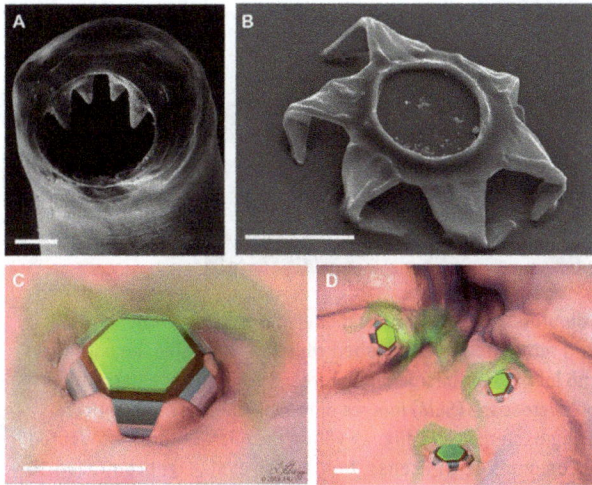

Fig. 2.32: Micro-gripper inspired by a hookworm. **(A)** Electron microscope image of the hookworm showing the sharp teeth by which the hookworm latches onto the epithelial layer of the intestine. **(B)** Electron microscope image of the biomimetic "theragripper" highlighting the sharp microtips. **(C–D)** Schematic illustration of the theragrippers embedded in the walls of the gastrointestinal tract (GI), injecting their drug cargo. Adapted with permission under the Creative Commons Attribution License 4.0 (CC BY), from Ghosh et al, Gastrointestinal-resident, shape-changing microdevices extend drug release in vivo, *Sci. Adv.* 2020; 6: eabb4133

The theragripper was made of a thin, shape-changing metal composite comprising gold and chromium, coated with a thermosensitive wax. Following rectal application of the devices, the wax is softened in the intestinal environment due to equilibration with the body temperature, giving rise to spontaneous folding, or contraction, of the microtips (Fig. 2.32B) and subsequent penetration through the mucosa (Fig. 2.32C-D). The GI-inserted theragripper then slowly releases its drug cargo into the GI tissue environment. A particular advantage of this biomimetic drug delivery platform lies in the intrinsic thermo-responsive mechanism, allowing controlled latching of the theragrippers specifically on the intestinal walls rather than immediately upon uptake.

Researchers have also harnessed microorganisms to perform mechanical work in biohybrid microrobots. Fig. 2.33 depicts a micromotor utilizing live bacterial cells as motion agents, akin to the ancient flour grinding devices operated by cows walking in circles. In that study, R. Di Leonardo and colleagues at the University di Roma, Sapienza in Italy, designed a microscale rotating unit containing microchambers that can accommodate swarming bacteria. The bacteria are directed to the microcompartments via the shaft structure (their trajectory is shown in the broken white line, Fig. 2.33a,b). Upon entering a microchamber (magnified image in Fig. 2.33d), the torque induced by a swimming bacterium results in the circular motion of the rotor. Importantly, the researchers fabricated an array of rotors (i.e. Fig. 2.33c) facilitating, in principle, macroscopic mechanical motion through operating large numbers of rotors.

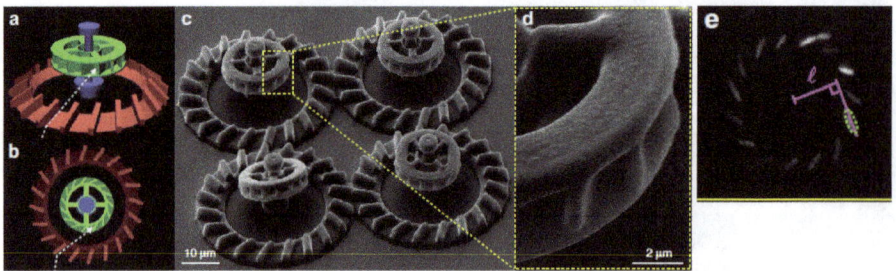

Fig. 2.33: Bacterially powered micromotors. (**a, b**) Schematic model of the micromotor, showing the ramp (red), motor axis (blue), and rotor (green). The dashed white line indicates the pathway of a bacterial cell guided by the ramp into a rotor microchamber, resulting in the circular motion of the rotor. (**c, d**) Scanning electron microscope (SEM) images of the micromotors. (**e**) Optical microscopy image of the rotor showing bacterial cells aligned within the microchambers. Adapted with permission under the Creative Commons CC BY license, from Ghosh et al., Light controlled 3D micromotors powered by bacteria, Nature Communications, 8:15974.

The fluorescence microscopy image in Fig. 2.33e confirms the experimental concept, depicting fluorescently labeled rod-like bacterial cells captured within all available microchambers in a rotor. Indeed, the simultaneous motion of the bacteria resulted in continuous motion. Furthermore, the bacteria selected in the experiment were genetically engineered to express a light-driven protein "pump", creating an ionic gradient across the cell membrane thereby enhancing cell motility. As such, the bio-hybrid micromotors could be optically modulated, both through tuning the light wavelength (the ion pump was triggered by green light) and through modulating light intensity.

In some instances, actual tissues have been incorporated within artificial or synthetic constructs to create interesting soft robots. Construction of biohybrid micro-actuators is a case in point. Muscle tissues constitute useful actuators, displaying high energy conversion efficiency, functioning in microscale environments (even a few muscle cells can contract and expand upon stimulation) and are relatively resilient in non-physiological conditions. Fig. 2.34 illustrates the construction of fast-moving muscle-powered microrobot employing insect dorsal vessel (DV) – the insect's heart tissue – as a bio-actuator. The research team of K. Morishima and colleagues at Osaka University, Japan, utilized the DV for powering the movement of the microrobot since such muscles are more robust and less prone to degradation in comparison to mammalian muscle tissues.

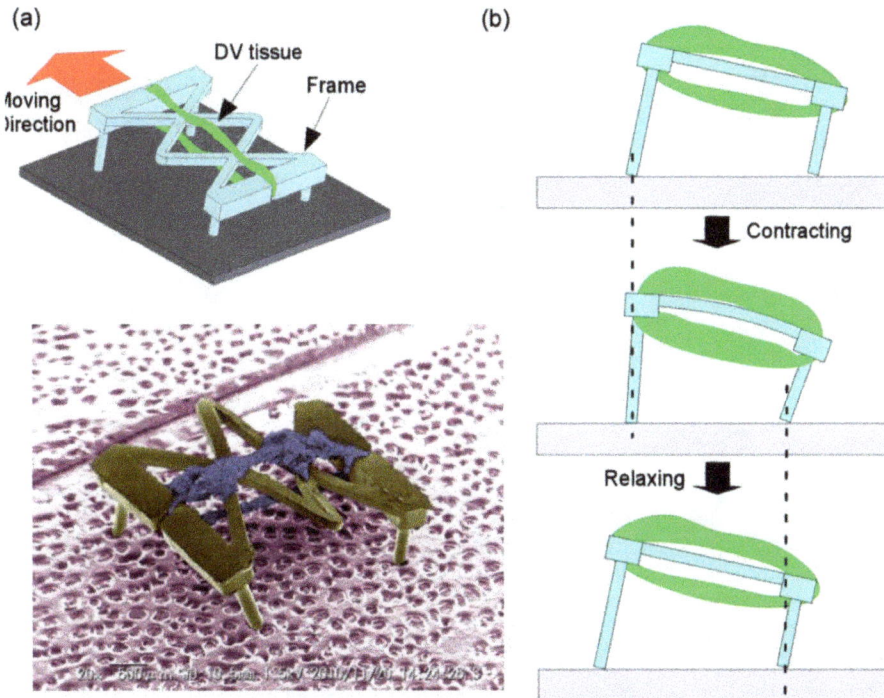

Fig. 2.34: Insect muscle-powered moving microrobot. **(a)** Design and forward motion of the microrobot, comprising a soft polymeric frame and the dorsal vessel (DV), contractible insect heart muscle. **(b)** Contract/release of the muscle induces strain to the polymer frame, affecting forward motion. Adapted with permission from Akiyama et al, Rapidly-moving insect muscle-powered microrobot and its chemical accelerations, *Biomed Microdevices* (2012) 14:979–986, Copyright (2012) Springer-Nature.

The muscle-powered microrobot design is shown in Fig. 2.34a. A flexible soft polymer four-legged frame comprised the microrobot "body", with the insect (moth larvae) DV muscle wrapped around the frame in the middle. Importantly, the two front legs were longer than the rear legs providing the movement directionality. As schematically outlined in Fig. 2.34b, when the muscle contracted, the friction between the surface and two front legs was larger than the corresponding friction of the hind legs, consequently forcing their forward movement. The opposite occurred upon muscle relaxation – stronger friction force applied to the rear legs compared to the front legs pushing them forward. The spontaneous contraction/relaxation of the DV muscle facilitated the continuous movement of the microrobot. Interestingly, the researchers have shown that the addition of a cardioactive peptide to the medium solution of the microrobot dramatically accelerated the muscle contraction/relaxation frequency and concomitant motion, achieving spontaneous movement of several millimeters by the microrobot.

3 Biomimetic surfaces

Surfaces are of foremost importance in natural systems and materials design alike, since a surface is where most physical, chemical, and biological processes and reactions take place. Attempts to modulate surface properties with the aim of mimicking natural phenomena are perhaps as old as civilization; from administering oil to the surface of wooden planks in ships to reduce friction with water to the deposition of poison on arrowheads, numerous inventions and innovations have aimed to endow surfaces with desired properties through modifying their structures, compositions, and molecular organizations. Contributions to this field have gone both ways: bio-inspired materials and synthetic interfaces in which the biological components confer useful properties, but also biological interfaces in which artificial, synthetic elements are designed to alter or enhance the biological properties.

Biological surface structures, compositions, and configurations have evolved for millions of years, conferring specific functionalities to the biological entities displaying them. In particular, *physically patterned* surfaces have attracted interest both from a fundamental scientific perspective but also as inspiration for synthetic surfaces exhibiting distinct properties. Examples of the unique roles of organized, patterned surface structures abound in nature: the anti-reflective properties of a moth's eye are ascribed to nanostructured surface "bumps", the grooved pattern of a shark's skin that minimizes flow resistance, and the lamellae on the gecko's feet that allow it to cling to smooth glass walls and rough, rocky surfaces.

3.1 Adhesion and wetting

The gecko's feet (Fig. 3.1) present perhaps the most celebrated biological-functional surfaces, inspiring many biomimetic designs. The adhesion interface on the gecko's foot consists of a hierarchical structure of lamellae—hair-like "setae" that are further split at their tips into miniscule "spatulae" 100 to 300 nm in diameter. This organization allows adaption to virtually any surface condition and roughness. The "pull-off force" increases with the number of contacts made between the spatulae and surface, bestowing the gecko with a sophisticated mechanism for attachment, detachment, and movement on smooth surfaces.

Even though the extraordinary stickiness of the gecko's feet have amazed humanity for millennia and attracted scientific research for more than a hundred years, the apparent breakthrough in understanding its molecular origins occurred less than 15 years ago, when R. Full and colleagues at the University of California, Berkeley appeared to decipher the micromechanical factors contributing to the gecko's feet adhesiveness. Specifically, they proposed that the extraordinary adhesion is traced to the van der Waals (hydrophobic) forces between a surface and the millions of tiny

https://doi.org/10.1515/9783110709490-003

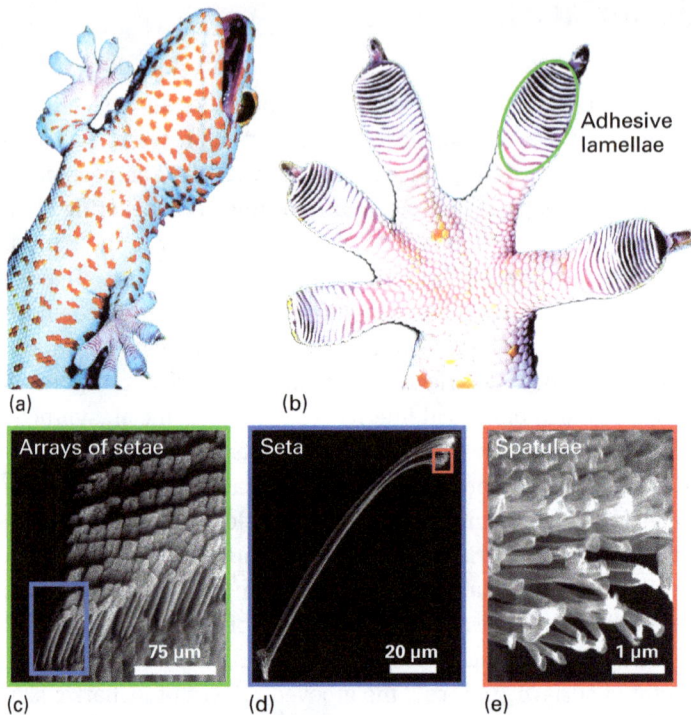

Fig. 3.1: Hierarchical structure of the gecko's foot. **(a)** Tokay gecko climbing on a smooth glass surface. **(b)** magnified gecko's foot showing the adhesive lamellae; **(c)** SEM image showing the array of setae comprising the lamellae; **(d)** single seta, and **(e)** nanoscale spatulae at the tip of the seta. Reprinted with permission from Hansen, W. R. and Autumn, *K. PNAS* **2005** *102*, 385–389. Copyright (2005) National Academy of Sciences, U.S.A.

setae coating the gecko's feet. Furthermore, slight modulation of the *orientation* of the setae induces significant reduction of the adhesive forces and consequent detachment. Intriguingly, the same parameters responsible for the highly effective attachment of the gecko to diverse surfaces have been shown to contribute to remarkable *self-cleaning* properties, e.g. removal of small particulates stuck to the gecko's setae.

Elucidating the hierarchical structures and multi-contact binding underlying the extremely effective surface adhesion of the gecko and other organisms has opened new biomimetic avenues for surface design. Indeed, it is now accepted that surface phenomena such as adhesion, de-wetting (water resistance), self-cleaning, and others are, on a fundamental level, manifestation of *physical forces* cleverly modulated by nature rather than originating from the presence of specific chemical substances. Accordingly, to reproduce these surface properties, the researcher or engineer basically needs to successfully mimic the relevant microscopic configurations, such as the multi-contact phenomena in the gecko's feet, without resorting to complex chemical formulations.

Building biomimetic surfaces designed to achieve effective adhesion through modulation of surface morphology has been widely pursued with varying degrees of success. K. Suh and colleagues at Seoul National University, Korea have devised, for example, a "dry" adhesive patch which, conceptually similar to the gecko's foot, employed a dense array of "micropillars" to achieve optimal adhesion to the skin (Fig. 3.2). The patterned surface was fabricated from soft polydimethylsiloxane (PDMS), which provided high intrinsic flexibility to the patch thus securing greater contact area with the skin and consequent enhanced adhesion. Fig. 3.2 also nicely shows the favorable distinction of the dry patch not just in maintaining strong adhesion while applied, but particularly the much smoother *peel-off* of the patch from the skin compared to a conventional patch. Indeed, this phenomenon is also intimately related to the micropillar structure for which slight tilting affects efficient disengagement from the surface.

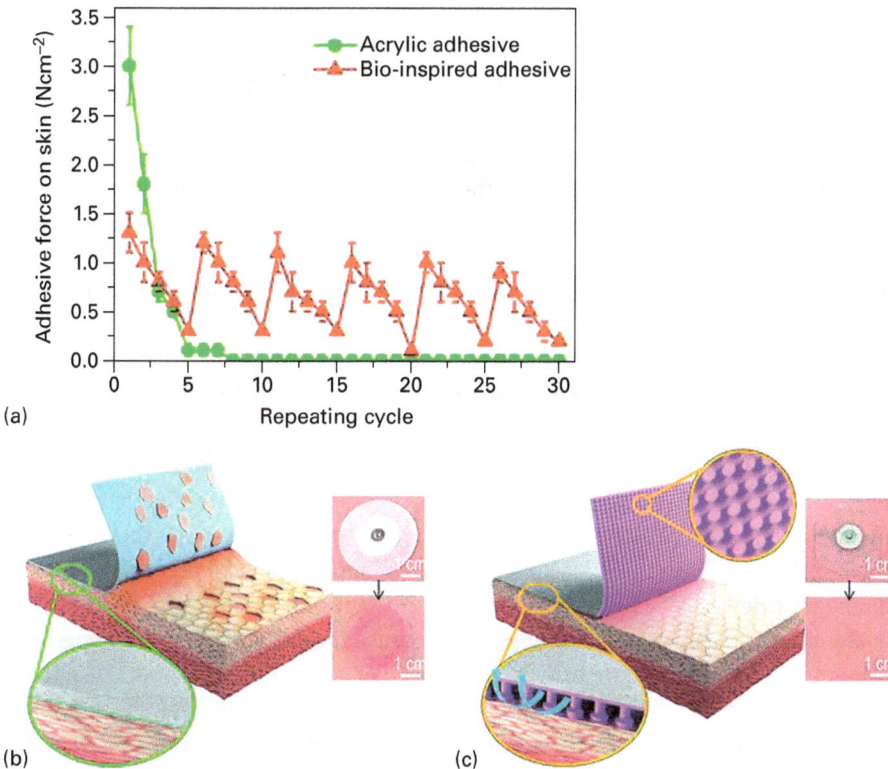

(a)

(b)　　　(c)

Fig. 3.2: Multi-contact adhesive patch. Multi-pillar surface morphology of an adhesive patch provides superior therapeutic properties. **(a)** Retaining of adhesive forces upon repeated applications of the adhesive, as compared to a complete loss of adhesive force for a conventional acrylic adhesive; **(b)** removal of acrylic patch results in skin damage and redness; **(c)** removal of the multi-pillar "dry" adhesive is smooth and less harmful to the skin. Reprinted with permission from Jeong, H.E. et al., *Adv. Mater.* **2011** *23*, 3949–3953. Copyright (2011) John Wiley and Sons.

Indeed, eliminating the need for *chemical additives* is among the most attractive practical aspects of synthetic surfaces whose morphological features are designed to imitate the strong physical adhesive forces recorded in various organisms. Possible applications of such biomimetic surfaces are many, for example in adhesive "band-aid"-type tapes, millions of which are administered every year for treatment of minor wounds, and which currently use polymer-based glues. Biomimetic tapes based upon physical adhesion could cause much less damage to fragile skin tissues during removal, and in principle should exhibit greater durability after application since no degradation of adhesive chemical substances is encountered.

While varied biomimetic designs have successfully reproduced strong surface adhesion in ambient conditions, designing surfaces tightly adherent in *aqueous environments* is more difficult due to the presence of water molecules. Several organisms can be perceived as "role models" for wet adhesion, for example mussels, which exhibit extremely tight binding to submerged surfaces. To make adhesives that work in water environments, researchers have often tried to mimic the action of specific *marine adhesive proteins (MAPs)* identified as the primary adhesive molecular constituents in mussels. These proteins contain *dihydroxyphenylalanine (DOPA)*, a derivative of the amino acid tyrosine which contains an addition hydroxyl moiety which helps attachment to almost any kind of surface mussels encounter under water, including metal, wood, rock, and plastic, through "crowding out" the water molecules and tightly binding polar regions. Researchers have synthesized DOPA-containing polymers to make waterproof adhesives that are self-healing and biocompatible. Since DOPA tends to spontaneously oxidize to dopaquinone, which mostly attracts to *inorganic* surfaces, there have been attempts to synthesize DOPA-mimetics that on the one hand exhibit chemical properties preventing oxidation, but on the other hand retain the extraordinary stickiness of native DOPA.

In contrast to biomimetic surfaces promoting strong adhesion, nature has also provided remarkable mechanisms for *preventing* adhesion, specifically related to reducing adhesion of *water* – i.e. limiting surface wetting. Creating synthetic surfaces that repel water would have broad technological implications for areas ranging from biomedical devices, fuel transport, window design, and "green" architecture. This goal, however, has been challenging, both because the exact physical characteristics of surface "wetting" and "de-wetting" have been elusive, and also since the technical means for producing reliable and effective water resistant surfaces have been lacking. In this field, however, more than many branches of modern materials science, the natural world – plants, insects, reptiles, among many examples – has provided extraordinary inspiration which has led to numerous water resistant surface designs in recent years.

The lotus plant is probably the most well-known example of a natural water-repelling and self-cleaning surface, technically defined as *superhydrophobic*. Man has have been fascinated for millennia by the fact that the lotus flowers and leaves appear clean and seemingly dry even in environments of muddy water. How the lotus performs this feat was unknown until quite recently, when close microscopic examination revealed that water resistance is rooted in quite a simple physical mechanism facilitated through the unique microstructure of the leaf surface. Specifically, the leaf is covered with tiny bumps called papillae (Fig. 3.3). The papillae effectively entrap air bubbles, and these air pockets prevent water droplets from establishing contact with the solid leaf surface (in physical-mathematical terminology, the extremely low surface adhesion of the water droplet is due to the spherical drop being in the "Cassie-Baxter regime"). The droplets then roll on the surface, collecting dirt and performing "surface cleaning" while they wash away. This effect is, in fact, not limited to the lotus plant or even the plant kingdom; similar phenomena have been found for insects (maintaining clean wing surface and preventing wetting, for example), birds, and mammals.

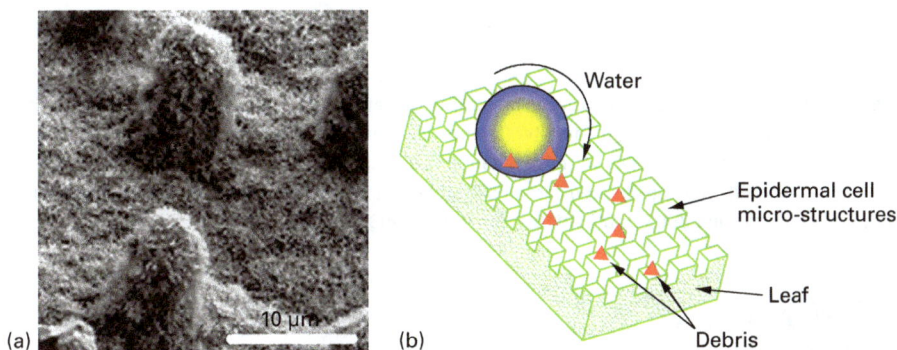

(a)

(b)

Water

Epidermal cell micro-structures

Leaf

Debris

10 μm

Fig. 3.3: The "lotus leaf" effect. **(a)** SEM image of the surface of a lotus leaf. Reprinted with permission from Sethi, S. et al., *Nano Lett.* **2008** *8*, 822–825. Copyright (2008) American Chemical Society. **(b)** Surface morphology of the lotus leaf. Trapped air repels water droplets, which subsequently slide on the surface and collect dirt and foreign substances.

Deciphering the physical core of the "lotus effect" has had broad technological significance, since it indicates that surface texture and morphology are the prominent parameters responsible for wettability properties, rather than the *chemical* characteristics of the surface. This realization has opened the door to broad technological advances and since the 1990s the lotus effect has been reproduced in many biomimetic artificial water-repelling surfaces through the fabrication of micro- and nanostructures capable of capturing and retaining air bubbles and maintaining a stable solid-air-liquid interface. Fig. 3.4 presents such a design in which a *carbon nanotube-based* surface developed by A. Dhinojwala and colleagues at the University of Akron achieved excellent water-repelling and self-cleaning properties.

Unlike the *isotropic* (uniform) distribution of microstructures on the lotus leaf, *anisotropic* textures on surfaces have important roles for many organisms. The butterfly wing (Fig. 3.5) is a striking example, displaying a *tilted* array of hierarchical

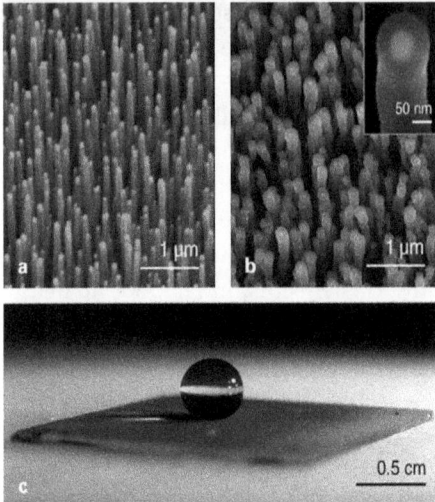

Fig. 3.4: Artificial water-repelling surfaces mimicking the lotus effect. **(a)** and **(b)** SEM images of the carbon nanotube forest before (a) and after (b) chemical treatment. Air pockets trapped within the protruding micropillars produced water-repellant and self-cleaning properties. **(c)** A spherical water droplet sitting on the micropatterned carbon nanotube forest. Reprinted with permission from Lau, K.K.S. et al., *Nano Lett.* **2003** *3*, 1701–1705. Copyright (2003) American Chemical Society.

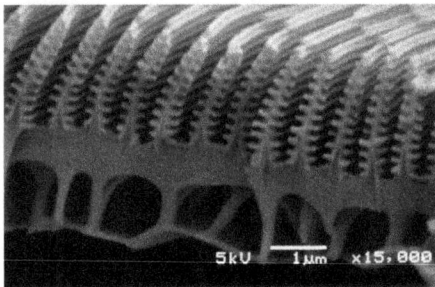

Fig. 3.5: Anisotropic surface morphology of a butterfly wing. SEM image depicting a cross section of the wing scale of *Morpho cypress* butterfly. Note the tilted orientation of the microscopic ridges on the upper part of the scale. Image courtesy of Prof. Shinya Yoshioka, Osaka University.

microstructures on its surface. This organization enables quick and efficient shedding of water droplets from the butterfly wing, critical for its survival. Anisotropic surface morphologies have been observed in other organisms and plants, and specifically mimicked in various synthetic schemes designed to produce self-cleaning and water resistant surfaces.

Anti-fogging is another intriguing phenomenon closely related to the microstructure of biological surfaces, and which continues to inspire functional surface design. Such surfaces, in addition to repelling water adhesion, minimize spatial and temporal accumulation of water. Their biological significance mainly lies in maintaining optical transparency of pertinent coatings such as on the eye. Design principles which follow the "lotus effect" might be insufficient in that regard; a surface morphology which comprises a coating of micro- or nanostructures might still cause microscopic-size water droplets to be entrapped within tiny cavities on the surface, consequently adversely affecting optical transparency. A tiny mosquito (the *C. pipiens* genus), in fact, offers a sophisticated mechanism to overcome this problem.

Fig. 3.6: Morphology of the mosquito eye. SEM images of the mosquito eye at different scales. **(a)** Single eye of the mosquito *C. pipiens*. **(b)** Hexagonal close packing of the microhemisphere (ommatidia) on the eye. **(c)** Two neighboring ommatidia. **(d)** "Nano-nipples" covering the surface of the ommatidia. Reprinted with permission from Gao, X. et al., *Adv. Mater.* **2007** *19*, 2213–2217. Copyright (2007) John Wiley and Sons.

The eyes of *C. pipiens* have extraordinary antifogging capabilities (essential from an evolutionary standpoint for a creature active in humid environments). The mosquito achieves this feat through a remarkable *hierarchical* organization (Fig. 3.6). The eye surface features a *nanoscale* configuration of non-close-packed hexagonal "pillars", which achieves repelling of microscopic water droplets through entrapped air bubbles (similar to the lotus effect). Furthermore, a tight hexagonal close-packing of the nanopillar-displaying nipples in the *microscale* prevents entrapment of the repelled water droplets, consequently ejecting the water rapidly and efficiently from the mosquito eye surface.

The hierarchical structure of the mosquito eye can be artificially reproduced through surface patterning techniques such as lithography (Fig. 3.7). In the scheme developed by L. Jiang and colleagues at the Chinese Academy of Science, the microscopic hexagonal pattern was reproduced by hemispheres of PDMS, a widely-used

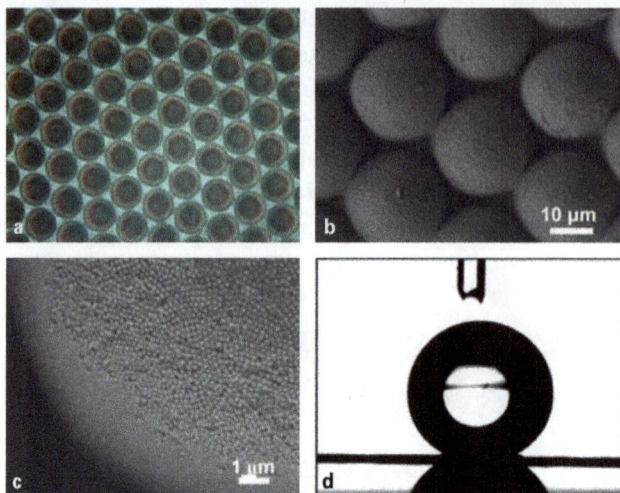

Fig. 3.7: Biomimetic anti-fogging surface morphology. Synthetic surface design mimicking the hierarchical organization of the mosquito eye. **(a)** Optical microscope image of the artificial compound-eye analog; **(b–c)** SEM images of the PDMS micro-hemispheres and silica nanospheres, respectively. **(d)** Spherical water droplet on the artificial surface; no adhesion occurs. Reprinted with permission from Gao, X. et al., *Adv. Mater.* **2007** *19*, 2213–2217. Copyright (2007) John Wiley and Sons.

transparent polymer, while a dense coating of silica *nano*spheres upon the PDMS hemispheres was employed for mimicking the nanoscale pillar organization. Together, this cleverly designed biomimetic surface demonstrated excellent anti-fogging properties.

Varied other approaches have been reported aiming to recreate the unique hierarchical organization responsible for the anti-fogging properties of mosquito eyes and other biological surfaces. A generic strategy is depicted in Fig. 3.8, in which a "two-scale" sphere-packing morphology is generated. The sphere *compositions* in such surface designs are quite flexible; reports have indicated the use of silica microspheres covered with gold nanoparticles, or polymers as the building blocks for the spheres; other materials were also used for constructing the spheres. While Fig. 3.8 might suggest that fabrication of a multiscale surfaces exhibiting anti-fogging properties is quite an easy undertaking, to the mosquito's and nature's benefit it should be emphasized that technical challenges are still often encountered for realizing the technological potential; it has been found, for example, that slight defects in surface structures significantly reduce their water-repelling performance.

Beside the diverse biomimetic surfaces displaying superhydrophobicity, nature has also inspired development of *omniphobic* surfaces – which repel both *polar* and *non-polar* liquids (Fig. 3.9). Such surfaces are intrinsically much harder to design since water and non-polar liquids like oil and organic solvents usually exhibit significantly different surface tensions, resulting in divergent attachment and spreading behavior on micropatterned surfaces. More specifically, non-polar droplets are much

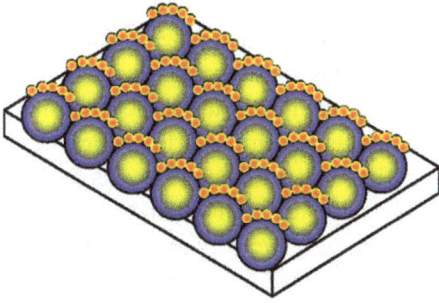

Fig. 3.8: Generic "two-scale" surface morphology for achieving anti-fogging properties. Synthetic surface comprising close-packed *micro*spheres coated with *nanoscale* spheres. The air trapped within the nanospheres prevents adhesion of small water droplets, while the tight packing of the microspheres prevents bigger droplets from staying on the surface.

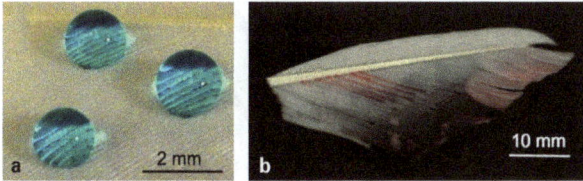

Fig. 3.9: Omniphobic surfaces. Water droplets **(a)**, and rapeseed oil (colored in red organic dye, **(b)** are both repelled from the surface of a duck's feather, which exhibits omniphobic properties. Reprinted with permission from Tutejaa, A. et al., *PNAS* **2008** *105*, 18200–18205. Copyright (2008) National Academy of Sciences, U.S.A.

more likely to disintegrate upon rough, micropatterned surfaces, thereby attaching to the surface features rather than being repelled in their original drop-form as encountered for *water* in conventional de-wetting surface scenarios.

Nature provides some avenues for omniphobic surface design. *Nepenthes* pitcher plants, for example, are insect-digesting plants maintaining a smooth highly-lubricated surface within their pitcher. The water-repelling surface blocks attachment of insects to the pitcher wall, leading to their entrapment and digestion. Similar to the lotus effect discussed above, water de-wetting here is accomplished through surface display of abundant microscopic cavities. Different from the lotus plant (or other plants and organisms displaying water-repelling surfaces), however, the surface cavities on the pitcher's surface do not entrap air, but rather a lubricating fluid responsible for the ejection of the non-polar liquids. Biomimetic synthetic omniphobic surfaces have been reported, primarily based upon nano- or microporous substrates comprising gel materials, cross-linked fibers, and similar designs. Inert fluids are injected into these cage-like networks and the matrix design ensures retention of the fluid.

While water-repellant surfaces have attracted huge scientific and technological interest, *water-attracting* surfaces might actually have a much more significant impact on humanity. This is due to the stark reality of growing water scarcity in many parts of the world and the constant search for technological means to remedy this situation. Indeed, the sophisticated mechanisms of plants growing in arid environments to capture and retain water have intrigued and inspired researchers in this field for many years. *Fog harvesting* mechanisms, in particular, are often related to the same surface morphology factors associated with the "lotus effect". Indeed, several plant species combine microscopic surface morphology designed to minimize water adhesion with a *macroscopic* three-dimensional organization aimed to entrap the non-adhesive water within the plant for further use. Fig. 3.10 schematically shows the water collection strategy of *Cotula fallax*, an alpine plant displaying an interspersed hair-like structure which captures the accumulated water.

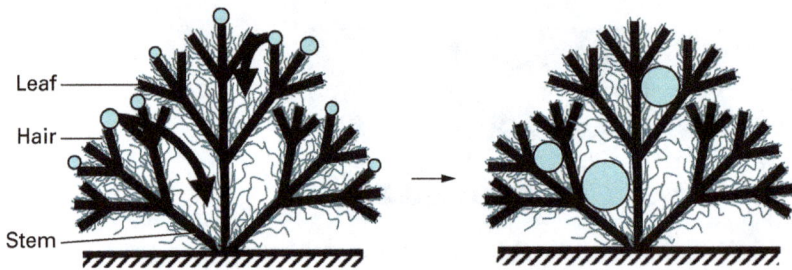

Fig. 3.10: Water-retention in plants based upon the lotus effect. Tiny water droplets are not adsorbed onto the surface of the *Cotula fallax* leaves, but rather are slowly accumulated closer to the stem, shielded by a mat-like structure formed by overlapping microscopic hairs. Reprinted with permission from Andrews, H.G. et al., *Langmuir* **2011** *27*, 3798–3802. Copyright (2011) American Chemical Society.

Recent studies have combined different biomimetic surface strategies to construct devices for environmental applications. Fig. 3.11 depicts a "water harvesting windmill" which integrates surface morphologies and structural features borrowed from different plants and organisms. The device, developed by Q. Zhang and colleagues at the Chinese University of Hong Kong, comprises of a fan [Fig. 3.11(a)] exhibiting bio-inspired nano-to-micro structural features and mechanical properties. Specifically, the surface of the fan blades consists of micrometer-scale corrugated features mimicking the directional grooves of rice leaves; this anisotropic surface topography induces accumulation and movement of water droplets towards the blade edges and eventually to the water collection vessel [Fig. 3.11(b)]. As shown in Fig. 3.11(c), the individual synthetic grooves further comprise of "nano-tips" designed to mimic *cactus* surfaces. The nano-tips constitute water deposition points, and the pyramidal structure facilitates ready transport of the droplets towards the groove bottoms. An

additional feature imparted on the fan blade surface was the application of a hydrophilic lubricant layer, mimicking the "pitcher" surface of *Nepenthes* plants and aiding
water adsorption and droplet transport [Fig. 3.11(d)]. Finally, the windmill design also
makes possible varying the orientation of the entire fan, mimicking the motion of butterfly wings in relation to the airflow aiming at achieving maximal centrifugal forces
for efficient, rapid water removal from the blade surfaces and towards the collection
vessel.

Fig. 3.11. Water-collecting biomimetic windmill. The design concept and specific surface features
of the windmill fans **(a)**. **(b)** Aligned microgroove morphology mimicking the surface of rice leaves
facilitating water droplet accumulation and directed movement of the droplets along the grooves.
Scale bar corresponds to 500 mm. **(c)** Pyramid-shaped nano-tips, mimicking cacti's surface,
displayed on the grooves for water capture and directing the droplets to the groove trough. Scale
bar corresponds to 500 mm. **(d)** Application of a hydrophilic lubricant at the surface, mimicking the
surface of a pitcher plant. **(e)** Tuning the orientation of the fan, mimicking the twisting motion of
butterfly wings, designed to maximize the centrifugal force. Adapted with permission from Wang
et al., Nature-Inspired Windmill for Water Collection in Complex Windy Environments, *ACS Appl.
Mater. Interfaces* 2019, 11, 17952–17959, Copyright 2019 American Chemical Society.

The system depicted in Fig. 3.11 nicely illustrates an integration of several biomimetic design concepts which together accomplish efficient water capture from
moisturized air and its collection. The device can be implemented, for example, for
extraction of water from foggy air. In particular, the bio-inspired water-collecting
fan can, in principle, be employed in field and home-use settings. The strategy outlined by the research summarized in Fig. 3.11 represents an important future direction of the biomimetic surfaces field in addressing the growing needs for water
extraction and purification associated with the impact of climate change. Indeed,
the diverse pathways evolved by plants and animals to capture and preserve
water from air will likely continue to inspire and aid development of innovative
technologies.

3.2 Color and photonics

Color is among the most abundant and enduring contributions of nature to mankind. While organisms use color as a fundamental evolutionary tool, humanity has used colors and pigments extracted from plants or animals primarily for aesthetic reasons, but also for diverse functional applications such as coatings, sensors, artificial photosynthesis, and others. Recent studies have also explored *photonics* applications of natural color phenomena and biological pigments. While there have been numerous examples of biological dye molecules extracted from natural sources, the discussion below will rather focus on *structural color* effects, i.e. colors produced through unique structural features of biological surfaces, which might be mimicked in artificial surface designs.

Perhaps surprisingly, structural color in nature is far from rare, encountered mostly in plants, but also in insects and other organisms. In most cases, colors in different wavelengths are generated from *periodic* arrays of surface features – very much like the colors produced by the facets of inorganic crystals via Bragg reflection of light. For example, the periodicity of the scales on the wing surface of the *M. cypris* butterfly, shown in Fig. 3.5 above, gives the butterfly wings their blue color. Indeed, the distances between the periodic features are the main structural parameter that determines the wavelength of the reflection and consequent color in other butterflies from that genus.

The analogy between structural color in biology and *gem* colors is, in fact, quite appropriate since, similar to many gems, some of the structural colors observed in the plant and animal kingdoms exhibit remarkable "glossiness" and "glare" which make them look highly attractive to the eye. The extraordinary color of the *Pollia condensata* fruit is a dramatic demonstration of this phenomenon, as recently discovered by U. Steiner and colleagues at the University of Cambridge (Fig. 3.12). The *Pollia* fruit does not contain pigments responsible for either the blue color, or the shiny, glossy appearance of the fruit (both color and glossy texture last for years even after the fruit is harvested!). Rather, the source of the unique fruit color and texture is the anatomy of the fruit surface. Specifically, periodic multilayer morphology of cellulose microfibrils (Fig. 3.12b) results in reflection of blue light; slight variations in microfibril orientations on the fruit surface produce the dramatic gloss and greenish-purple hues giving the fruit their unusual appearance.

The *Pollia condensata* fruit exhibit other remarkable features. The *thickness* of the corrugated multilayer structure at the fruit surface was found to vary between adjacent fruit coating cells, subsequently producing slight spatial variations in the reflected light – responsible for the "pixel"-like surface color appearance. Another intriguing and surprising observation has been *polarization-dependent* light reflection by the fruit surface, ascribed to progressive "rotation" of the cellulose microfibril orientation (Fig. 3.12b). This unusual phenomenon might be "accidental" since no pertinent evolutionary advantage has been identified. Understanding the underlying physical and structural factors responsible for the unusual and beautiful color of *Pollia condensata,* like other species in nature, could lead to diverse practical applications – artificial colorings on surfaces, cosmetics, and even photonic devices.

Fig. 3.12: Structural color of the *Pollia condensata* fruit. **(a)** Photographs of the *Pollia* fruit. Left: Single fruit from dried specimen collected in 1974. Right: Infructescence (cluster of fruits) from alcohol-preserved specimen collected in 1974. The diameter of each fruit is about 5 mm. **(b)** TEM image of the cellulose microfibril-formed ridge structure that constitutes the external fruit surface. Note both the periodic wavy structure as well as the curved directionality of the individual microfibrils. Reprinted with permission from Tutejaa, A. et al., *PNAS* **2012** *109*, 15712–15715. Copyright (2012) National Academy of Sciences, U.S.A.

Nature endows a plethora of surface architectures affecting varied photonic properties. Butterfly wings, for examples, display diversity of colors, generated in large part via periodic sub-micrometer structures on the wing surface. The multiscale morphologies of butterfly wings (observed also in other organisms and plants) yield distinct light absorbance and reflection profiles that together determine the color, degree of reflectivity, pattern formation, and other physical/optical parameters. Such surfaces have provided inspiration to varied photonic applications. Fig. 3.13 illustrates a functional surface (employed in photocatalysis) in which the microscopic structure, based upon butterfly wing architecture, facilitates efficient light absorbance. In the experiments, T. Fan and colleagues at Shanghai Jiatong University constructed an "artificial butterfly wing" designed to achieve maximal light harvesting, employed for photocatalytic "water splitting" – a prominent reaction in clean energy (i.e. "carbon neutral") applications by which hydrogen generated from breaking water molecules is used as energy-rich fuel.

The researchers examined the wing surface morphology of the butterfly genus *Papilio helenus Linnaeus* which appears black, accounting for highly efficient light absorbance (i.e. minimal reflected light). Interestingly, the virtually complete light absorbance has been traced to the hierarchical architecture of the wing, comprising

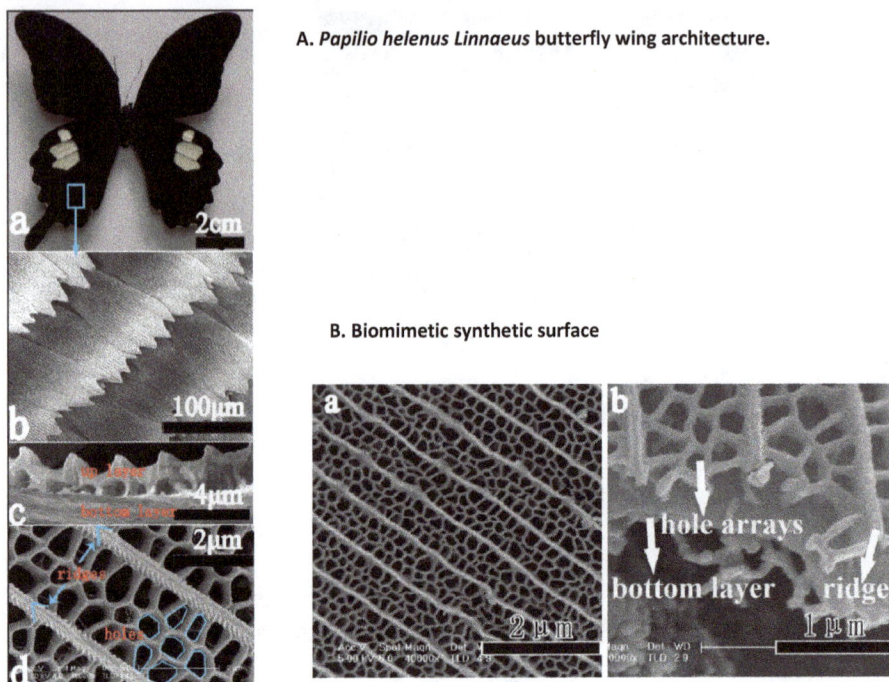

A. *Papilio helenus Linnaeus* butterfly wing architecture.

B. Biomimetic synthetic surface

Fig. 3.13: Hierarchical wing surface morphology of a butterfly wing facilitating efficient light absorbance. A. Hierarchical architecture of the butterfly wing tiles depicting the microporous morphology giving rise to significant light absorbance (and corresponding black appearance). B. Synthetic TiO2 surface mimicking the butterfly wing morphology designed to achieve efficient light absorbance. Adapted with permission from Liu et al, Hydrogen evolution via sunlight water splitting on an artificial butterfly wing architecture, *Phys. Chem. Chem. Phys.*, 2011, **13**, 10872 -10876, Copyright 2011 The Royal Society of Chemistry.

micrometer-scale "roof tiles" (Fig. 3.13A). Each tile is made of microporous arrays separated by elongated "ridges". Optical simulations indicated that the combination between the hole-array and parallel ridges account for the significant anti-reflection effect and concomitant efficient light absorbance. Fig. 3.13B depicts the "artificial wing" architecture consisting of titanium oxide (TiO$_2$) microporous "tiles". TiO$_2$ was selected in the biomimetic surface since this material is a prominent photocatalyst for water splitting. Indeed, the researchers observed that this biomimetic butterfly wing architecture accomplished excellent light harvesting and hydrogen generation, confirming the occurrence of water splitting. Notably, the porous TiO$_2$ microstructure further allowed doping the framework with platinum nanoparticles, which promote gas desorption thereby accelerating the splitting reaction.

Mimicking the intricate and often complex morphology of butterfly wings is not trivial. Accordingly, actual butterfly wings have been also used as *templates* for innovative light absorption platforms. In such systems, researchers have exploited the hierarchical micro- and nanoscale organizations of wings' surfaces as templates

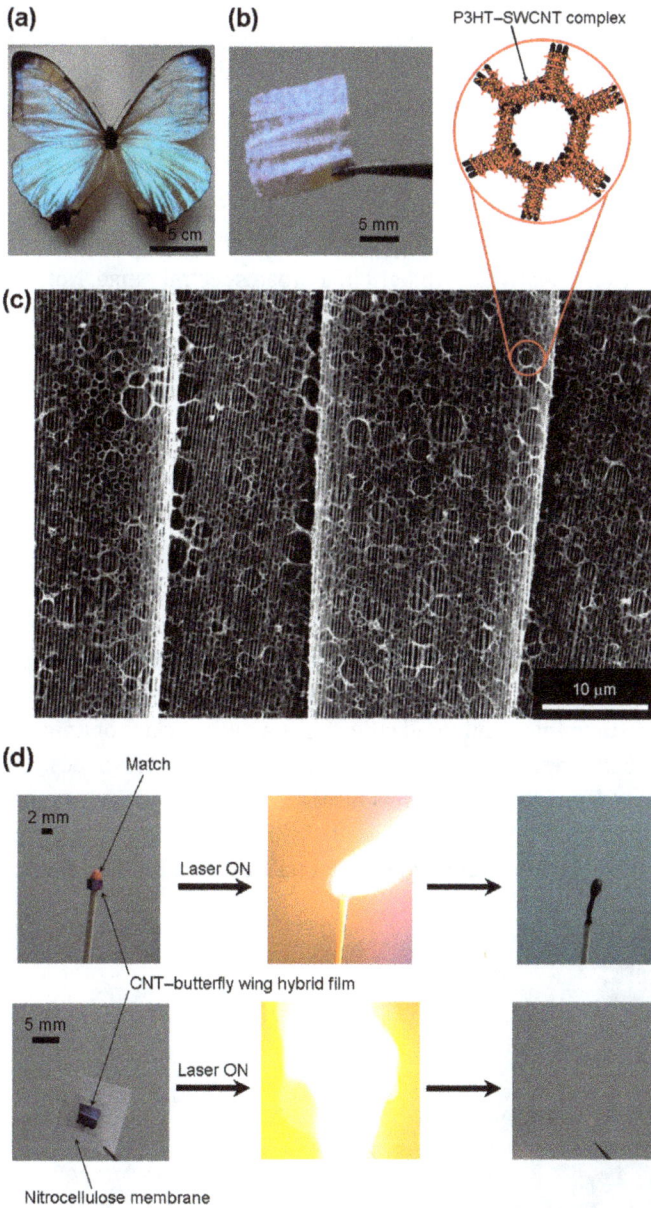

Fig. 3.14: Butterfly wing as a template for efficient light absorption. *Morpho sulkowskyi* butterfly (a) and a fragment of its wing used as a template for a composite polymer (P3HT)-single wall carbon nanotubes (SWCNTs) microporous film (c). (d) Photothermal effect of the templated matrix. Efficient absorbance of light and conversion to heat gave rise to flame generation. Adapted with permission from Miako et al, Self-Assembled Carbon Nanotube Honeycomb Networks Using a Butterfly Wing Template as a Multifunctional Nanobiohybrid, *ACS Nano* 2013, *10*, 8736-8742. Copyright (2013) American Chemical Society.

upon which non-biological architectures exhibiting distinct functionalities can be constructed. Fig. 3.14 illustrates this strategy. In that study, E. Miyako and colleagues at the Health Research Institute, Osaka, Japan, utilized the wing of the *Morpho sulkowskyi* butterfly genus as a substrate for assembling a polymeric film exhibiting remarkable photothermal properties.

Specifically, the researchers deposited on the butterfly wing a thin "honeycomb"-shape film comprising a polymer framework embedding single-walled carbon nano-tubes (SWCNTs), known to efficiently absorb light in a broad spectral range. Notably, the researchers observed that the *Morpho*-templated honeycomb-shaped film featured excellent absorbance of infrared (IR) light, generally challenging in light harvesting film technologies, which was then dissipated as heat. In a visually striking demonstration, the photographs in Fig. 3.14(d) show that brief IR illumination of *Morpho* wing-templated SWCNT honeycomb films resulted in ignition – accounting for the significant amount of generated heat. This result is important, because chemical systems capable of efficient light absorbance, particularly in the IR spectral region which is a prominent part of sunlight irradiation, are highly sought for solar cell designs, sensors, and photothermal devices.

To exploit natural photonics for possible practical applications we might not have to limit ourselves to mimicking surface structures, but rather use actual organisms. *Diatoms*, in that case, might be excellent candidates. Diatoms, described in more detail in Chapter 5 (Biomimetic mineralization), are a diverse class of unicellular algae displaying a broad range of fascinating microscopic structures, shapes, and patterns. In the context of *biophotonics,* diatoms have attracted interest due to their mostly transparent and intricately patterned silica shell structure. These "glass-like" shells often display remarkable appearance when interacting with light (Fig. 3.15).

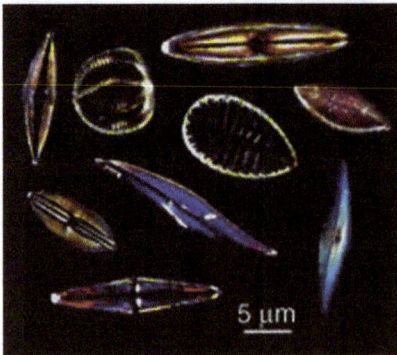

Fig. 3.15: Structural colors of diatoms. Light microscopy images of *pennate* diatom species illuminated with white light. Reprinted from Gordon, R. et al., The glass menagerie: diatoms for novel applications in nanotechnology, *Trends Biotechnology,* **2009** *27*, 116–127. Copyright (2009), with permission from Elsevier.

The range and beauty of diatom structural colors is related to diffraction and interference occurring when light interacts with the silica framework, which often exhibits a high degree of periodicity. Furthermore, the organized shell structures of diatoms could, in principle, enable propagation of only specific wavelengths passing through – the exact phenomenon occurring in *photonic crystals* (accordingly, diatoms have been referred to as "living photonic crystals"). Another intriguing optical property observed for some diatom species is *photoluminescence* – prolonged light emission following illumination with ultraviolet light. This phenomenon, observed also in synthetic porous silicon frameworks, opens the door to many practical applications, such as ultrasensitive vapor sensing (adsorbed gas molecules within the diatom pores modulate the photoluminescence, which is associated with dissipation of the irradiation energy within the silica framework).

The horizon of possible diatom-based photonic applications is wider still. Since the biological properties of diatoms, like all organisms, are ultimately determined by their genetic information, *genetic engineering* might be applied to tune photonic properties through modulating the structure of the diatom shell surface. The porous surface framework of the diatom might also function as a host matrix, encapsulating photoactive (phosphorescent, fluorescent) molecules. Furthermore, the broad range nano- to microscale dimensions of the pores in different diatom species and the transparent shell structure might enable placement of actual nanoscale optoelectronic components upon the diatom surface.

3.3 Biosensing

Biosensing is a huge area of research with considerable commercial significance (mostly beyond the scope of this book). *Biomimetic surfaces* have been implemented as core components in numerous biosensors, due to the fact that surfaces play a crucial role in enabling *molecular recognition* of tested analytes, thereby triggering the sensor response. Biomimetic surfaces need to satisfy the fundamental yardsticks of biosensors: sufficient *sensitivity* and *specificity*, as well as adherence to parameters such as stability, fidelity, and acceptable cost. In addition, the recognition event occurring at the surface should be able to trigger an adequate signal (mechanical, electrical, other), to be processed (and amplified) ultimately producing a sensor response.

A large body of work has been devoted to designing surfaces that would satisfy the biosensor design criteria described above. A common approach has been the immobilization of biomolecules upon a transducer surface, which act as the recognition modules. This scheme has often been difficult to implement, however, since surface immobilization should not adversely affect the binding capabilities of the recognition molecules towards the analyte. In particular, retaining the functionalities of *protein receptors* or similarly large biomolecules which serve as recognition elements can be a formidable task, limiting the broader applicability of such sensing designs. The binding between an analyte and the biological recognition element attached to the surface must result in detectable physical transformation which can be recorded.

Antibody-epitope recognition has been among the most widely used biomolecular binding schemes exploited on biosensor surfaces. This process, in which an antibody recognizes and tightly binds a small fragment, usually short peptides, is a fundamental event central to activation of the immune system. Other triggering events occurring at biomimetic sensing surfaces include *protein-ligand* interactions, in which small molecules bind at specific sites of protein molecules thereby leading to structural changes and functional activation of the proteins, and *enzyme-substrate* binding, giving rise to biocatalysis following docking of the substrate at the active site of the enzyme molecule.

3.4 Cell-surface interactions

Surfaces play prominent roles in the behavior of **cells** attached to them. The interest in cell-surface interactions also stems from the fact that cells in the body either reside on rigid surfaces (such as organ linings, bones, vascular pathways), or respond to cues provided by interaction with external surfaces – either surfaces of organs and tissues or surfaces of other cells. Indeed, modifications of cell-surface interactions through surface manipulations are considered a primary tool in diverse biomedical and research fields. Below I will discuss synthetic and semi-synthetic surfaces designed to mimic physiological environments for cell proliferation and manipulation of cell properties, with particular emphasis on the burgeoning field of stem cell biology.

Surface engineering for studying and modulating cell adhesion and cell transformations has traditionally been pursued through two main routes – *mechanical* modification of surface features, and *chemical* functionalization. Both strategies have provided a wide variety of surface configurations and adhesion features; surfaces produced through a combination of the two approaches have attracted interest as well. Physical (e.g. mechanical) patterning of surfaces, in particular, has been a burgeoning research track in recent years, moving hand in hand with the significant technical progress in lithography and nanolithography capabilities. Furthermore, elucidating mechanisms of cell response to parameters such as surface elasticity or stiffness and topography has contributed to our understanding and control of cell biology as affected by the cell-surface interface.

Numerous studies have analyzed the effect of surface *chemistry* upon cell structure, metabolism, motility, and overall viability and proliferation, and the scope of this chapter (and book) is too limited for providing a comprehensive review of this vast field. Engineered surfaces exhibiting *gold nanostructures* have been a popular means for controlled cell adhesion (Fig. 3.16). The use of gold in such arrangements is preferred due to the many chemical derivatization routes available, enabling linkage of diverse residues (mostly through *thiol* moieties) onto the gold surface. Among the more popular surface-displayed motifs has been the arginine-glycine-aspartic acid (RGD) cell-adhesion tripeptide (either bound to gold, or displayed by other chemical means). In the many instances in which discreet regions, i.e. patterns, for cell adhesion are intended, a common strategy has been to attach gold micro- and nanoparticles onto the surface which are subsequently derivatized with cell-binding elements such as RGD, while surface coating with *biologically-inert* molecular units, such as polyethyleneglycol (PEG), is carried out to prevent cell binding in the areas *outside* of the gold domains (Fig. 3.16).

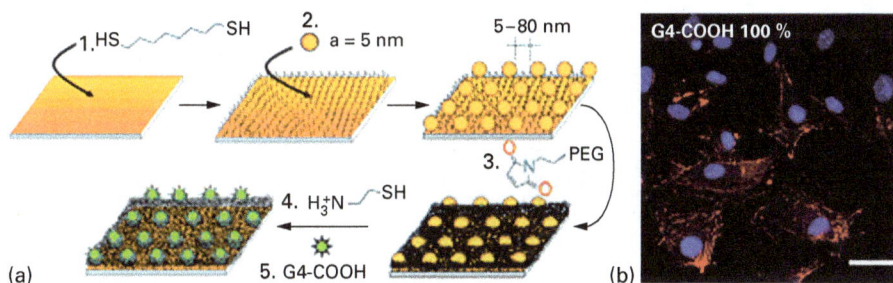

Fig. 3.16: Chemical modification of surfaces for cell adhesion. **(a)** Scheme showing the stages in surface preparation. **1.** Formation of an alkylthiol monolayer; **2.** immobilization of Au nanoparticles (NPs); **3.** malemide-functionalized PEG is added, coating the areas around the attached Au NPs with biologically-inert PEG moieties; **4.** the NP template is reacted with cysteamine for subsequent immobilization of the cell-attachment functional groups **5.** (presented on *dendrimer* particles – G4-COOH – in this particular study). **(b)** Confocal microscopy image of vascular endothelial cells attached to the G4-COOH – derivatized surface. Reprinted with permission from Lundgren, A. et al., *Angew. Chemie* **2011** 50, 3450–3455. Copyright (2011) John Wiley and Sons.

Another cell-recognition peptide often used for promoting cell patterning on surfaces is the vascular endothelial growth factor (VEGF) sequence. VEGF is a highly potent signaling peptide promoting blood vessel proliferation (e.g. "angiogenic" factor), and its high-density covalent display on synthetic (biocompatible) surfaces has been shown to significantly aid the proliferation of endothelial cells – a process that is crucial in medical emergencies such as ischemic damage.

Since adhesion mechanisms of cells to surface-displayed macromolecular ligands are believed to be *non-specific*, techniques have been introduced to achieve effective and/or more specific attachment of cells using synthetic rather than biological ligands. *Dendrimers,* highly branched multivalent ligands, have been successfully tried as substitutes for peptide ligands to promote patterned cell adhesion (Fig. 3.16, above). Dendrimers present several advantages, primarily the synthetic capability to produce a variety of surface-displayed molecular units exhibiting diverse structures, which could be made accessible for cell adhesion. Dendrimer assemblies can be grown directly on surface gold domains, producing patterns for cell adhesion.

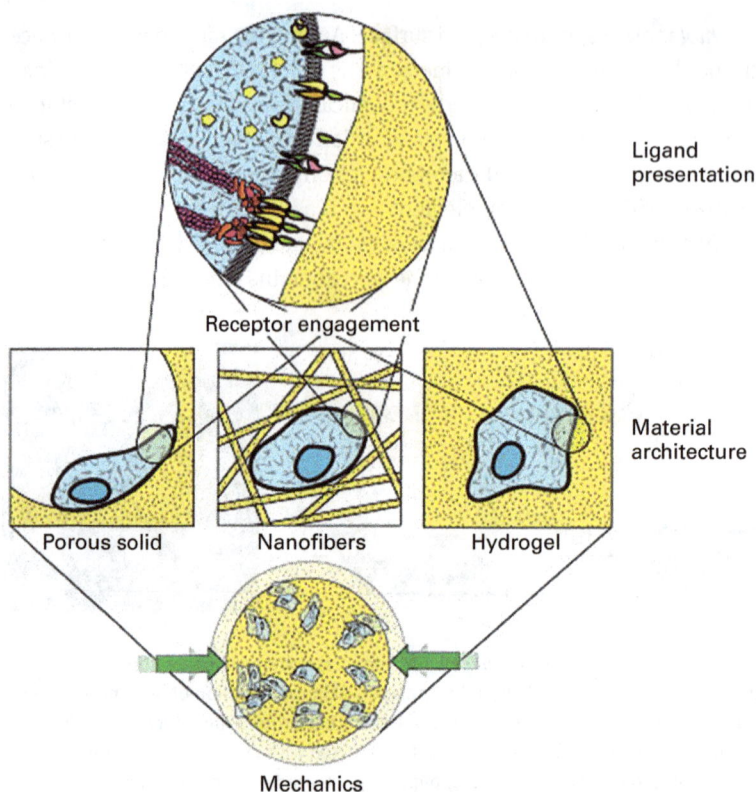

Fig. 3.17: Surface interactions affecting stem cell properties. Design parameters of biomimetic surfaces affecting stem cell phenotypes. See text for details. Reprinted from Saha K. et al., Designing synthetic materials to control stem cell phenotype, **2007** *11*, 381–387. Copyright (2007), with permission from Elsevier.

Surfaces and surface modulation are particularly relevant to the burgeoning field of **stem cell** biology. Stem cells are undifferentiated cells capable of self-renewal and differentiation into varied cell species. Accordingly, stem cells are considered as among the most promising therapeutic avenues for numerous diseases, pathological conditions, and injuries. The notion of replacing defective or injured cells with a "fresh start" through stem cells that can subsequently differentiate might indeed revolutionize medicine. Other promising applications for stem cells include technologies for cell-based diagnostics. The fundamental requirement for using stem cells, which can be exceedingly difficult in many instances, is being able to intimately control differentiation of the stem cells into a desired cell type – such as muscle, skin, blood vessels, bone or neurons. Importantly, the structural features and chemical composition of surfaces were found to exert significant effects upon stem cell differentiation in particular, and cellular processes in general.

Stem cell self-renewal and differentiation pathways are highly dependent upon the *microenvironment* in which the cells reside (niche) – providing a powerful tool for controlling the fate of the stem cell and the eventual target tissue arising from the cell. Indeed, the topographical features of the niche – from the macro- to the nanoscale – closely influence stem cell behavior and functionality by altering cell morphology, adhesion, motility, proliferation, protein expression, and gene regulation. Inside the niche, stem cells are furthermore exposed to complex, spatially- and temporally-controlled biochemical stimuli. Together, the environmental and chemical cues generated at the stem cell-substrate interface are considered primary factors affecting differentiation pathways and growth profiles.

Fig. 3.17 depicts the main external factors impacting stem cell phenotype, properties, and ultimate differentiation through the physical interface between the cell and the surface. *Ligand identity* and *density* are among the most important determinants for stem cell fate through interactions with receptors on the cell surface and the consequent effects on intracellular signaling pathways. *Spatial configuration* of the surface supporting cell growth is also of prime importance as the presence of flat, curved, corrugated, or constrained surfaces closely influence the cell microenvironment. The *mechanical properties* of the surface, particularly its *elasticity* and *microviscosity,* play additional roles in modulating stem cell processes. Indeed, a central tenet of the research into biomimetic surfaces for stem cell applications has been the realization that composition and structure of the substrate essentially transmit distinct signals to the cells which consequently "decode" them through different biological pathways. The still largely unrevealed "secret" is to decipher the exact "codes" directing stem cells through their surface interactions.

Periodic patterns have been shown to provide intimate guidance for stem cell adhesion, growth direction, and morphologies. Among the first impressive demonstrations of the relationship between surface patterns and stem cell growth properties was the observation by the American biologist R. G. Harrison (more than 100 years ago!) that embryonic stem cells cultured on spider webs followed the fibers rather than grew randomly. Recent years brought numerous examples of aligned and elongated stem cells growing on surfaces displaying groove and spatially-oriented structures (Fig. 3.18). The effect of surface patterns is in fact more nuanced; in some cases such surfaces modulated cell differentiation not directly through interacting with the cells themselves but rather through affecting the organization and density of microtubules – the intracellular matrix framework maintaining cell structure.

Periodicity of microstructures and nanostructures was found to not only affect organization of cells on surfaces, but also induced distinct differentiation profiles of stem cells. Fig. 3.19 presents interesting results recorded by P. Colpo and colleagues at the Institute of Health and Consumer Protection, Italy, demonstrating that neuronal stem cells cultured in the presence of a micropatterned polymeric surface were specifically attached to an array of cationic polymer domains deposited on a surface through

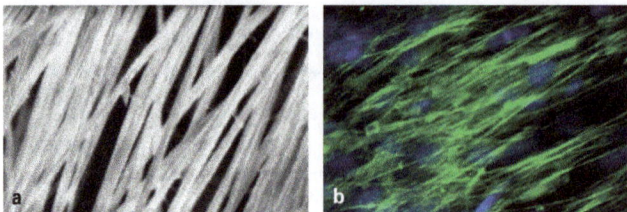

Fig. 3.18: Stem cells grown on aligned surface patterns. **(a)** Scanning electron microscopy (SEM) image of a bundle of aligned polysaccharide (chitosan) and collagen microfibers used as a substrate for stem cell growth. **(b)** Confocal fluorescence microscopy image of highly aligned human mesenchymal stem cells (hMSCs) grown on the microfiber surface. Reprinted with permission from Dang, J. M. and Leong K. W. *Adv. Mater.* **2007** *19*, 2775–2779. Copyright (2007) John Wiley and Sons.

microcontact printing. Furthermore, the experiments depicted in Fig. 3.19 showed profound dependence of *cell morphology* upon surface geometry; in a condition of high cell concentration the cells maintained round shape crowding upon the protruding polymer "islands", while low cell density enabled cell flattening and formation of axon-like cell extensions linking with adjacent microstructures. These cell outgrowth features also served as conduits for cell migration between the elevated polymer domains.

While the importance of external physical signals for stem cell differentiation is widely accepted, establishing direct links between external factors and differentiation pathways and thus controlling such pathways has been difficult. Yet, major efforts are directed towards identifying and utilizing new biomaterials as substrates for stem cell growth, and modulating surface architectures for controlling differentiation without adversely affecting cell viability.

Supplanting the *topographical* patterns designed to affect cell attachment and properties, researchers aim to develop *chemical* strategies aimed at directing cell growth on surfaces. A common detriment, however, for utilization of biomimetic surfaces for cell proliferation is the occurrence of *biofouling* – the non-specific adsorption and accumulation of proteins onto surfaces subsequently rendering them inactive. Several schemes have been introduced to overcome biofouling in the context of cell adsorption and viability. An interesting approach for assembling biocompatible antifouling surfaces involved coating patterned surfaces with metal nanoparticles (NPs) exhibiting different surface charges; such functionalized surfaces were found to diminish protein adsorption and were thus more accommodating for cells.

Stem cells, like other living systems from the cell to the organism level, respond to electric fields and electrical currents in various ways. Electric stimuli can be coupled to the cell's physical environment. An example of such an approach is the use of *electro-active polymers (EAPs)* as scaffolds for stem cell growth. Many types of EAPs undergo pronounced morphology changes induced by electrical currents; this environmental modulation, together with the actual electric field or currents applied on the cells, could significantly influence cell properties in a manner which diverges from just a change of physical environment.

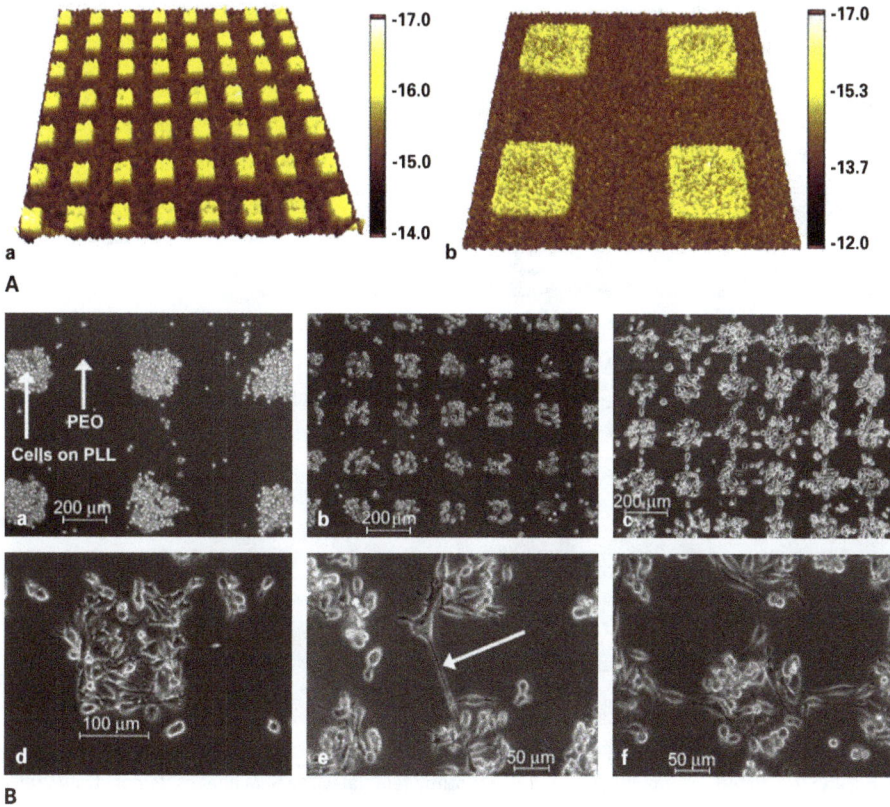

Fig. 3.19: Growth of neuronal stem cells on periodic microstructures **A.** Thickness maps showing the periodic poly-L-lysine (PLL) stamped protrusions. The heights are in relation to the silicon surface. Size of the top image 1500 µm x 1690 µm; bottom image 380 µm x 390 µm. **B.** Microscopy images of HUCB neuronal stem cells incubated for 1 day on the patterned surfaces. **(a)–c)**: High cell density; the cells are crowded on the PLL protrusions and adopt round shapes. **(d)–f)**: Low cell density; cells flatten and show extended projections to adjacent PLL domains (e.g. arrow in **e**). Reprinted from Ruiz, A. et al., Micro-stamped surfaces for the patterned growth of neural stem cells, **Biomaterials** **2008** 29, 4766–774. Copyright (2008), with permission from Elsevier.

Direct electrical stimuli in some instances lead to pronounced transformations of cell adhesion, migration, and morphologies. This observation has opened new research avenues designed to endow cell-growth substrates with electrical conductivity. The burgeoning field of nanotechnology has significantly contributed to progress in this field. Specifically, studies have shown that inclusion of metallic NPs or conductive nanostructures (such as carbon nanowires or nanotubes) yield composite materials that provide useful means for controlling stem cell behavior via application of electrical currents. Fig. 3.20, for example, depicts neuronal cells aligned along carbon nanotube "tracks". The alignment of the cells on the nanotubes is still not fully understood; however, the electrical conductivity of the carbon tubes is believed

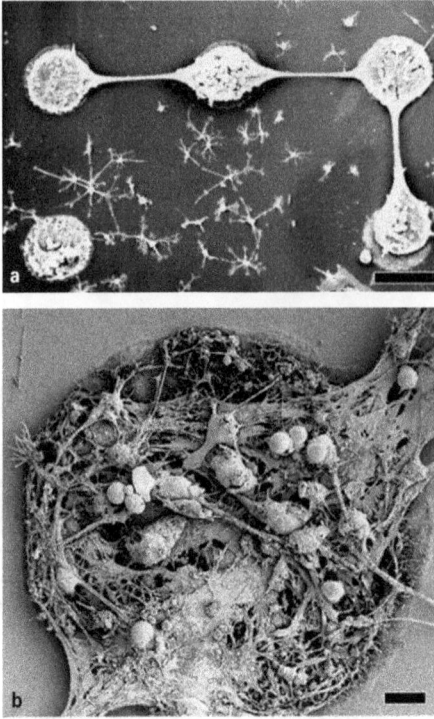

Fig. 3.20: Restricted cell growth on a carbon nanotube template. The SEM images demonstrate that the cells are confined to surface areas displaying the nanotubes **(a)**. Scale bar 100 μm. **(b)** Higher magnification of the nanotube-coated surface region, showing that the nanotube area boundary indeed restricted cell growth, except where extended to adjacent neuronal cells (bottom left and top right of the circular nanotube domain). Images courtesy of Y. Hanein, Tel Aviv University, Israel.

to play a role in promoting adsorption and growth of the neuronal cells (which exhibit sensitivity to electrical currents in physiological scenarios).

Distinct stem cell induced-proliferation methods employ three-dimensional stem cell "clumps" called "embryoid bodies" (EBs). Indeed, researchers have found that embryoid bodies constitute important junctures in the differentiation pathways of stem cells, and varied methods have been devised to manipulate the growth and environments of the aggregates. While EBs generally grow in suspensions, surface attachment can contribute to EB poliferation and subsequent differentiation. In particular, in some instances it has been found that surface *elasticity* intimately affects EB properties and ultimate differentiation pathways. Other studies reported that *hydrophobic surfaces* promote assembly of uniform (mono-dispersed) EBs, which can be further differentiated into specific cell lineages. These results are important in light of the availability of a wide range of commonly-used hydrophobic biocompatible materials and synthetic techniques for fabricating hydrophobic surfaces.

While most studies of surface interactions of stem cells (or other types of cells for that matter) consider the surface as an overall *flat* environment (albeit not necessarily *smooth*), the effects of *confined spaces* which present high surface curvatures are significant. Indeed, cells often behave differently in restricted, crowded environments compared to more open and accessible spaces. This realization has initiated attempts to mimic cell confinement in reduced-dimensionality artificial systems. Such efforts take advantage of recent advances in micro- and nano-fabrication for construction of nanoscale porous scaffolds and templates, which in many cases include microsystems designed to recreate as faithfully as possible the physiological conditions experienced by cells in such scenarios.

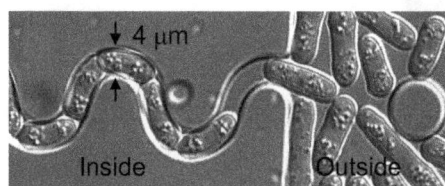

Fig. 3.21: Cell shape-change induced in a microfluidic channel. Fission yeast cells populating highly-curved, narrow channels, undergoing a pronounced shape alteration to accommodate the channel structure. Reprinted from Terenna, C.R., et al., Physical mechanisms redirecting cell polarity and cell shape in fission yeast, *Curr. Biol.* **2008** *18*, 1748–1753. Copyright (2008), with permission from Elsevier.

Interestingly, recent elegant experiments have demonstrated that microfluidic channels can actually force shape changes upon cells grown internally. Fig. 3.21 depicts experimental results obtained in the laboratory of P. Tran at the University of Pennsylvania, which nicely show that constrained, curved space induced "bending" of encapsulated yeast cells. In essence, the cells had to adapt to the constrained environment, consequently adopting curved structures which allowed their snug fit into the channel. Such synthetic channels not only demonstrated cell shape-adaptability to highly constrained environments, but were also used to study fundamental questions of cell movement and molecular transport into individual cells in the context of engineered tissues.

Interfaces between cells and abiotic surfaces play roles in intriguing functional biomimetic systems. Figure 3.22 illustrates an artificial construct comprising a living cell encapsulated within a rigid nanoparticle-assembled "shell". The "SupraCell", reported by J. Brinker and colleagues at the University of New Mexico, was based upon rapid mixing of mammalian cells and NPs, in which formation of an external abiotic skeleton (exoskeleton) was facilitated by co-addition of small molecule ligands mediators. In particular, the synthesis scheme relied upon the fast ligand-enabled complexation (within seconds) between the NPs – important for preventing NP uptake by the cells (through common endocytosis processes).

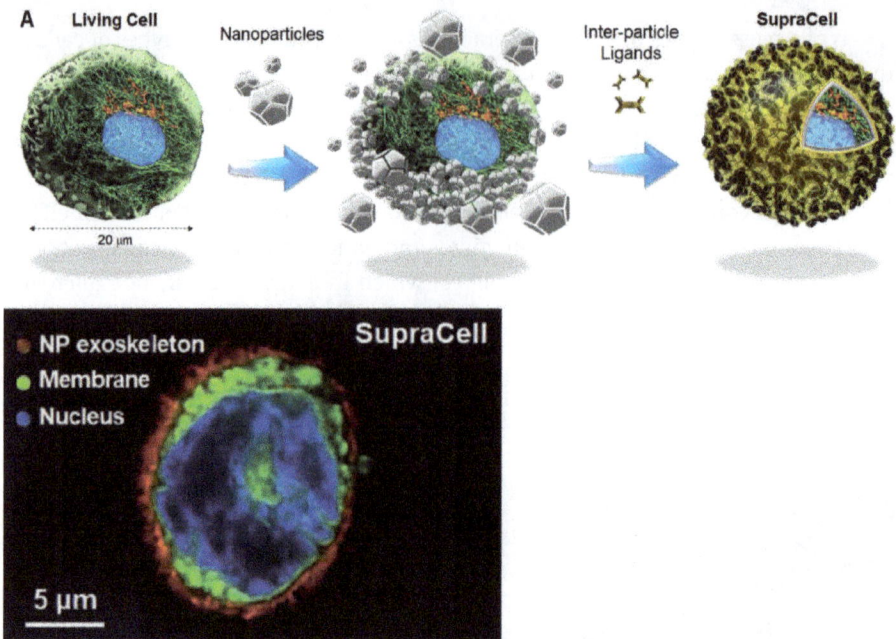

Fig. 3.22: Supracells. Top: Scheme depicting the formation mechanism of the supracells. A live mammalian cell is coated with nanoparticles (NPs), addition of small ligands induces linking of the NPs, preventing their uptake by the cells, effectively forming an exoskeleton around the cell. **Bottom:** Confocal fluorescence microscopy image showing the embedded cell within a relatively homogeneous NP exoskeleton (stained in red). Adapted with permission from Zhu et al, SupraCells: Living Mammalian Cells Protected within Functional Modular Nanoparticle-Based Exoskeletons, *Advanced Materials*. 2019, *31*, 1900545, Copyright (2019) John Wiley and Sons.

The SupraCell concept was shown by the researchers to be highly generic – accomplished using diverse classes of NPs and different cells. Remarkably, the NP exoskeleton constituted an effective protective shield to the encapsulated cell, blocking cell damage from external stress factors such as heat, UV exposure, acidity or oxidative stress. In effect, the entombed cells could be perceived as existing in a "spore-like" state, in which viability is maintained while cellular processes such as replication or spatial migration are inhibited. Importantly, however, the cells retained their functionalities upon dissolution of the NP coating, indicating that interactions between the cells and NP shell did not adversely affect overall cell viability and that cell functions were likely sustained in the SupraCell state. As such, the SupraCell biohybrids may open the way for varied applications, including cell storage in extreme environments, cell-based sensors, tissue engineering and others.

4 Artificial Organs and Tissue Engineering

Historically, development of tissue engineering technologies has aimed to address two distinct albeit related goals. A primary, highly rewarding objective albeit still elusive, is generating functional organs (or organ components) *outside* of the body, e.g. on the "laboratory bench-top", that could be subsequently implanted and/or substituted for damaged or defective organs. Another important field of research and development is the construction of functional tissues for testing of drug candidates and examination of biological processes and phenomena in environments which best mimic actual physiological and cellular conditions. This latter field of research partly derives from the enormous costs and uncertainties embedded in drug design and discovery which dictates a constant search for cost-effective biomimetic drug discovery platforms. These two distinct topics will be discussed below.

The challenges and practical difficulties inherent in regenerative medicine and its sub-discipline tissue engineering appear daunting. Indeed, the barriers to successful implementation put this field as a fine example for the intertwined hurdles and enormous potential of *biomimetics*. The fundamental requirement of any organ or tissue created in an artificial environment is that it will morphologically resemble its native counterpart and will be able to perform similar biological functions. Specifically, engineered tissues need to be physiologically compatible, i.e. integrate well and satisfactorily perform their functions after implantation into the body. More than that – such tissues need to respond and sometime transform following external stimuli in the same way as native tissues – for example lengthening and thickening of a muscle tissue following extended physical exercises. Overall, successful tissue engineering can be broadly defined as the structural and functional reconstitution of tissues in which the cells, biomaterials, and biological signals are combined and fully mimic their physiological settings.

4.1 Artificial organs

The quest for creating artificial body parts – organs, limbs, bones, joints, and others – is perhaps as old as medicine itself, and touches the core of the interface between medicine and materials science. Until the present, however, growing viable, complete organs in the laboratory has been often delegated to the realm of science fiction books or movies (e.g. "Frankenstein"). Successful incorporation of synthetic, foreign assemblies in the human body poses two fundamental challenges. First, the embedded component has to adhere to specific requirements in terms of functionality, long-term stability, and robustness. Second, and perhaps more crucial and often problematic is the issue of "host rejection" of the foreign object.

https://doi.org/10.1515/9783110709490-004

To become an adopted therapeutic solution, artificial organs need to be superior to conventional donor-extracted organs in several ways. One crucial parameter is *time*. Since artificial organs belong in large part to the synthetic realm of materials science, in principle they can be made to order more quickly than a donor organ is found. In situations where the organs are grown from a patient's own cells, they also do not require immunosuppressant drugs to prevent rejection, often the case with organ donation, which can be harmful to the body by themselves and occasionally ineffective. Many research avenues are currently being pursued to create artificial organs, from the relatively simple and already available as products, such as the commercial production of human skin for burn therapy, towards goals that might be realized only in the distant future, such as growing heart, kidney, liver and similarly complex organs.

Assembling whole organs at the laboratory-bench is a formidable task. To construct a kidney, for example, one has to engineer a complex structure comprising 15 (!) cell-types, all functioning in tandem. This is undoubtedly a huge challenge that might never be fully realized. More conceivable and realistic goals currently pursued focus on regeneration of organ constituents, ultimately embedding them as substitutes for damaged or defective counterparts within the native organ. An alternative to partial organ regeneration is to use a "hierarchical" approach in which core elements of the organ are synthetically constructed, while more peripheral parts are produced in more conventional manners. As an example of such schemes, among the foci of recent research in kidney regeneration has been the reconstruction of the renal *tubular* framework, upon which the appropriate cell population and blood vessel network could form.

Similar efforts center on mimicking the *functionality* of major components within the target organ. Again, in the context of artificial kidneys, P. Dankers and colleagues at the University of Groningen created artificial renal *membrane-like layers* which constitute a platform for seeding and growth of kidney epithelial cells on one side, endothelial cells (for regeneration of blood vessels) on the other side (Fig. 4.1). The basic idea in systems such as that depicted in Fig. 4.1 is to fabricate the core physical scaffold of the organ (consisting of a fibrillar biocompatible polymer in the system presented in Fig. 4.1), endowing it with the appropriate physical/chemical properties (such as mechanical flexibility, high surface area, and permeability), and inducing growth of functional cell monolayers on the scaffold. The process culminates in integrating the resultant product within the actual organ or surrounding tissue through implantation.

In some cases, artificial organs have been created and utilized not as actual substitutes to damaged human organs but were implanted in non-human targets for investigating biological properties, toxicity, effects of therapeutic treatments and drugs, and overall integration of foreign tissues within a recipient organism. An interesting example has been the implantation of artificial human organs in mice – creating "humanized" mice that allow examination of varied aspects pertaining to the artificial organ's functionality and systemic response of the mouse host. In particular, such mixed human/animal models could offer significant advantages in preclinical studies of drugs over *in vitro* analyses utilizing just cells or tissues, or studies involving sole animal models. S. Bhatia and colleagues at MIT have shown, for example, that a tissue-engineered liver constructed with human hepatocytes (primary liver cells) can be implanted in mice and utilized for analysis of cell metabolism, drug effects upon liver tissues and drug degradation, and gene expression induced by added molecules.

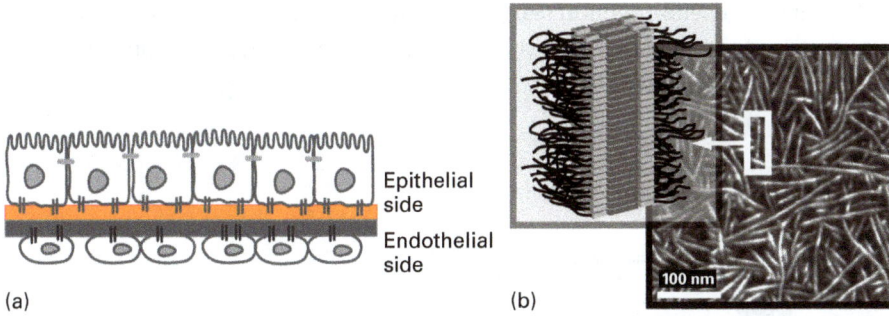

Fig. 4.1: Engineered kidney membranes. **(a)** Artificial cell layers for kidney-tissue engineering. **(b)** Macro-scale organization of the polymer scaffold employed for cell growth; the SEM image shows the long nanofibrillar stacks which provide a very high surface area. Reprinted from Dankers, P.Y.W. et al., From kidney development to drug delivery and tissue engineering strategies in renal regenerative medicine, *J. Controlled Rel.* **2011** *152*, 177–185. Copyright (2011), with permission from Elsevier.

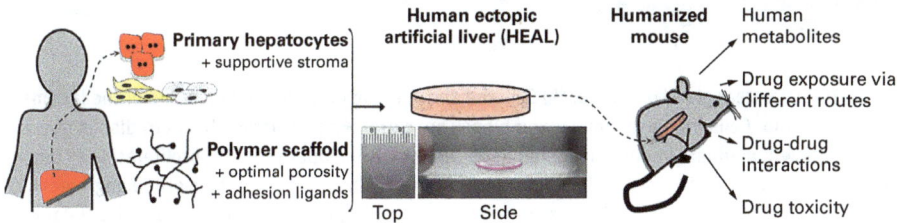

Fig. 4.2: Human organ implantation in mice. Technique for creating artificial liver, implanting the organ in mice, and using the "humanized" mice for biological and pharmacological studies. Primary hepatocytes are cultivated on a polymer scaffold together with liver endothelial cells, creating an artificial liver that is subsequently implanted in a mouse (i.e. "ectopic" liver). The "humanized" mouse model enables examination of varied drug effects and biological processes associated with the implanted liver. Reprinted with permission from Chena, A. A. et al., *PNAS* **2011** *108*, 11842–11847. Copyright (2011) National Academy of Sciences, U.S.A.

4.1.1 Organoids

Organoids, sometime referred to as "synthetic mini-organs", constitute assemblies of cells grown in laboratory settings. The key feature of organoids is the possibility to mimic biological functions of actual tissues, while obviating the need to generate in the laboratory the complex composition and architecture of an entire tissue (or organ). This goal can be naturally difficult to accomplish, and a variety of creative approaches have been introduced for organoid design and applications. Fig. 4.3 illustrates the conceptual framework of organoid construction and an example of a practical utilization. In that study, H. Clevers and colleagues at the Hubrecht Institute, The Netherlands, created an artificial "tear gland" (lacrimal gland) for studying the molecular and genetic origins of tear generation; furthermore, the research may open new avenues for producing synthetic substitutes for protective eye fluid addressing harmful conditions such as the "dry eye syndrome".

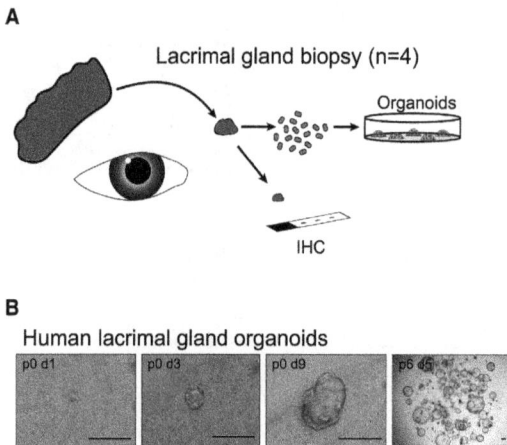

A

Lacrimal gland biopsy (n=4)

Organoids

IHC

B

Human lacrimal gland organoids

p0 d1   p0 d3   p0 d9   p6 d5

Fig. 4.3: Human tear gland organoid. (A) Scheme showing the preparation procedure of the lacrimal gland organoid. Cells from a lacrimal gland biopsy are extracted and plated on a petri dish in specialized growth medium. Immunohistochemistry (IHT) is carried out to confirm proliferation of functional gland cells. (B) Microscopic images showing the growth of functional tear-producing organoid within 9 days after cell seeding. Scale bars correspond to 200 μm. Reprinted from Bannier-Helaouet et al., Cell Stem Cell, 28, 1-12 Exploring the human lacrimal gland using organoids and single-cell sequencing, Copyright (2021), with permission from Elsevier.

As depicted in Fig. 4.3a, the lacrimal gland organoid was prepared by incubating surplus cells from lacrimal gland biopsies in a rich medium, containing varied cell growth agents and signaling molecules. Molecular profiling of the resultant proliferating cells (via immunohistochemistry, IHC) confirmed their identity as tear-producing epithelial cells found in the human tear glands, underscoring that the lacrimal gland organoid may display the desired physiological function as a source of tears. Optical microscopy analysis (Fig. 4.3b) reveals striking organization of the cells into a circularly shaped aggregate mimicking an actual physiological lacrimal gland. Importantly, the researchers employed the lacrimal gland organoids for genetic screening, identifying expression of core proteins which regulate formation and differentiation of the tear-gland cells. As such, this study underlies the potential of organoid-based research in furnishing possible therapeutic avenues for lacrimal gland dysfunction and pathologies.

Fig. 4.4 presents another striking demonstration of a successful organoid technology for creating a functioning human tissue. The research, carried out by T. Seto and colleagues at Keio University, Japan, focused on recreating the epithelial lining of the small intestine. The small intestine is the primary organ for nutrient absorption in the digestive system, accomplished via the extensive surface area maintained by the protruding "villi" of epithelial cells lining the intestine. Several bowl diseases, such as short bowel syndrome (SBS) and inflammatory bowel disease (IBD), induce significant damage to the small intestine epithelium and villi functionality thus disrupt nutrient absorption. The intestine restoration strategy depicted in Fig. 4.4 aimed to construct functional small intestine organoids which may be implanted in disease-inflicted patients, offering a possible remedy.

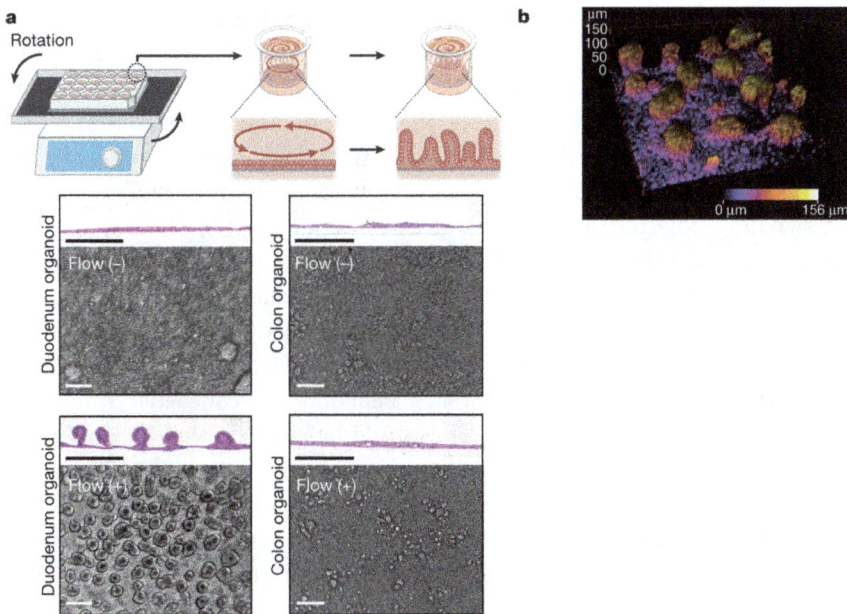

Fig. 4.4: Small intestine organoid. **(a)** Scheme showing the organoid growth procedure. The microscopy images demonstrate that the rotational motion of the well-plate creating water flow is crucial for producing the villi. Left images: duodenum (top intestine) organoid surface without rotation (top image; flat surface without villi) and with rotation (showing circular villi). Right images: colon organoid not forming villi upon rotation, as expected. **(b)** Multiphoton depth-coded three-dimensional image depicting the villi formed in the organoid. Reprinted by permission from Nature Publishing, Sugimoto et al., An organoid-based organ-repurposing approach to treat short bowel syndrome *Nature, 592,* 99-101. Copyright 2021.

The intestine organoids were generated through placing of epithelial cells (obtained from volunteers) in conventional laboratory well plates (Fig. 4.4a). Remarkably, the researchers observed that villi developed in the organoids only upon the application of plate rotation, which induced continuous flow of the medium overlaying the cells (Fig. 4.4a, left; Fig. 4.4b). This result is important, since it attests to the significance of the digestive solution flow – a characteristic feature of the small intestine which continuously undergoes mechanical contraction/release – as a critical environmental cue affecting the physiological organization of the intestine epithelial cells. Notably, medium flow did not induce villi formation in *colon* organoids (Fig. 4.4a, right) as the colon experiences nearly stagnant flow of the digestive liquid. Notably, the study further confirmed the biomimicry of the intestine organoid, as triggering and upregulation of varied genes associated with properly functioning small intestine epithelial cells were observed.

Stem cells have been often utilized in organoid construction. Stem cells constitute an attractive platform for organoid applications since the cells can differentiate, yielding distinct mature cell populations and cell compartmentalization, defining features of most organs. The primary challenges in stem-cell-originated organoids (and stem cell technology, in general) have been the identification and proper control of cell differentiation, its dynamic profile and ultimate formation of a functioning organoid/tissue. This is generally carried out through stimulation with varied molecular cues, usually signaling molecules and gene regulators.

Fig. 4.5 portrays the fabrication of a kidney organoid employing human pluripotent stem cells (hPSCs). Pluripotent stem cells constitute "master" stem cells, capable of differentiating into almost all cell types in the human body. As such, they are considered effective cell source as they can be transformed into different cell / tissue targets. In the work depicted in Fig. 4.5, J.C.I. Belmonte and colleagues at the Salk Institute, US, showed that kidney organoids could be created using hPSCs. Importantly, the researchers modulated cell differentiation within the organoid through induction and tuning several cell-signaling pathways. Indeed, the study outlined in Fig. 4.5 represents a major accomplishment since the complex architecture of the nephrons, including the distribution of blood vessels (e.g. vascularization) could be recreated in the organoid.

The cartoon in Fig. 4.5 illuminates the usage of the kidney organoid for studying nephron development and properties, pathophysiology phenomena associated with kidney malfunction, and possible therapeutic avenues. Specifically, through analysis of individual cells within the organoid, the researchers identified specific molecular pathways triggered by distinct kidney cells that are responsible for generating the vascular network critical for kidney viability. Remarkably, implantation of the organoids in mice kidneys resulted in their maturation into functional units performing effective blood filtration and solute re-adsorption. Moreover, the kidney organoids could be used for screening drug candidates against cyst-forming kidney disease.

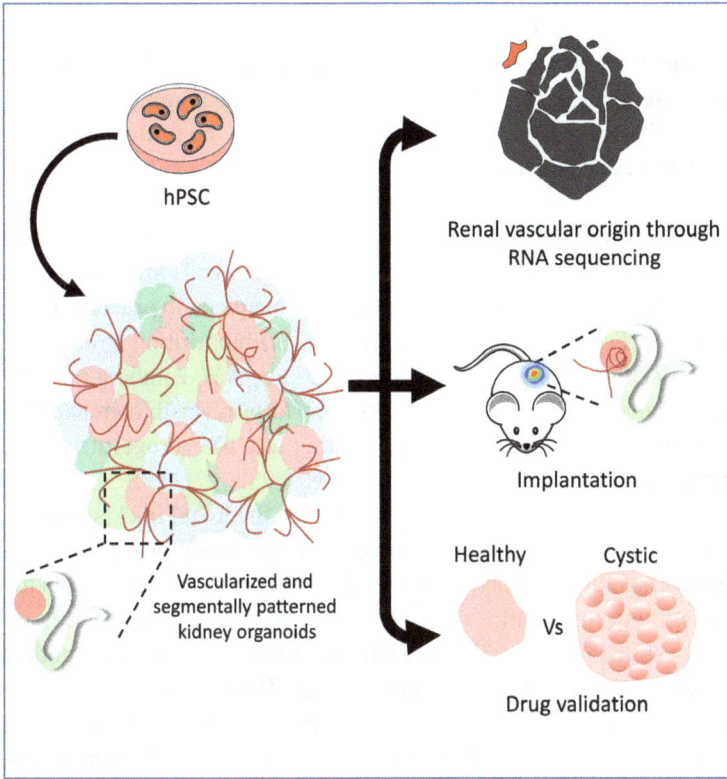

Fig. 4.5: Kidney organoid generation from stem cells. **Left column:** Scheme showing organoid development. Starting from human pluripotent stem cells (hPSCs), subsequently transforming to kidney (nephron) cells. Importantly, blood vessel formation (vascularization) was accomplished in the organoid. **Right column:** applications of the organoid. Identification of cell compartment responsible for vascularization; maturation of the organoid into functional unit upon implantation in mice; drug validation for kidney diseases. Reprinted from Low et al., *Cell Stem Cell, 25*, 373–387, Generation of Human PSC-Derived Kidney Organoids with Patterned Nephron Segments and a De Novo Vascular Network, Copyright (2019), with permission from Elsevier.

4.1.2 Three-dimensional organ printing

The remarkable progress in three-dimensional printing (3D printing) technology in recent years has contributed to the advancement of artificial organ development, possibly taking "organ mass production" from the realm of science-fiction literature into the real world. Specifically, introduction of new materials and 3D assembly techniques, availability of high precision and high-fidelity instrumentation and development of computational tools have made possible dramatic accomplishments in artificial organ research. Furthermore, this field has been aided by the expansion of "personalized medicine" research, as imaging techniques such as computed tomography (CT) and magnetic resonance imaging (MRI) have been harnessed to create sophisticated "replicas" of whole (damaged) organs from patients, potentially using the replicas as actual organ substituents. Such applications face significant technical challenges, some of which are linked to the difficulties in 3D printing of soft materials that are the predominant framework constituents in most organs.

Varied biologically applied 3D printing techniques have been introduced (often referred to as 3D *bio*printing) aimed at both overcoming the technical barriers imposed by soft (biological) materials, as well as maintaining, in principle or in practice, physiological functionalities. Fig. 4.6 presents an innovative 3D bioprinting strategy for artificial organ manufacturing. The technique, developed by A.W. Feinberg and colleagues at Carnegie Mellon University, US, termed "freeform reversible embedding of suspended hydrogels" (FRESH) was based on creating soft 3D hydrogel structures inside a biocompatible host matrix acting as a template.

Fig. 4.6a illustrates the principles of the technique. A "hydrogel ink" is injected into gelatin slurry serving as the support template for the assembled organ. The gelatin matrix allows assembly of embedded three-dimensional guest construct in the desired structures; without the gelatin scaffolding, such soft structures would naturally collapse. Moreover, the gelatin scaffold is biocompatible and can accommodate cells that could migrate and seed the constructed organ. The gelatin matrix can be readily melted after formation of the embedded structures and removed upon heating slightly above 37°C, releasing the templated organ without harming the embedded cells. To demonstrate applicability of the 3D organ bioprinting method, the researchers utilized organ images obtained via conventional bioimaging techniques (i.e. CT, MRI) in conjunction with simple computer-aided design (CAD) software to successfully produce a human femur (Fig. 4.6, middle row).

As indicated above, one of the main challenges in 3D printing of human organs concerns accomplishing reconstruction of complex architectures of elastic domains and internal cavities, for example, as present in the human heart. This requirement has been tackled both through development of new 3D methodologies and instrumentation, as well as selection of appropriate biomaterials. Significant progress has been achieved in both fronts, making possible the tantalizing possibility of "organ-by-request" fabrication. Fig. 4.7 illustrates a remarkable feat by the research team of A.W. Feinberg at Carnegie Mellon University, reporting the construction of an artificial heart by the FRESH 3D printing technique.

The main achievement exemplified in Fig. 4.7 is the feasibility to construct an entire heart, including its internal anatomical structure. Specifically, the researchers utilized alginate, a biocompatible soft polymer approved for human use, as the printing medium. Importantly, alginate exhibits similar mechanical properties as native cardiac tissues, particularly elasticity and reversible morphological transformations. Furthermore, the material can be injected and molded using commercial 3D bioprinters. The images in Fig. 4.7a–d depict CAD of the heart based upon a MRI scan

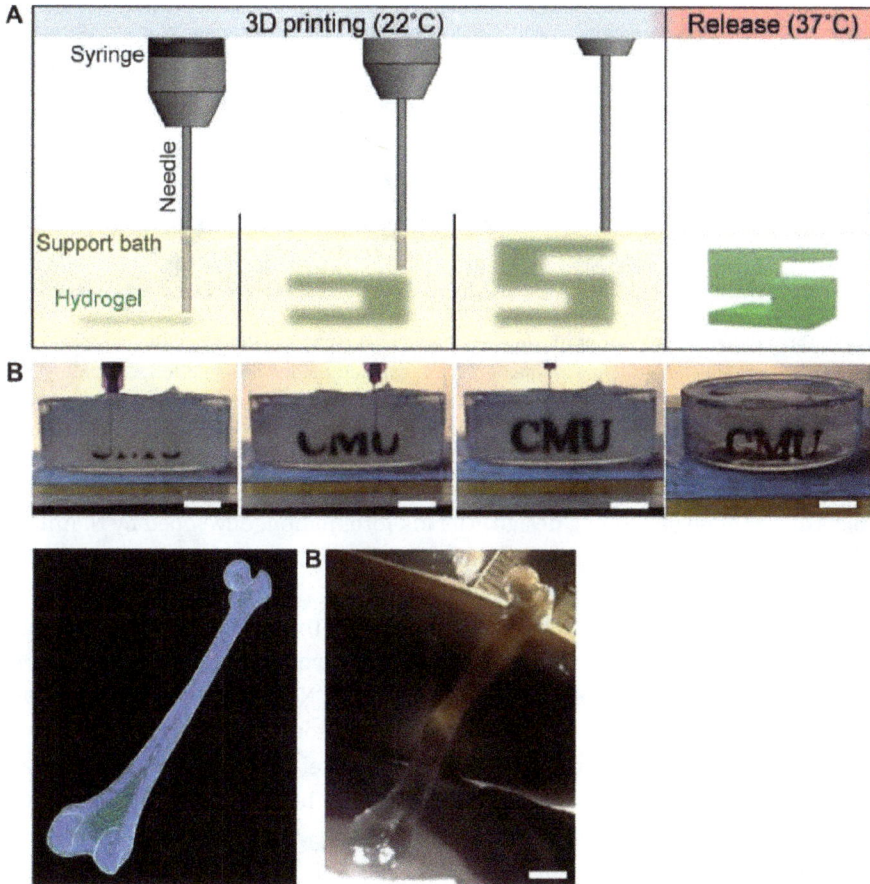

Fig. 4.6: Three-dimensional organ printing using the freeform reversible embedding of suspended hydrogels (FRESH) technique. **Top row:** (a) Schematic depiction of the technique, in which the hydrogel precursor is slowly injected into the support bath comprising gelatin at low temperature and subsequent release at a higher temperature. (b) Photographs showing the process. **Bottom row:** a computer-generated human femur based on computed tomography (CT) imaging data (left) and the 3D-printing femur using the FRESH method (right). Adapted from Hinton et al., *Science Advances*, 2015;1:e1500758, Three-dimensional printing of complex biological structures by freeform reversible embedding of suspended hydrogels, under Creative Commons Attribution License 4.0, Copyright (2015).

obtained from a patient. The full-size heart in real-life dimensions has been produced (Fig. 4.7d), underscoring also the intricate internal accomplished.

Among the most significant challenges to the progress and implementation of 3D printing of tissues and organ has been the selection of appropriate "bioinks". Bioinks need to adhere to several requirements, including biocompatibility, printability, post-printing stability and resilience, and particularly appropriate mechanical profile which could allow usage in mimicking soft tissues. Materials research has thus become an important and vibrant component of 3D organ printing R&D. Fig. 4.8

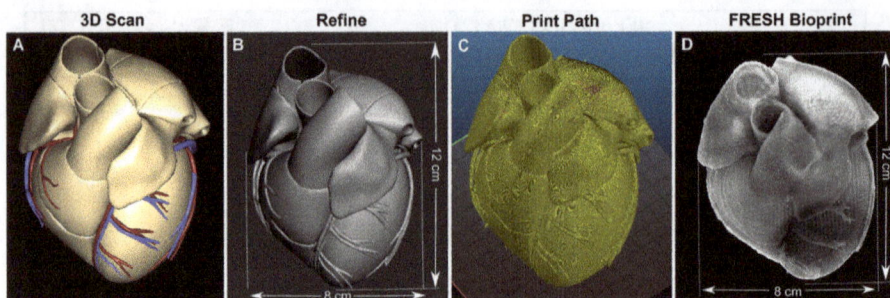

Fig. 4.7: Three-dimensional printing of a human heart model. Stages in the 3D printing process, starting a 3D scan, further structural refinement using computer-aided design (CAD), computer-aided print pathway design, 3D-printed heart from alginate, a biocompatible polymer. Adapted with permission from Mirdamadi et al., FRESH 3D Bioprinting a Full-Size Model of the Human Heart, *ACS Biomater. Sci. Eng.* 2020, 6, 6453–6459, Copyright (YEAR) American Chemical Society.

depicts an example of a bioink design. The ink formulation, developed by N. Annabi and colleagues at UCLA, comprised mixed soft polymers, including chemical derivatives of the well-known food ingredient gelatin and the mechanically stretchable human protein elastin. This mixture could be injected through the printer nozzle and spatially manipulated, and, importantly, exhibited favorable mechanical properties following maturation and polymerization (Fig. 4.8a). Notably, the researchers optimized the properties of the final printing product through tuning parameters such as printing speed, extrusion pressure and temperature. Fig. 4.8b shows objects fabricated through the technique, including a slice of a human heart displaying high fidelity of the inner morphology.

While varied multifunctional materials have been developed for 3D bioprinting applications, a core requirement which needs to be addressed is the co-incorporation of the *live components,* that is, cells, for construction of viable tissues and organs. This aspect of 3D bioprinting is challenging, since cell integrity needs to be maintained in occasionally harsh, non-biological conditions employed for synthesis of the biomimetic matrixes. Fig. 4.9 illustrates incorporation of living yeast cells within a 3D-printed gel framework, further facilitating their proliferation and biochemical functionalities. The experiment relied on mixing yeast cells with the gel precursor which maintains a liquid phase at low temperatures, solidifying at room temperature. This feature allowed 3D printing of complex shapes, such as the cube depicted in Fig. 4.9, in which viable yeast cells could be embedded.

4.2 Tissue engineering

Constructing artificial tissues in the laboratory and subsequently implanting them in the body is naturally more feasible compared to recreation of entire organs, and vast R&D efforts are currently underway in this field. Considerable challenges, however,

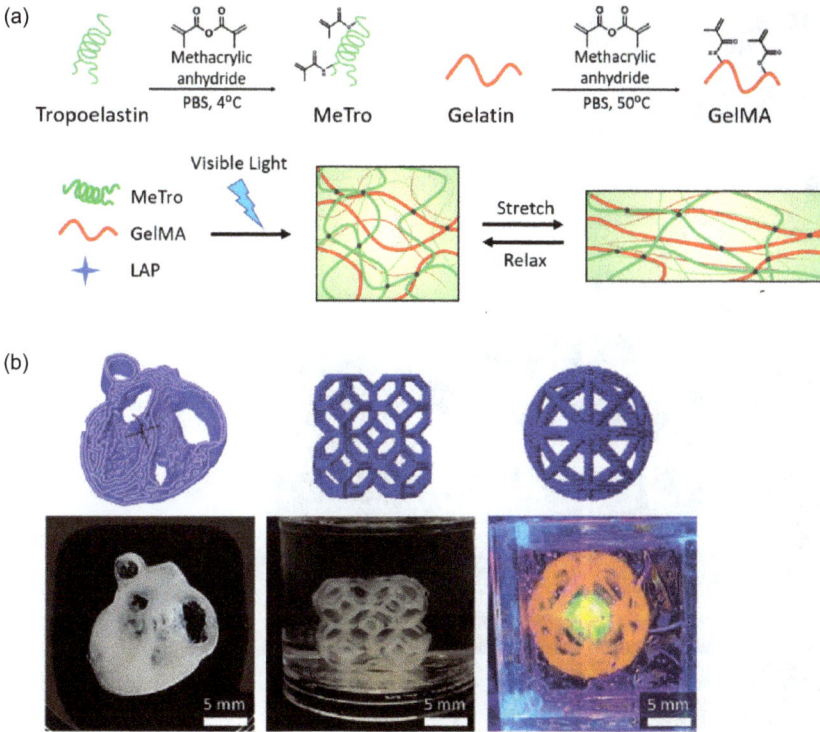

Fig. 4.8: Three-dimensional printing using elastic bioink. **(a)** Scheme showing the chemical composition and synthesis pathway for bioink fabrication. **(b)** Complex soft structures prepared from the bioink, including a human heart model (left). Adapted with permission from Lee et al., Human-Recombinant-Elastin-Based Bioinks for 3D Bioprinting of Vascularized Soft Tissues, *Advanced Materials*. 2020, 32, 2003915, Copyright (2020) John Wiley and Sons.

are encountered in the successful implementation of tissue engineering (TE) technologies. Difficulties in this field vary, depending upon the type of tissue being recreated in the laboratory, properties of the scaffold materials used (in particular their stability in the physiological environment and biocompatibility), and ultimately – the viability and integration of the produced tissues in their target locations in the body.

A representative case underscoring the fundamental and practical requirements of TE is the quest towards construction of *cardiac* tissues. Such tissues, if successfully engineered, would be enormously beneficial for patients suffering from heart diseases or those experiencing myocardial infarction or heart failure. The ideal artificially-created cardiac tissue should mimic the functional and morphological characteristics of a native heart muscle. These involve the distinct mechanical, electrical, and metabolic properties of heart tissue which are ultimately responsible for efficient systolic and diastolic performance. Accordingly, the engineered heart muscle tissue should be capable of efficient contraction and expansion, respond to electrophysiological signals, exhibit long-term mechanical stability, develop an effec-

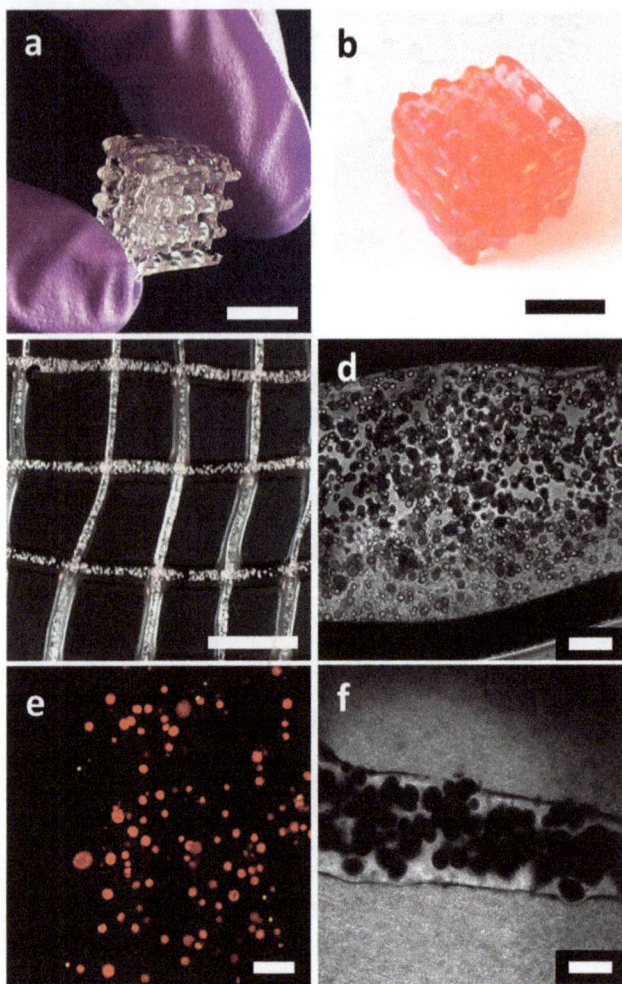

Fig. 4.9: Live yeast cells embedded in a soft hydrogel. (a–b) Shape and appearance of the gel cube comprising of a hydrogel material embedding live yeast cells. (c) Visual appearance (c) optical microscopy image (d) and confocal microscopy image (e) of the cube components depicting the embedded yeast cells, collected after 3 days. (f) Optical microscopy image after 15 days, showing viable yeast cells. Scale bars: **(a)** 15 mm **(b, c)** 10 mm, and **(d, e, f)** 200 μm. Reprinted with permission from Saha et al., Additive Manufacturing of Catalytically Active Living Materials, *ACS Appl. Mater. Interfaces* 2018, 10, 16, Copyright (2018) American Chemical Society.

tive vascular (i.e. blood vessel) network after implantation, and should not be rejected by the body. Satisfying all these conditions constitutes a formidable task.

4.2.1 Cell seeding and growth in engineered tissues

TE research can be divided into three main components: growth and proliferation of **healthy and functional cells**, proper design of **scaffolds** upon which the cells will assemble, and achievement of effective **biological signaling** within the artificially-created tissue. The obvious starting point for the tissue engineer is the *cell*. Generally, the type of tissue one desires to construct will in most instances determine the cell species that needs to be grown. Selecting and growing the appropriate cells have been a significant impediment to progress in practical TE applications – mostly due to limited availability of desired cells which often had to be carefully matched to the individual in whom the tissue will be grafted in order to minimize tissue rejection. Advances in stem cell technology, however, have made a significant contribution in overcoming this limitation, see below. Another factor to consider when working with cells in engineered tissues concerns the risk of adversely affecting cell properties. In principle, creating healthy cell populations on laboratory-fabricated scaffolds does not entail modification of intrinsic cell properties, such as signaling pathways, gene expression, and metabolic processes. However, in many instances the close interplay between a cell and its environment significantly affects the biochemical profile of the cells, producing another layer of complexity.

The advent of *stem cell* technology in recent years has dramatically altered the field of tissue engineering (among other biomedical disciplines upon which stem cells have had considerable impact). Stem cells can be made to differentiate into varied cell types, generally using specific molecular cues that have been thoroughly explored. Accordingly, such molecular inducers (usually different *transcription factors*) dictate which cells will be generated from the stem cells. This is particularly powerful in the context of TE, since stem cells extracted from the bone marrow, for example, might be made to evolve into specific tissues through the use of appropriate transcription factors and physical scaffolds. Importantly, after the engineered tissue is placed inside the body, the body itself will assist in the regeneration process and forego tissue rejection since the tissue itself was produced from the stem cells of the individual in whom the tissue was implanted.

Seeding and proliferation of cells in engineered tissues is, in fact, quite a complex process. Recent studies point to intriguing scenarios involving cells seeded in artificially-produced tissues that are subsequently implanted; surprisingly, J. Roh and colleagues at Yale University have discovered that stem cells embedded and grown within laboratory-engineered blood vessels gradually *disappeared* after implanting the vessels around the heart (the specific vessels were targeted as a therapy for severe heart malformation in children; the experiments were conducted in mice). This unexpected observation stands squarely against the conventional wisdom in TE which assumes that seeded cells proliferate and subsequently populate the entire engineered tissue.

This interesting phenomenon appears to point to a previously unknown role for seeded cells (Fig. 4.10): recruitment of the host immune cells to the engineered tissue and subsequent rapid formation of new blood vessels – a critical prerequisite to integration and functionality of the implanted TE vessel. This hypothesis might significantly expand TE and even artificial organ applications; it implies that to successfully introduce an engineered tissue into the body one does not necessarily need to laboriously embed specific cells and promote their proliferation, but rather employ "standard" cell lines whose role would be to trigger the immune system to induce vascularization within the artificial tissue which would enable growth of the *endogenous* cells.

4.2.2 TE scaffolds

A fundamental distinction between proliferating colonies of human cells in a Petri dish or upon a specifically-designed surface, and an actual *tissue* is the three-dimensional organization of cells in a tissue. Indeed, perhaps the most intensive R&D efforts in TE focus on facilitating appropriate *scaffold configurations* in order to recreate the actual organization and environments of the tissues within the body. In this context, controlling the microscopic and macroscopic architecture of the scaffold has an utmost importance in realizing the biomedical potential of TE. A variety of three-dimensional TE scaffoldings have been reported, including arrays of biological and natural fibers, porous gel matrixes, polymer *tubes* as vehicles for production of blood vessels, and others.

A critical precondition for effective tissue growth is the need for the scaffold material to be both *biocompatible* and *biodegradable*. These stipulations are understandable, since the tissue grown in the laboratory is likely to be implanted back into the body along with the scaffold upon which it is constructed. Accordingly, a broad area of scientific activity aims to identify new biocompatible and biodegradable materials, which would also exhibit specific physical properties and architectures allowing their use in TE designs.

Porous materials, in particular, have attracted great interest as scaffoldings for TE (Fig. 4.4). Such materials can host cells in their pore structure, and, furthermore, their structures and surface chemistry can be synthetically controlled. The pore structure yields large internal surface areas for cell attachment, proliferation, and physical support for the tissue formed. In addition, the pore configuration is conducive to an efficient transport of molecules to and from cells, enabling the vital metabolic pathways, signaling, and cell-cell communication processes taking place in viable tissues. A range of permeable biomaterials have been explored as three-dimensional TE platforms, including polymers, fibrous materials, hydrogels, and others (Fig. 4.11).

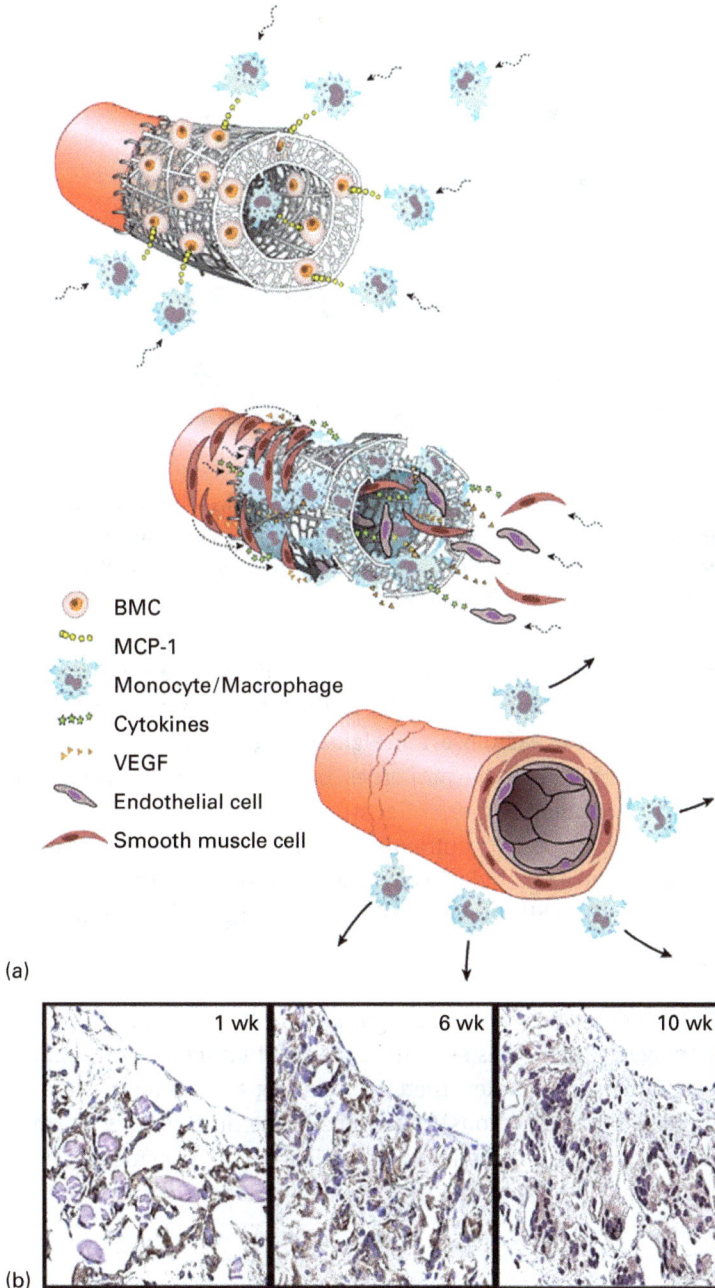

BMC

MCP-1

Monocyte/Macrophage

Cytokines

VEGF

Endothelial cell

Smooth muscle cell

(a)

(b)

1 wk   6 wk   10 wk

Fig. 4.10: Tissue regeneration through seeding cells in blood vessels. **(a)** Possible mechanism of engineered blood vessel through tissue engineering based upon cell seeding. The immune system's monocytes and macrophages are initially recruited onto the biodegradable TE scaffold via chemical inducers. The seeded monocytes induce attachment and proliferation of endothelial cells (ECs) and smooth muscle cells (SMCs) which populate blood vessel walls. The ECs and SMCs organize on the

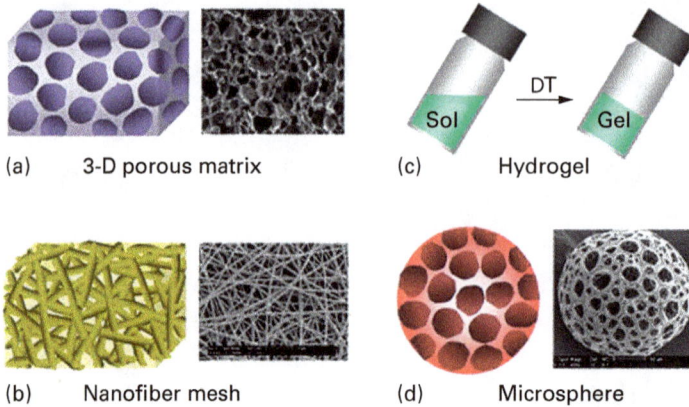

(a)　　3-D porous matrix　　　　(c)　　　　Hydrogel

(b)　　Nanofiber mesh　　　　　(d)　　　Microsphere

Fig. 4.11: Porous scaffolds used in tissue engineering. Examples of porous matrix configurations for cell growth in TE applications. Reprinted from Chung, H.J. and Park, T.G., Surface engineered and drug releasing pre-fabricated scaffolds for tissue engineering, *Adv. Drug Deliv. Rev.* **2007** *59*, 249–262. Copyright (2007), with permission from Elsevier.

Several polymers, both naturally available and synthetic, have been utilized as TE scaffolds. Among the most common are poly lactide glycolic acid (PLGA), and polysaccharides such as chitosan, cellulose, and alginate (Fig. 4.12). These branched polymers produce scaffolds with suitable properties for TE applications, specifically interconnecting networks of pores both for effective migration, growth and attachment of cells, as well as transport of nutrients and cell secretions. Other useful properties for such porous hosts include mechanical stability and inert framework materials unreactive with cells while biodegradable through digestion by proteolytic enzymes.

Hydrogels have been popular substances as substrates for tissue proliferation and assembly. The advantages of hydrogels stem from the fact that their porous matrix can be perceived as immersed in water, thereby providing a convenient platform which mimics the physiological (hydrous) conditions in the human body. Furthermore, hydrogel matrixes can be produced from biological macromolecules, particularly proteins, which then allow exploiting the protein gel framework to act both as

Fig. 4.10 (continued) luminal surface of the scaffold, forming mature blood vessel structure following scaffold degradation. **(b)** Immunohistochemical vascular endothelial growth factor (VEGF)-staining of the cell-seeded scaffolds. Images were taken at the indicated times after implantation, and show clear proliferation of the VEGF peptide (stained in brown) which is associated with vascular network formation. Image magnification was 400×. Reprinted with permission from Roh, J.D. et al., *PNAS* **2010** *107*, 4669–4674. Copyright (2010) National Academy of Sciences, U.S.A.

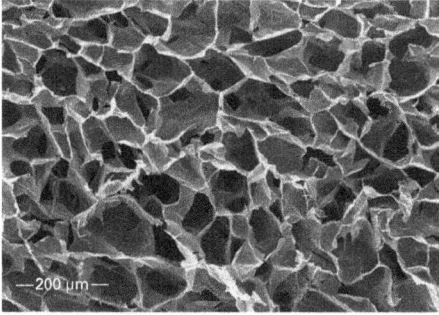

Fig. 4.12: Porous *alginate* scaffold. Scanning electron microscopy (SEM) image showing an alginate matrix utilized as a host for cells in TE applications. Image courtesy of Professor S. Cohen, Ben Gurion University, Israel.

the scaffolding, but also as means to generate biological signals to the attached cells. Indeed, the variety of bioactive peptide sequences and our ability to manipulate protein structures open the way to use biological hydrogels for "fine tuning" of tissue growth patterns, growth initiation mechanisms, and overall cell manipulation.

Fig. 4.13 illustrates an innovative application of hydrogel technology for nerve regeneration. Neuronal tissue damage following nerve injury is often irreversible. Indeed, the development of nerve healing technologies has been elusive and highly challenging. The system developed by Q.-D. Shen and colleagues at Nanjing University, China, utilizes important properties facilitated by hydrogels – electrical conductivity and mechanical flexibility. In the hydrogels reported, the researchers combined two popular, biocompatible polymers – polyaniline (PANI) and polyacrylamide (PAM) – forming a flexible, transparent matrix termed "conductive polymer hydrogel" (i.e. CPH) which could serve as an effective connective tissue between two damaged peripheral nerve termini.

The CPH exhibited important properties that were exploited for nerve cell regeneration. The hydrogel was adhesive, allowing attachment onto the irregularly shaped surface of the exposed nerve termini. The transparency of the hydrogel was also important functionally since the penetration of near-infrared (NIR) light into the material was necessary for cross-linking of the polymeric units comprising the hydrogel. Notably, the electrical conductance of the CPH, attained through seamless ion diffusion, is an essential requirement for nerve physiology. Additional benefits furnished by the CPH matrix include stretchability and resilience, both essential for nerve healing. Furthermore, the microscopic porosity of the hydrogel facilitates cell penetration and diffusion of biological molecules critical for effective regeneration of the injured nerve tissue.

Hydrogel matrixes have been studied not just as potential TE scaffolds for therapeutic uses, but also as useful three-dimensional *models* for cell migration and proliferation. For example, researchers have constructed internal *gradients* of signaling peptides, such as vascular endothelial growth factor (VEGF), within the porous frame-

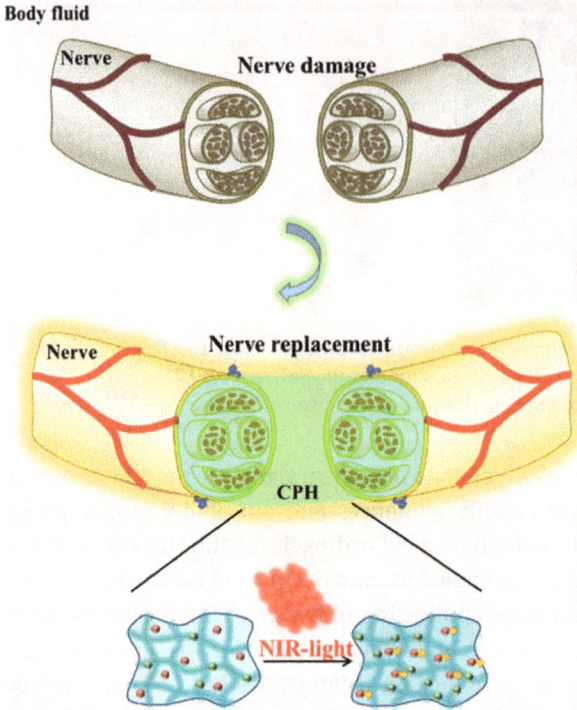

Fig. 4.13: Conductive flexible hydrogel for nerve damage repair. Scheme depicting a flexible conductive polymer hydrogel (CPH) placed in between the termini of a frayed nerve. The orange spheres within the CPH account for ions diffusing within the hydrogel matrix. Fusion of the hydrogel is carried out through irradiation with near-infrared (NIR) light. Reprinted with permission from Dong et al., Conductive Hydrogel for a Photothermal-Responsive Stretchable Artificial Nerve and Coalescing with a Damaged Peripheral Nerve, *ACS Nano* 2020, 14, 12, 16565–16575, Copyright (2020) American Chemical Society.

work of hydrogels – designed to facilitate *spatial guidance* of cell growth, particularly endothelial cells (ECs) lining blood vessel walls. ECs are critical for delivering nutrients and metabolites into or out of formed tissues. Specifically, covalently attaching peptides such as VEGF in "strategic" locations within the hydrogel matrix provides a powerful tool for directing the cells within the porous matrix.

Biodegradable TE scaffolds are usually synthesized through relatively simple chemical protocols, including fiber weaving, solvent casting, particulate leaching, membrane lamination, and melt molding, which contribute to their broad applicability. A problem encountered with several fabrication technologies, primarily involving *solvent casting* methods, is the extremely long and often insufficient removal of the chemical precursors and solvent molecules from the scaffold interior – which might adversely affect cell viability and usefulness of the scaffold. To address this deficiency, scaffold synthesis techniques relying upon textile processing concepts which do not require solvent encapsulation have been developed in recent years.

Rapid prototyping (RP) technologies constitute another interesting scheme for producing TE scaffolds, particularly in cases where the scaffold shape needs to closely mimic the actual tissue, for example in hard-tissue regeneration such as bone and cartilage. RP techniques, also referred to as *solid free form (SFF)* fabrication and aided by advances in "three-dimensional printing" are capable of producing complex rigid structures directly from computer models. Unlike conventional synthesis-based techniques, RP fabrication produces a 3D scaffold through a layer by layer deposition of thin horizontal cross sections directly from the computer-generated model. In the context of TE applications, RP can be implemented for development of manufacturing processes to create porous scaffolds that both mimic the *internal* microstructure of living tissue but also the exact desired shape and morphology of the tissue (or organ) to be implanted in the individual patient.

Three-dimensional (3D) "organ (or tissue) printing" technologies offer particular hope for regeneration of hard tissues such as *bones*, for which both the external and internal structures of the rigid matrix constitute intrinsic, critical parameters for maintaining healthy cell populations and tissue functions. For example, the interconnected porous architecture of bone, necessary for osteogenic (i.e. bone) cell recruitment, blood vessel formation, and the flow of oxygen, nutrients, and other chemicals can be easily recreated via this technology. Importantly, tissue printing procedures can accomplish cell deposition and seeding that is *concurrent* with fabricating the template (Fig. 4.14), different to most conventional TE schemes in which cells are seeded only *after* making the scaffold. This distinction provides a powerful means for cell organization within the tissue graft, but also imposes constraints upon the printing conditions, which need to be mild and should not adversely affect the seeded cells (e.g. temperature, solvents, etc).

In some cases, biocompatible scaffolding has been prepared not so much to support tissue growth *per se* but rather as an actual replacement of damaged *soft tissues*. Delicate soft tissues, such as the cheeks, nose, and other fibrous tissues, often pose significant difficulties for reconstruction through surgery. Accordingly, development of materials that on the one hand can be "sculpted" to the desired shapes and on the other hand will be biocompatible and integrate within the physiological tissue environments would be a great biomedical boon. Research has already yielded promising materials, mostly involving common biocompatible polymers already approved for human use (such as hyaluronic acid and other fatty acids) and polyethylene glycol. The advantage of such materials is that they can be sculpted *in situ* (within the site of implantation within the body) and set in place through clever external control – such as light-induced polymerization.

Artificial skin is an example of soft-tissue engineering in which the scaffold design plays a crucial role in determining the outcome of the TE procedure. Fig. 4.15, for example, shows that inclusion of the natural protein *collagen* in a polysaccharide polymer scaffold is essential for promoting cell attachment and proliferation within

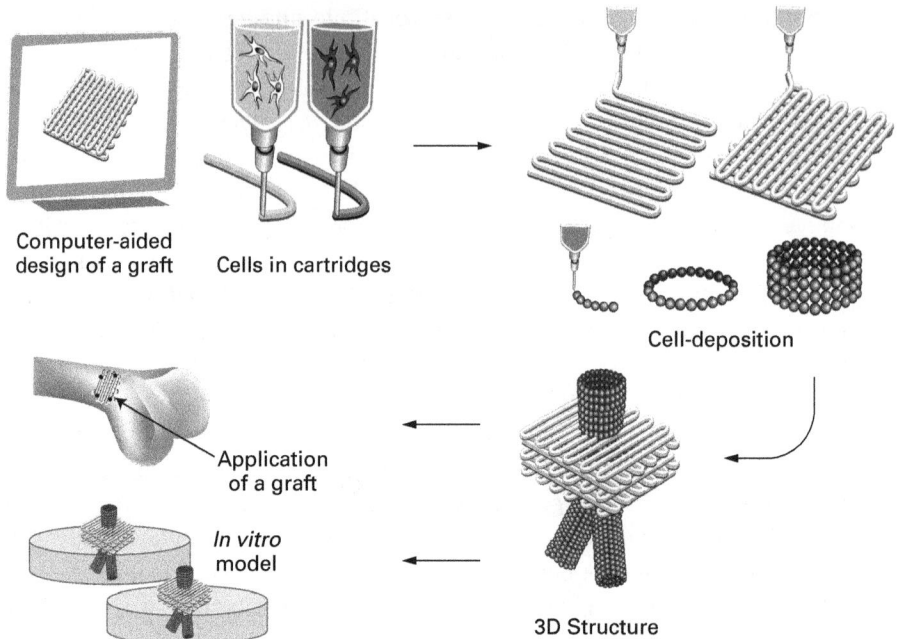

Fig. 4.14: Three-dimensional tissue printing. Generic scheme for computer-aided fabrication of tissue scaffold concurrent with the embedded cells. Reprinted from Fedorovich, N.E. Organ printing: the future of bone regeneration?, *Trends Biotech.* **2011** *29*, 601–606. Copyright (2011), with permission from Elsevier.

the scaffold. Indeed, G. Chen and colleagues at the National Institute for Materials Sciences, Japan, demonstrated that only when collagen was added to the polymer fiber mesh, did the skin fibroblast cells succeed to form a continuous graft. This effect, encountered in many other TE systems, is ascribed to both the physical organization of the collagen fiber network, but also to putative biological cues provided by this protein, which are absent in case of the non-biological (even though biocompatible) polymer scaffolds.

Similar to soft tissue regeneration, several TE applications focus on adjusting the *mechanical* properties of the artificial constructs in order to make possible their effective substitution with the target organs or tissues. Replacement of damaged **intervertebral discs** (IVDs) is an interesting case in point. Lower back and neck pain are among the leading physical conditions for physician visits in the US, and both have huge economic costs. Persistent back pain is usually associated with damaged or diseased IVDs. Accordingly, identifying satisfactory biomaterials and developing TE technologies for artificial IVD substitutes could have major benefits in terms of public health and quality of life for millions of people. The significant challenge, however, is to design a biomaterial assembly that would both display a balance between mechan-

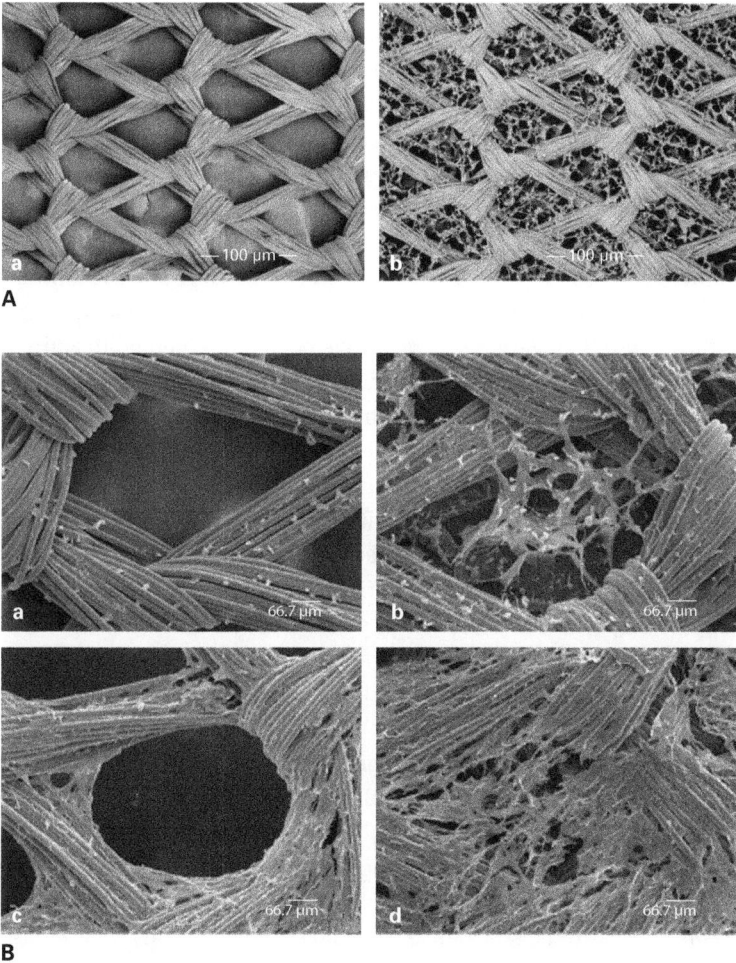

Fig. 4.15: Scaffold effects on artificial skin growth. Skin fibroblasts proliferated only in a polymer mesh scaffold that additionally contained collagen. SEM images of: **A.** the polymer scaffold poly-DL-lactic-co-glycolic acid (PLGA) mesh **(a)** without collagen; **(b)** with embedded collagen. **B.** Cell proliferation: **(a)–(b)**: only PLGA; **(c)–(d)**: PLGA-collagen hybrids. **(a), (c)**: 30 minutes after cell seeding; **(b), (d)**: 5 days. Dense cell coating observed only in the PLGA-collagen scaffold. Reprinted from Chen, G. et al., Culturing of skin fibroblasts in a thin PLGA–collagen hybrid mesh, *Biomaterials* **2005** *15*, 2559–2566. Copyright (2005), with permission from Elsevier.

ical flexibility and rigidity inherent to physiological discs, as well as exhibit biocompatibility with the surrounding tissues and spine.

While several artificial IVD designs based upon composite polymeric materials have been tried, such devices mainly aim to conform to the mechanical requirements of the discs; however they often exhibit poor biocompatibility resulting in inflammation. The key features in research aiming to overcome biocompatibility constraints and maintain mechanical properties has been to reproduce the actual tissue hierarchy of the disc, specifically the proteoglycan- and collagen-II-rich nucleus (denoted *nucleus pulposus,* or NP) surrounded by a fibrocartilage region exhibiting high content of collagen-I and proteoglycans (the *annulus fibrosus* or AF, Fig. 4.8).

A particular challenge in biomimetic IVD design is the fact that major structural and mechanical determinants of the tissue are related to *anisotropic* microscopic morphology, specifically the annular and circumferential organization of the collagen fibers comprising the AF (Fig. 4.8). This configuration poses a significant technical challenge since the tissue engineer needs to successfully recreate the annular organization within the artificial disc, in addition to mimicking the molecular and cellular composition of the IVD. Varied schemes have been implemented to address this anisotropy condition, mostly focusing on generating aligned collagen fibers in macroscopic architectures that form circumferential morphology. The collagen fibers in such engineered IVD systems constitute "templates" for the cells populating the AF, overall yielding the desired annular structure. An example for the sensitivity of cell growth to template directionality is provided in Fig. 4.9. In that study, D.L. Kaplan and colleagues at Tufts University showed that stem cells adopted the prefabricated alignment of patterned silk fibers. Particularly dramatic was the observation that *changing* the alignment of the silk fiber scaffold in the midst of the growth period resulted in modulated cell growth tracing the *modified* fiber directionality.

A particularly important aspect for construction of biomimetic scaffolds is the need to re-establish the rather complex physiological environment of the native tissue. Beside the inclusion of varied bioactive peptides, growth factors, and other biochemical signaling molecules, this requisite also involves maintaining a delicate balance of *physiological ions* inside the scaffold. These include notable ions such as Na^+, K^+, Mg^{2+}, and CO_3^{2-}. Specific tissues need the presence of additional ions; bone tissues, for example, require furnishing of Sr^+ and SiO_4^{4+}. Silicon, in fact, poses particular challenges in TE scaffolds, since its leakage might create highly toxic conditions for growing cells.

Naturally-occurring biomaterials that do not have physiological human roles have occasionally inspired the design of extraordinary TE scaffolds. The ligneous structures found in wood, for example, exhibit morphology and semi-porous organization similar to *bone* tissues. A. Tampieri from the National Research Council, Italy, pointed out the remarkable resemblance of the microchannel structure in *Rattan* wood to human bone (Fig. 4.10). Such structural similarities underscore the fact that wood frameworks exhibit other mechanical properties sought after in engineered bone tissues, including withstanding heavy loads, microscopic stretching, and shape flexibility. Indeed, from a bioengineering standpoint the unique hierarchical microstructural architecture of wood confers an attractive combination of high strength, stiffness and toughness at relatively low material density.

Carbon nanostructures, e.g. nanotubes and nanocomposite materials have attracted interest in recent years both because of the unique properties of carbon – a

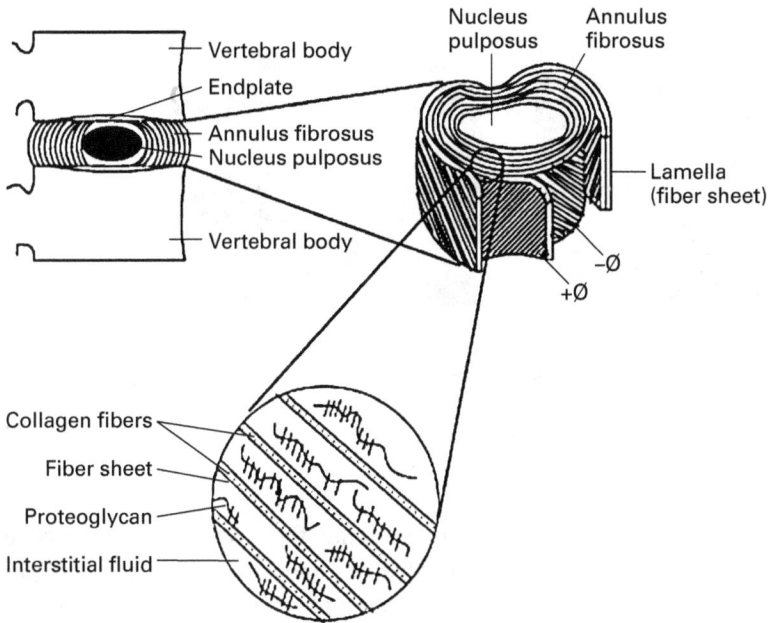

Fig. 4.16: Intervertebral disc (IVD) structure. Location of the IVD between vertebrae (top left), and the layers of the *annulus fibrosus* (AF, top right). The internal structure of the AF comprising collagen fibers in alternating layers (bottom). Reprinted from Ngwa, M. and Agyingi, E., A model of fluid injection into the spinal disc, *Appl. Mathem. Model.* **2012** *36*, 2550–2566. Copyright (2012), with permission from Elsevier.

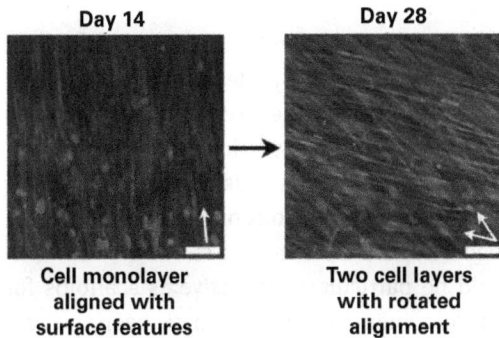

Fig. 4.17: Cell alignment determined by directionality of surface groove patterns. Microscopy images of mesenchymal stem cells (MSCs) grown on surfaces patterned with aligned silk fibers. The patterned surface was rotated after 14 days. The arrows indicate groove fiber direction. Reprinted with permission from Tien, L.W. et al., *Macromolecular Biosciences* **2012** *12*, 1671–1679. Copyright (2012) John Wiley and Sons.

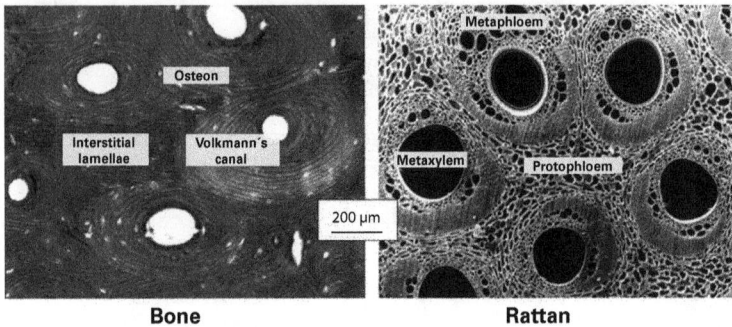

Fig. 4.18: Morphological similarities of wood and bone as tissue-engineering scaffolds. Microstructure of a bone cross-section (**left**); Rattan wood (**right**). Reprinted from Sprio, S. et al., Biomimesis and biomorphic transformations: New concepts applied to bone regeneration, *J. Biotech.* **2011** *156*, 347–355. Copyright (2011), with permission from Elsevier.

generally biocompatible material in its pure elemental form, relatively non-reactive with biological molecules, and exhibiting diverse nano- and microscale configurations. In particular, carbon nanotubes have been proposed as potentially useful scaffolds for regeneration of *heart tissues*, specifically due to the *aligned arrangement* of the elongated nanotubes which mimics the natural tissue pattern in the heart. The *electrical conductivity* of carbon nanotubes is another parameter mimicking the physiological conditions of the heart. Indeed, V. Baron at the National University of Ireland demonstrated that electrical stimulation of carbon nanotube scaffolds gave rise to dramatic alignment of stem cells in the direction of the nanotubes (Fig. 4.11).

Other biocompatible materials were explored as candidates for TE scaffolds. *Silk* fibers have been considered a particularly promising system. Silk, discussed in detail in Chapter 2, has long attracted great interest in the chemistry and materials science communities due its unique physical properties, particularly the extraordinary mechanical strength of its fibers. Several studies have demonstrated that silk proteins exhibit favorable biological properties, including biocompatibility, slow biodegradation, low immunogenicity, and low inflammatory response.

Biomimetic fibers such as silk could be particularly attractive as scaffolds for *blood vessel* regeneration. Interesting developments in this field have been the synthesis of bioactive peptide-polymer conjugates that form fibers upon injection into the injured site or regions in which blood vessel regeneration is needed. A recent demonstration of this concept has been the inclusion of a VEGF-mimetic peptide as a component of fiber-forming amphiphilic molecules designed to induce blood vessel growth (i.e. "vascularization") in engineered tissue locations (Fig. 4.12). VEGF is a key biological constituent, promoting formation and growth of new blood vessels. While many schemes have been proposed to embed VEGF in tissue environments where vascularization is desired, such approaches have often failed *in vivo*, most likely because of the rapid diffusion of the protein out of the impacted areas. The novel VEGF-mimic amphiphilic peptides developed by S. Stupp and colleagues at Northwestern University appear to exhibit superior vascularization-inducing properties. This effect likely corresponds to the covalent display of the VEGF-mimic modules upon the fiber surface, consequently acting for longer time periods to recruit endothelial cells before biodegradation of the polymer framework.

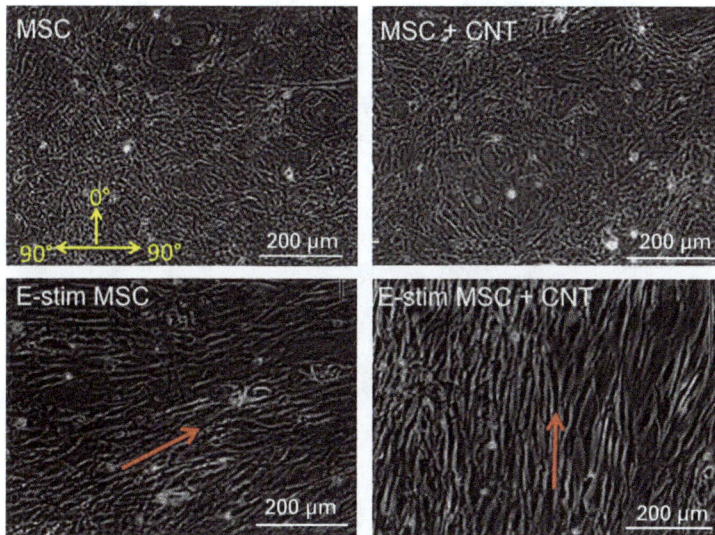

Fig. 4.19: Alignment of cells on an electrically-stimulated carbon nanotube scaffold. Effects of a carbon nanotube scaffold and electrical stimulation (duration: 14 days) upon the morphology of seeded mesenchymal stem cells (MSCs). Pronounced elongation and cell alignment were clearly linked to application of electrical stimulation through the carbon nanotube cell-growth scaffold (panel at bottom right). Reprinted from Mooney, E. et al., The electrical stimulation of carbon nanotubes to provide a cardiomimetic cue to MSCs, *Biomaterials* **2012** *33*, 6132–6139. Copyright (2012), with permission from Elsevier.

Endothelial biomaterials, in which materials used for tissue grafts are artificially coated with a layer of endothelial cells (ECs), have been used for decades for treating vascular defects and related diseases. In TE applications, deposition of ECs upon scaffolds' surfaces could aid blood vessel proliferation within the engineered tissue. Endothelial biomaterials have been produced either by seeding ECs inside a TE scaffold prior to implanting, or by relying on the infiltration of the host's ECs onto the walls of the implanted biomaterial (Fig. 4.13). The primary requirement from such materials is that they serve as efficient substrates for EC growth. The constituents of endothelial scaffolds may be biological in nature, such as extracellular matrix proteins, synthetic peptides, or biocompatible polymers. In addition, scaffolds targeted for EC coating are further modified to introduce abundant cell binding sites to expe-

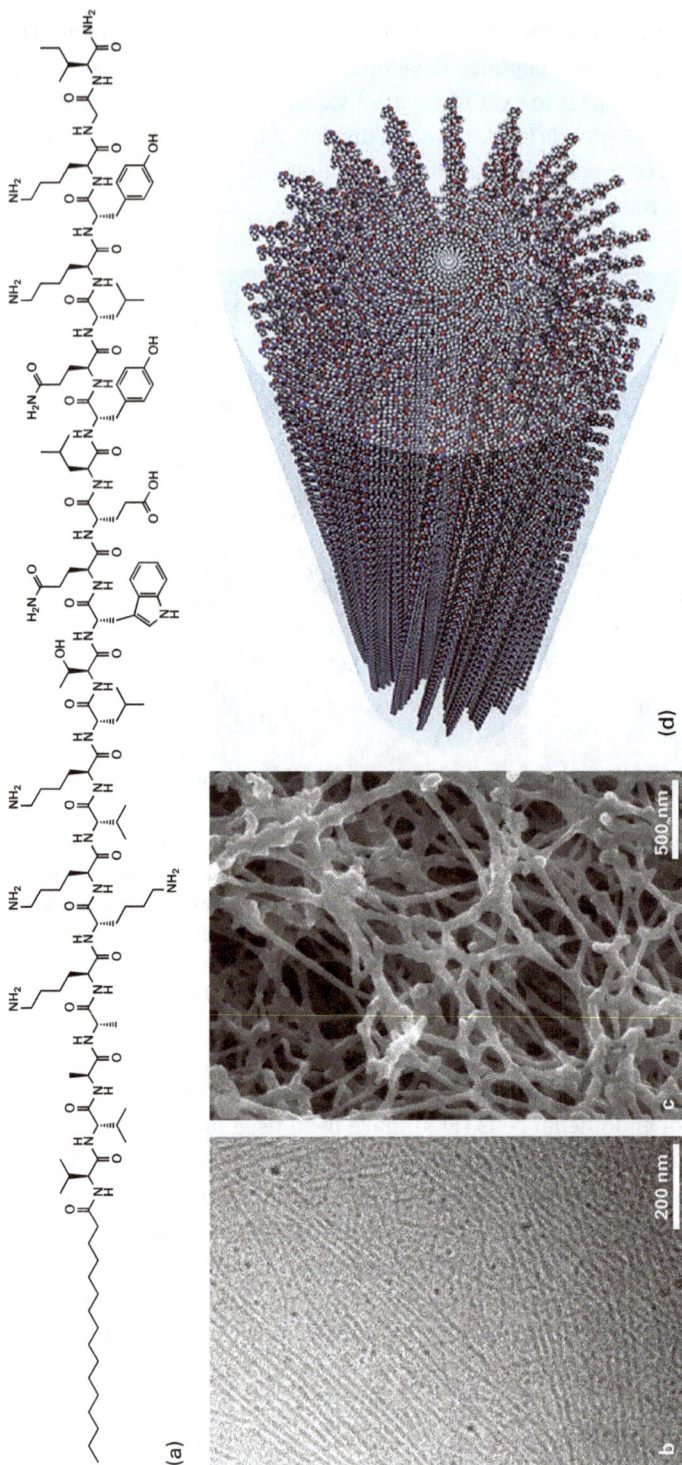

Fig. 4.20: VEGF-mimic peptide fibers for blood vessel regeneration. **(a)** Chemical structure of the VEGF-mimic peptide amphiphile. The amphiphilic region is shown at the left, while the peptide sequence mimicking VEGF is specifically outlined; **(b)** the VEGF-mimic peptide assembles into nanofibers shown in the transmission electron microscopy (TEM) image; **(c)** SEM image showing the porous gel structure formed by the peptide fibers; **(d)** schematic depiction of the peptide amphiphile fiber. Reprinted with permission from Webber, M.J. et al., *PNAS* **2011** *108*, 13438–13443. Copyright (2011) National Academy of Sciences, U.S.A.

dite cell adhesion, either through covalently displaying adhesion peptide sequences at the scaffold inner surface (such as the arginine-glycine-aspartic acid cell binding motif), or through physically imbuing the scaffold with common adhesion factors such as laminin or gelatin.

Achieving effective vascularization is indeed among the more significant challenges in TE applications. While cells in the body are usually within a short distance (less than 100–200 μm) from blood vessels providing essential nutrients, in artificial 3D TE scaffolds the comparable distances to nutrient sources can be orders of magnitude greater, leading to cell death (i.e. necrosis). Beside *chemical* means designed to promote inner-tissue vascularization described above (for example through displaying the VEGF sequence or VEGF-mimics), there are efforts to develop *physical* (or bioengineering) approaches to achieve the same goal. An innovative scheme introduced recently by C. Chen and colleagues at the University of Pennsylvania relies on the construction of a sugar-based microtubular network as the scaffold for cell seeding (Fig. 4.14). The selection of the *carbohydrate* polymer cast is crucial – it is biocompatible, and what is particularly important – it can be easily dissolved in the cell growth media, leaving behind a network of microscopic channels that allow blood flow. The engineered tissue subsequently comprises a microcapillary blood vessel organization enabling molecular transport to and from the entire cell population. Furthermore, this synthetic concept is intrinsically "multiscale", since the carbohydrate scaffolding can be constructed in diverse configurations and sizes.

Research into achieving optimal tissue scaffolds has become highly sophisticated, as scaffold design has evolved from the construction of "passive" surfaces upon which cells just proliferate, into "active scaffoldings" which intimately participate in tissue development. Such matrixes direct cell organization through slow release of biological and chemical molecules, affecting selectivity of cell adsorption in different scaffold locations, or facilitate mechanical stimulation through dynamic changes to the three-dimensional structures of the scaffold matrix. Promising approaches focus upon careful optimization of the scaffold surface chemistry, producing molecular cues that the cells respond to. The availability of sophisticated chemical synthesis methodologies greatly aids these efforts. Examined systems go all the way from the display of generic adhesion and signaling species, such as the cell-adhesion RGD module, to facilitating within the scaffold complex inter-cellular communication networks which intimately affect cell growth, migration, and overall tissue organization.

Conjugation of *growth factors* to the scaffold has been a prominent component in "active scaffold" research. Growth factors are fundamental molecular inducers for cell growth and differentiation. In the real cellular world, however, growth factors are secreted to the extracellular space and the molecules can diffuse freely within tissues; mimicking this scenario in engineered tissues would make it very difficult to harness growth factor effects at specific areas in the scaffolds. To address

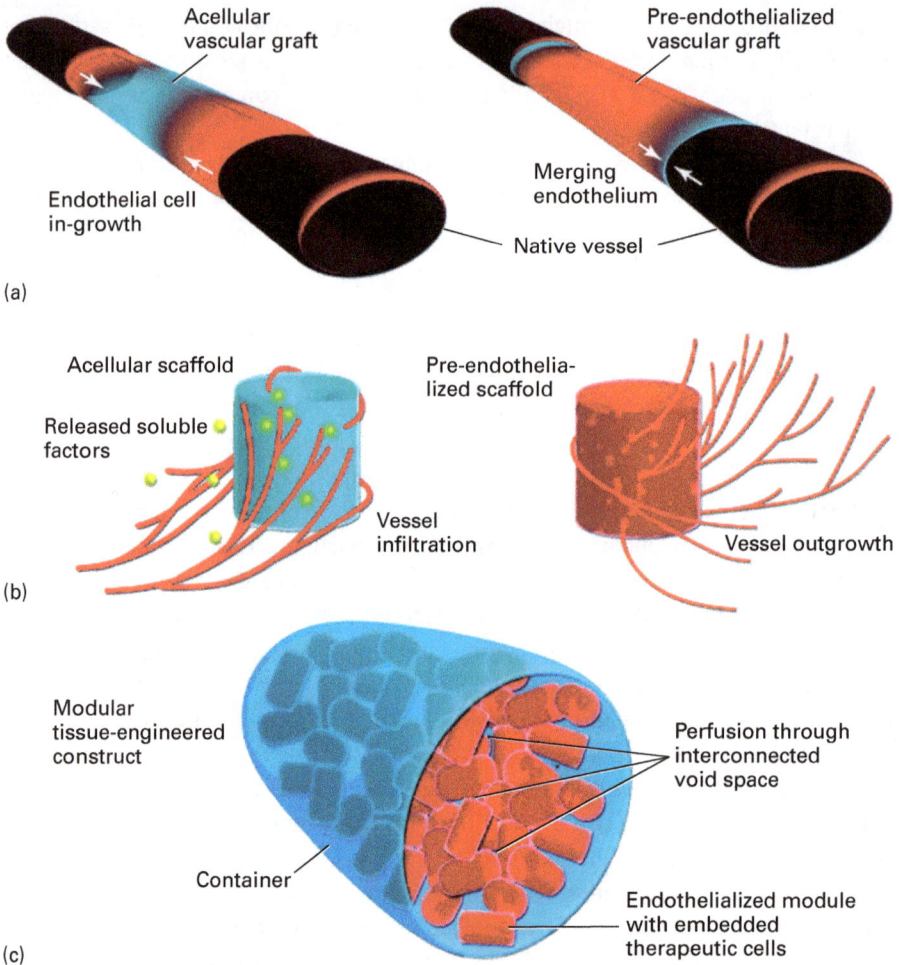

Fig. 4.21: Endothelial biomaterials. Common strategies to create endothelial scaffolds (summarized by M. V. Sefton, University Toronto). **(a)** Synthetic vascular grafts gradually coated with endothelial cells (ECs). The endothelial cells spread and cover the graft surface either prior to implanting or inside the body. Cell layer formation is additionally promoted through release of factors enhancing cell attachment and proliferation. **(b)** Vascularization (blood vessel formation) of endothelial scaffolds, essential for proper functioning of the tissue. For vascularized constructs for tissue engineering, rudimentary vascular network can be assembled prior to implantation (right), or the process can be initiated at the implanted site (left). **(c)** *Modular* tissue engineering. Small scaffolds pre-seeded with ECs can be implanted directly at the target site, or delivered in a larger container. The ECs form an interconnected network, ultimately producing a viable EC layer enabling blood transport. Reprinted from Mooney, E. et al., The electrical stimulation of carbon nanotubes to provide a cardiomimetic cue to MSCs, *Trends Biotech.* **2012** *26*, 6132–6139. Copyright (2012), with permission from Elsevier.

this issue, immobilization (e.g. covalent binding) of growth factors onto the scaffold matrix can result in local sensitization of the cells, producing synthetic niches for cell patterns.

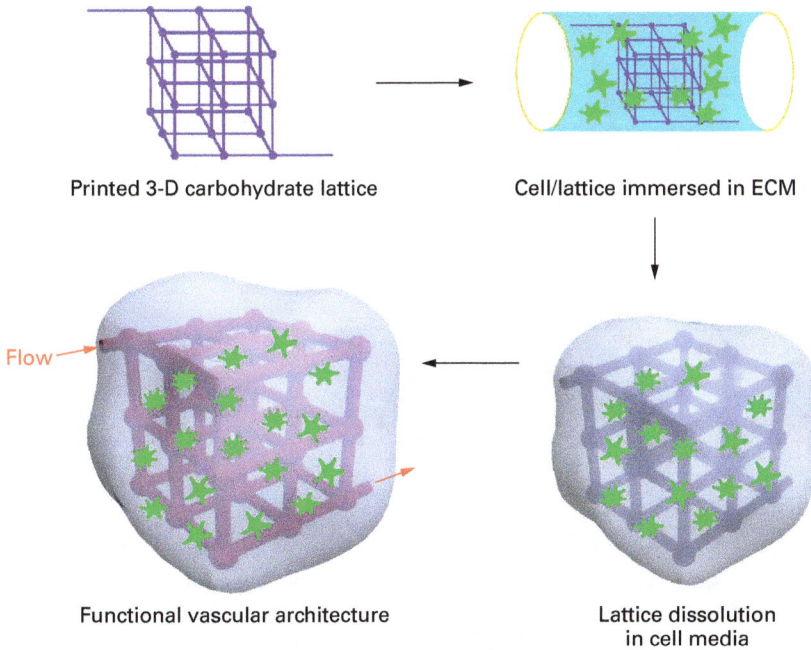

Printed 3-D carbohydrate lattice

Cell/lattice immersed in ECM

Flow

Functional vascular architecture

Lattice dissolution in cell media

Fig. 4.22: Vascular network via microtubular carbohydrate lattice. Experimental scheme for construction of a vascular network through physical means. A 3D carbohydrate-glass cast is printed and immersed in extracellular matrix (ECM) (cell environment). The carbohydrate lattice is then dissolved in cell media, leaving behind an interconnected vascular network tracing the dissolved lattice.

A "chemical tailoring" scheme for spatial activation of the ubiquitous TGF-β signaling pathway is depicted in Fig. 4.15. In the elegant experiment carried out by L. Kiesling and colleagues at the University of Wisconsin, a synthetic peptide exhibiting a TGF-β binding domain was covalently attached to a surface, subsequent binding of the TGF-β growth factor resulted in activation of the TGF-β signaling pathway in attached cells, indicating that the TGF-β factor adopted its biologically-active conformation while attached to the scaffold surface.

Growth factor display approaches such as the one shown in Fig. 4.15 provide an additional important advantage. Many growth factors elicit their action through *multimeric* (multivalent) assembly formation. Accordingly, the positioning of the molecules in specific areas within a scaffold results in multimer formation and effective initiation of the signaling mechanisms. Moreover, by using a method in which synthetic peptides are used for growth factor immobilization through molecular recognition, essential processes such as internalization of the growth factors into the recruited cells (usually occurring through endocytosis) can still take place. Overall, the growth factor scaffold display approach has the potential for accomplishing signaling control, affecting cell fate decisions on a highly local level.

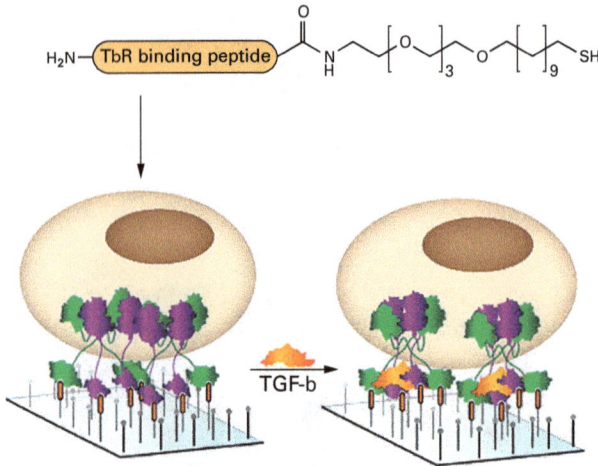

Fig. 4.23: Immobilization of growth factors on TE scaffold. Strategy to display the growth factor TGF-β on a scaffold. A synthetic peptide containing a recognition element for TGF-β is attached onto a surface through the alkanethiol unit. The TGF-β peptides are displayed upon specific binding to the recognition units, activating the TGF-β signaling complex in the cells. Reprinted with permission from Lia, L. et al., *PNAS* **2011** *108*, 11745–11750. Copyright (2011) National Academy of Sciences, U.S.A.

A paramount requirement in *nerve cell regeneration*, among the most sought after albeit difficult goals in regenerative medicine, is to attain cell growth *directionality*. This issue is particularly acute since the ultimate action of a nerve cell is determined by its connectivity with adjacent cells. Different biocompatible scaffolds have been developed which aim to "guide" a developing neuron within an engineered tissue. Specifically, scientists have focused on creating microenvironments within synthetic scaffolds containing controlled distributions of molecular inducers deigned to "attract" growing nerve cells (or significant components of the cells, such as the axons).

While biochemical signals play crucial roles in developing successful neuronal TE applications, surface topography has been also found to exert a significant effect in such systems. Similar to regenerated heart muscle tissues discussed above, *fibrous* scaffolds induced alignment of the cytoskeleton and nuclei of Schwann cells (which are the primary cells in the peripheral nerve system) along the fiber axes. Such alignment is likely related to the physiological orientation of neurons within body parts (such as arms, legs, etc.).

Nanotechnology has opened new avenues for TE research. Specifically, manipulating scaffold dimensions on scales that are pertinent to biological systems – proteins, subcellular organelles, and whole cells – gives new opportunities for design parameters and tissue control. The extracellular matrix (ECM) is a case in point when considering potential contributions of nanotechnology in TE. The *nanofiber* mesh structure of the ECM is physiologically important because it provides physical support to the cells and intimately influences cell behavior via cell-ECM interactions. Furthermore, the ECM functions as a natural "reservoir" for a wide range of biological factors, activating the molecules through sophisticated mechanisms involving cell-cell and cell-ECM communication. Recent advances in nanotechnology have enabled fabrication of biomimetic microenvironments which mimic the features and nanoscale organization of the ECM. In particular, networks of natural and synthetic fibers have been synthesized, further containing varied bioactive molecules such as cell-binding ligands, signaling molecules, growth factors, and others (Fig. 4.16).

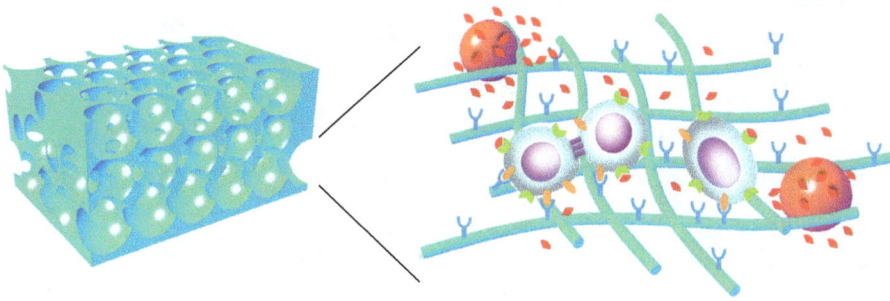

Fig. 4.24: Artificial extracellular matrix (ECM). Schematic illustration of a generic biomimetic scaffold comprising intertwined synthetic nanofibers. The scaffold provides both the physical infrastructure for accommodating cell growth and tissue formation, as well as a platform for facilitating biomolecular interactions of the cells with bioactive agents. Reprinted from Zhang, Z. et al., Nanofiber-based delivery of bioactive agents and stem cells to bone sites, *Adv. Drug. Deliv. Rev.* **2012** *64*, 1129–1141. Copyright (2012), with permission from Elsevier.

Nanotechnology has contributed to TE method development in other areas. Recent studies have shown that embedding *gold nanoparticles* (Au NPs) within TE scaffolds can endow important advantages for the resultant hybrid materials. Specifically, the *photothermal* properties of Au NPs (particularly gold *nanorods*) – the feasibility of local heating of the Au NP environment through irradiation with near-infrared light – makes the integration of the NPs within the scaffold a powerful tool, for example initiating adhesion of the scaffold or the encapsulated cells onto desired target regions in the body through local scaffold melting.

A TE system in which *Au nanorods* have played an intriguing role is depicted in Fig. 4.17. In that work, D. Kohane and colleagues at the Harvard Medical School have shown that placing Au nanorods within a porous alginate scaffold enhanced proliferation of cardiac cells, leading to a functioning cardiac patch. In particular, the inclusion of the electrically-conductive Au nanorods within the scaffold facilitated *electrical communication* between cells in adjacent pores – a critical parameter for proper functioning of a cardiac tissue. The concept highlighted in Fig. 4.17 demonstrates that the Au nanorods, which are essentially non-biological entities, were singularly responsible for inducing the transformation of nominally separate cells into a "unit" functioning in tandem as part of a beating heart muscle.

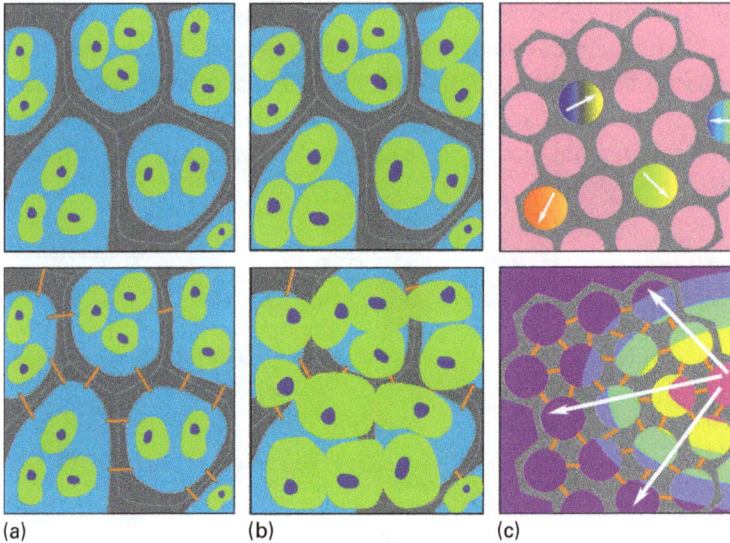

(a)　　　　　　　　　(b)　　　　　　　　　(c)

Fig. 4.25: Alginate and Au nanorod composite scaffold for engineering cardiac tissues. "Synchronized beats" produced upon embedding cardiomyocyte cells within an alginate scaffold further implanted with Au nanorods. *Top row:* pristine alginate; *bottom row:* alginate-Au composite. **(a)** Isolated cells are cultured within each scaffold; **(b)** the cardiomyocytes in the pure alginate matrix (top) form small clusters and beating occurs only within each cluster, whereas synchronized beating can occur in the alginate-Au system; **(c)** localized beating clusters assemble in pure alginate, while the presence of the Au nanorods promote formation of organized biomimetic cardiac tissue. The arrows and contour lines indicate the progression of the synchronized beating signal. Figure adapted from Dvir, T. et al., *Nature Nanotech.* **2011** *6*, 720–725.

4.2.3 Tissue-engineered cartilage: a case study

Efforts for growing artificial *human cartilage* through TE techniques are a good case study, highlighting both the enormous potential of this undertaking as well as the difficulties which might explain why "bench-to-bedside" engineered cartilage products have yet to be introduced. Cartilage is a highly specialized tissue essential for attaining low friction for adjacent bones and for maintaining efficient load bearing and distribution. The major constituents of cartilage are the chondrocytes, cells embedded in a highly hydrated and organized extracellular matrix (ECM). The interplay between chondrocyte density and their three-dimensional organization is the main factor affecting the unique physiological properties of cartilage, and this factor needs to be reproduced to create a fully-functional artificial cartilage. Particularly challenging is the significant variability in cell phenotypes, and matrix composition and organization in different regions of the cartilage tissue, reflecting its varied biological and mechanical roles.

Most current approaches for repairing or replacing damaged cartilage involve extracting host cells, growing them in laboratory settings, and implanting them back into the damaged site. However, a more broadly-applicable TE approach (yet to be realized) would be to create from basic building blocks an entirely synthetic cartilage in the laboratory, which will mimic the intricate mixture of the fluid environment and rigid ECM matrix, providing the distinct viscoelastic and mechanical properties necessary for efficient cartilage function. This scaffold should be combined with suitable constituent cells which properly respond to mechanical and biological stimuli to produce a fully functional cartilage tissue that could be subsequently implanted.

Chondrocytes, usually extracted from the patient undergoing cartilage repair, have been to-date the most abundant *cell source* for cartilage regeneration. The problems often encountered with using chondrocytes in engineered tissue have been the practical need for cell donation for each treatment, as well as the loss of *cell phenotype* (i.e. physiological cell morphology and functions) occurring through the tissue growth processes. Recent promising avenues involve the use of human *stem cells*, which might be successfully induced to differentiate into healthy pools of cartilage cells. Particular interest has been directed to *mesenchymal stem cells* (MSCs), which can be more readily programmed to differentiate into specific cell lines, such as chondrocytes (indeed, some researchers hypothesize the MSCs are in fact part of an inherent "tissue repair" mechanism in the human body). A challenge often encountered when employing stem cells in cartilage reconstruction (and other TE applications) concerns the delivery of *signaling molecules* to the cells. This is particularly pertinent to stem cells, which require specific growth factors to initiate differentiation into the required cartilage cells, e.g. chondrocytes. Accordingly, the engineered cartilage *scaffold* needs to allow transport of such signaling molecules to the cells in a controlled manner.

Scaffolds are indeed critical for effective cartilage TE. The ideal scaffold should exhibit the following physico-chemical properties: (i) three-dimensional, accessible pore network for maintaining cell growth and unhindered transport of nutrients, signaling molecules, and metabolic waste; (ii) surface chemistry that would facilitate cell attachment, proliferation, and differentiation; (iii) mechanical features, e.g. rigidity on the one hand, flexibility on the other, that would mimic endogenous cartilage;

(iv) biocompatibility and biodegradable properties – efficient breakup and desorption once the artificial cartilage functions properly in its physiological site. Numerous scaffold materials have been examined with various degrees of success in preliminary laboratory investigations. These include natural protein-based assemblies (fibril, collagen, gelatin), carbohydrate-based polymers (e.g. agarose, alginate, chitosan), artificial polymers (Teflon, carbon fibers, others), and hybrid materials combining different materials from these groups.

Perhaps the most promising, and already utilized, scaffold constituent in cartilage TE applications is **collagen**, the fibrous structural protein which is, in fact, a prominent structural element of natural cartilage. Collagen, discussed in Chapter 2, forms resilient fibers comprising three helical strands wrapped around each other (Fig. 4.18a). In cartilage, the collagen fiber network provides the extracellular matrix (ECM) scaffold and a platform for the three-dimensional matrix of the growing cells. Collagen has been shown in many instances to support and promote growth and organization of chondrocytes or stem cells in artificial cartilage tissues.

Sensitivity to *mechanical stimulation* is a particularly important property engineered cartilage tissue needs to exhibit. Indeed, flexibility in response to longitudinal pressure and torsion are fundamental physiological functions of cartilage. Furthermore, mechanical stimulation also affects the biochemical profiles of cells; it has been shown for example that mechanical effects modulated the *differentiation* of embryonic stem cells into cartilage-building chondrocytes. These observations are relevant to efforts towards fabrication of artificial cartilages through the use of stem cells. As a matter of fact, methods have been introduced to integrate the application of mechanical loads as part of the process of cartilage tissue growth.

Scaffolds that would satisfactorily reproduce both stiffness and mechanical flexibility are not the only precondition for cartilage regeneration. Sufficient *angiogenesis*, e.g. growth of blood vessel networks, is a fundamental requirement for the viability and functionality of the artificial cartilage tissue. Stimulation of angiogenesis through molecular cues, generally growth factors, embedded within the porous cartilage scaffold has been a common approach. Several strategies have introduced the use of *magnetic* scaffolds for generation of effective vascular networks in TE cartilage. Magnetic scaffolds can be manipulated by easily-applied external magnetic fields to direct angiogenic factors. This concept can be further extended to the inclusion of *magnetic nanocarriers* and magnetic nanoparticles in the porous scaffolds – which could be directed by an external magnetic field to deliver molecular inducers to desired areas within the artificial cartilage. Importantly, since the magnetic field is externally controlled, the methodology described above provides the physician with a powerful tool for adapting and tailoring the cartilage growth to specific conditions and personal needs of the patient.

While "conventional" *in vitro* TE (e.g. tissue production in the laboratory, and subsequent implantation of the "ready-to-use" tissue at the target location inside the body) is quite developed and has greatly increased our understanding of cellular

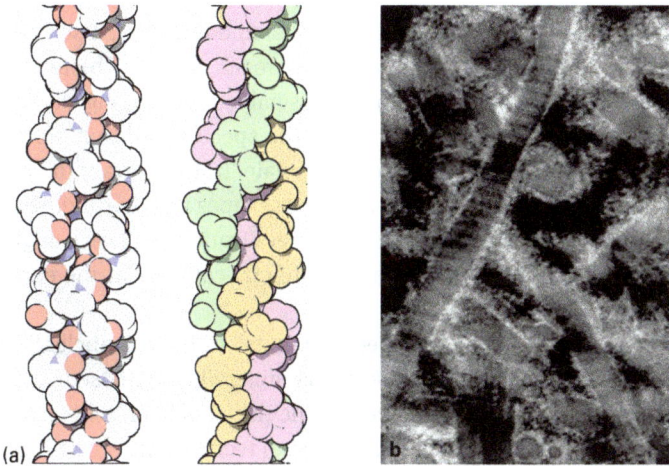

Fig. 4.26: Collagen fiber network in cartilage. **(a)** Collagen triple helix; **(b)** the porous network formed by collagen fibers in cartilage. The typical cross-striated fibrils are apparent. Reprinted from Aigner, T. and Stove, J. Collagens—major component of the physiological cartilage matrix, major target of cartilage degeneration, major tool in cartilage repair, *Adv. Drug. Deliv. Rev.* **2003** *55*, 1569–1593. Copyright (2003), with permission from Elsevier.

behavior and cell-scaffold interactions, this methodology has not been successful for regenerating tissues such as cartilage in which the internal and external morphology and hierarchical organization critically affect the physiological functions. Accordingly, alternative *in vivo* TE strategies have been introduced, whereby the triad of cells, growth factors, and scaffolds are directly implanted at the damaged tissue location for "on-site" regeneration.

An impetus for the expanding popularity of *in vivo* TE approaches for cartilage regeneration and as a general approach has been the hope (and recurring observations) that the vascularization process would occur more effectively in the target location compared to *in vitro* settings. Furthermore, it appears that populating the tissues with the appropriate cell types – a critical requirement in conventional *in vitro* TE systems – is in fact less of a hurdle in *in vivo* methods. This outcome is most likely related to the fact that appropriate combinations of endogenous molecular cues produced at the site of tissue growth can stimulate growth of *local* cell populations and ultimate formation of functional tissues, a scenario that is much more difficult to achieve in artificial settings. *In vivo* TE may specifically offer advantages for cartilage regeneration (and other musculoskeletal tissues) since the engineered tissue is formed at the specific physiological location in which recruitment of autologous cells (i.e. cells from the treated patient) can be carried out and rejection is thus significantly minimized.

An interesting *in vivo* TE approach for cartilage regeneration is to try to harness natural processes as the means for constructing a functional tissue. This is particularly the case with the use of *biomineralization* processes as "conceptual platforms" for construction of artificial cartilage and bone tissues. Biomineralization, discussed in more detail in Chapter 5, is a classical "bottom-up" approach, in which hierarchical three-dimensional structures are spontaneously assembled, exhibiting diverse properties and functionalities that are determined by the specific building blocks employed and their interaction with the (biological) template. It is important to note that highly organized biomineralized assemblies in diverse shapes and morphologies are encountered in nature, indicating that *in vivo* biomineralization pathways might provide a controlled avenue towards construction of artificial cartilage at the exact target location in the body.

Ceramic/biological constituents used as "seeding points" to initiate nucleation and crystal growth *in vivo* have been among several biomineralization-based cartilage regeneration strategies explored in recent years. The thrust of such schemes is that a cartilage tissue scaffold will slowly assemble through recruitment of cartilage-forming cells onto the biomineralized seeds, ultimately forming a functional tissue. Practically, such techniques aim to utilize synthetically introduced nucleation seeds for mimicking critical cartilage tissue environmental cues. Specifically, the engineered biomineralization promoting sites might mimic the column-like rigid structures believed to be an essential factor in shaping the distinct porous, flexible cartilage matrix (Fig. 4.19). Overall, drawing inspiration from natural biomineralization phenomena can open new horizons to materials scientists and bioengineers for developing innovative devices and strategies for regeneration of cartilage and other hard tissues.

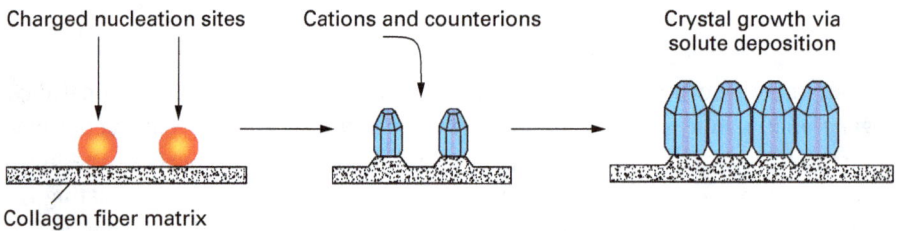

Charged nucleation sites   Cations and counterions   Crystal growth via solute deposition

Collagen fiber matrix

Fig. 4.27: Biomineralization in cartilage formation. Sequential biomineralization model in cartilage tissue formation.

4.3 Tissue engineering and drug design

The introduction of successful drugs or new molecular therapies faces formidable scientific, technological, and financial barriers. Indeed, numerous drug candidates that show great promise in "test-tube" settings, e.g. biological assays, cell cultures, and even laboratory animals, ultimately fail due to inadequate activity or even total lack of biological effects in the environment of an actual tissue inside the human body. While it is naturally inconceivable and unethical in most cases to perform drug effectiveness tests on humans, it would certainly be advantageous to carry out drug studies in the context of actual tissues *outside* of the body. Most current experiments along this line, however, generally require whole animals or fresh body parts extracted from animals, which are costly, exhibit significant variability, and might raise ethical issues. TE could nicely remedy many of these limitations, since functional tissues (and organs) artificially assembled outside of the body can be perceived as reliable surrogates for drug testing and analysis. Indeed, a crucial distinction between a conventional tissue culture and an *engineered* tissue in the context of the pursuit of pharmaceutical targets is the "systemic effect" – the impact of the tested molecule not on a single cell level or cell population, but rather on the entire target tissue.

Among the more innovative recent approaches towards development of TE scaffoldings for drug screening applications have been the attempts to construct tissues or even entire organs upon silicon chips. A primary goal of such schemes is to create the *smallest* tissue-like unit in which cells behave according to their "natural" functionalities in actual physiological tissue environments. A central premise here is the considerable degree of spatial control available in chip manufacturing technologies; extremely fine details can be etched upon a chip and complex micro- and nanofluidic networks fabricated. Such designs can be employed for directing cell growth and localization and to trigger cell processes in desired times and tissue areas. In this context, such "tissue-on-chip" systems could be ideal for drug discovery and testing applications, since the effect of drug molecules could be monitored on the chip and even potentially pinpointed in both temporal and spatial spaces.

Microfluidics (and *nano*fluidics by extension) underlies an active field integrating TE and drug design. Microfluidic cells can be fabricated from biocompatible materials through conventional methods, such as photolithography, etching techniques, laser ablation, and others, rendering such devices amenable for production in large quantities. Microfluidic devices could host up to millions of cells and down to single cells, depending upon the specific design and application, thereby providing much greater flexibility compared to conventional well plates. The structural features of microfluidic devices, specifically channels, membranes, and other physical components, provide a high degree of control over cell behavior and interactions between cells and cell within populations. Particularly important, a fundamental property of microfluidic devices is the available modulation of fluid flow, an important physiological condition not addressed at all in conventional, static culture dishes. Controlled flow dynamics in microfluidic assemblies could generate gradients of oxygen, growth factors, nutrients, and signaling molecules which intimately affect cells and tissues. The flow conditions also enable the application of mechanical forces to cells, and could interfere with cell-cell communication.

Organs on chips are still very much a "work in progress", but advances in culturing cells and manufacturing nanomaterials mean that such futuristic-sounding devices could eventually supplant animal studies. Efforts are already underway for producing "mini-organs" on chips – associations of specific cell types on carefully designed substrates mimicking physiological environments which maintain essential ingredients of a functional organ. In that regard, the increasingly sophisticated engineering tools of nanotechnology can be employed for fabrication of highly complex scaffoldings, complete with diverse microenvironments, nanofluidic channels, and reservoirs of biological and chemical substances needed for maintaining organ viability and functionality. The challenges in this field, however, are formidable; one has to mimic not only three-dimensional cell organizations, but in many cases several cell types need to be assembled together in appropriate spatial organizations, and the entire tissue should often exhibit distinct *mechanical* properties depending on the organ mimicked.

A recent remarkable feat has been the construction of "artificial lungs" on silicon chips. The artificial lung created by D. E. Ingber and colleagues at Harvard University (Fig. 4.20) was literally not "on chip"; however, the microfabricated "lung" elegantly recreated lung-tissue environments, including distinct configurations of the epithelial cells lining the lung walls, immune cells which protect the lung lining, complete with capillaries for fluid transport mimicking blood flow as well as miniscule air channels connected to a vacuum chamber designed to simulate the mechanical forces associated with passage of air in and out of the miniaturized lung. The configuration depicted in Fig. 4.20 underscores the importance of *functional mimicry* as a core feature in artificial organ design; as an example, the engineered cell layer of the "lung" could modulate its air transport capabilities and maintain cell functions throughout the "stretching" and "compression" processes akin to breathing motion.

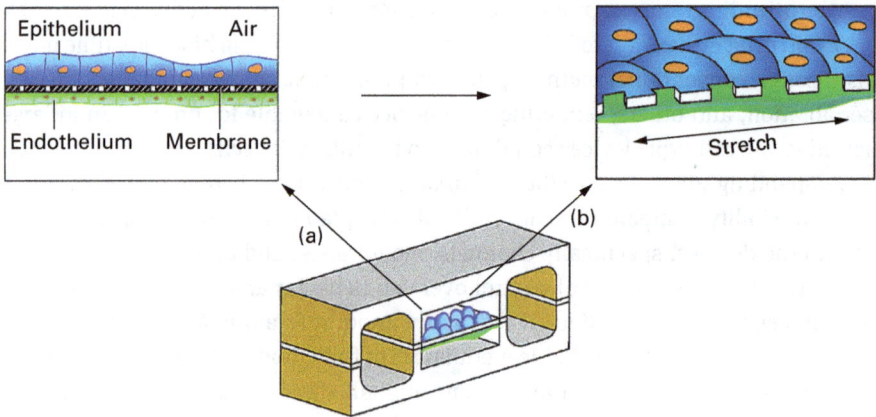

Fig. 4.28: Artificial lung. Schematic drawing of the vacuum microchamber which recreates human breathing. A layer of epithelial cells is assembled on a thin, porous flexible polymer which can stretch and shrink in response to the vacuum conditions in the microchambers. **(a)** Initial pre-inhalation stage, the epithelial cells comprising the lung epithelium are compressed together, forming a barrier to air passage; **(b)** following inhalation the biomimetic film is stretched (induced through application of vacuum in the two side chambers), allowing air passage through the intercellular microchannels.

"Lungs-on-chip" such as the design shown in Fig. 4.20 could facilitate detailed analysis of stretching of the lung lining upon air passage, the effect of environmental pollutants such as small dust particulates and gases, and therapeutic potential of drugs. Furthermore, different types of cells can be seeded within the artificial lung, for example epithelial cells associated with lung diseases such as asthma, making possible analysis of specific factors affecting cell- and lung-tissue properties, and examining potential therapeutic avenues for these cells.

The tissue-on-chip concept can be expanded beyond the construction of single tissues or single organs towards fabrication of more complex tissue arrangements – even possible "organism-on-chip" scenarios. In such an approach one could envision several microcompartments (e.g. microbioreactors) connected through microfluidic channels, with each compartment containing cells and tissues representing specific organs or body parts, and/or performing distinct physiological tasks. Such an artificial microscale "organism", if designed appropriately, might mimic complex systemic processes which are impacted by the interplay among molecular events in different parts of the body.

Like other new and innovative platforms for drug screening, the barriers to practical applications of the "tissue-on-chip" concept are significant. This is expected because the systems envisaged are highly complex, both from a biological as well as an engineering standpoint. For one thing, the chip design has to maintain optimal "blood flow" into and out of the artificial "organs" grown on the chip. This is particularly acute for organs (or tissues) that require a delicate balance of nutrients, biological effectors, and released metabolites (such as lung-lining cells described above). Furthermore, cell-cell interactions and communication (both between the same cell types and between different cells in a particular tissue) need to be carefully optimized in an artificial environment. In addition, the "regular" TE constraints need to be addressed – primarily cells which exhibit pronounced sensitivity to growth conditions particularly when grown in artificial and "unconventional" settings, such as the surface of an etched silicon chip. It is expected, however, that the intense interest in this field combined with greater understanding of cell behavior in artificial environments and advances in bioengineering will usher in an era in which tissue-on-chip platforms will constitute a powerful tool for drug discovery and biological research.

5 Biomineralization

Biomineralization encompasses broad and diverse natural phenomena that have long fascinated scientists and lay people, and has been harnessed for human use since the early days of civilization. Indeed, the concept in which organisms use biomolecules and biological substances to assemble organized inorganic scaffolds has been exploited by mankind from prehistoric times, including the attraction to gems (natural and synthetic), tissue engineering applications involving hard tissues like bone, development of custom-made functional biomaterials containing inorganic substances, and others. Considering the wide scope of biomineralization research, this chapter focuses primarily on *biomimetic* aspects of this field.

One of the most striking aspects of biomineralization is the considerable variety of structures, morphologies, and shapes of materials produced. This diversity is related to the underlying molecular mechanisms of biomineralization phenomena. Specifically, the combination of directed crystal growth occurring at biomolecular interfaces, recruitment of various organic and inorganic building blocks by many organisms, and the specific conditions of the aqueous solution environments, all entail distinct cooperative properties which result in the broad range of biomineralized structures observed in nature.

Biomineralization exhibits attractive features from a materials science standpoint. Most biomineralization processes are based on *peptides* and *proteins* which initiate and direct crystallization of inorganic structures. This implies that such processes generally take place in *mild* conditions (in terms of pH, temperature, solution composition, etc.) in which most proteins function. Accordingly, mimicking biomineralization events could open new materials synthesis routes precluding the need for elevated temperatures and pressure or usage of environmentally-harmful solvents. In addition, the extraordinary diversity of protein sequences, structures, and hierarchical organizations confers a rich source of templates for inorganic material construction.

Gaining a thorough understanding of biomineralization *mechanisms* has been a key issue in this field and its practical implementation. Indeed, shedding light on the parameters contributing to the enormous diversity of biomineralized structures, their physical properties, and biological functions have been key factors contributing to progress in this field. While a thorough analysis of the extensive research aimed at elucidating the molecular mechanisms of biomineralization is naturally beyond the scope of this book, it is generally accepted that distinct protein interfaces present *nucleation sites* for crystal growth. More specifically, nucleation sites are believed to have pivotal roles in biomineralization as they aid overcoming energy barriers to molecular organization (i.e. crystal growth). Indeed, the biological matrix itself, providing the interface upon which crystal nucleation and growth occur, is also intimately involved in lowering the energy barriers to mineral assembly.

https://doi.org/10.1515/9783110709490-005

5.1 Protein- and peptide-associated biomineralization

Design criteria for artificial biomineralized systems are usually conceptually similar to their biological counterparts. Natural biomineralization processes are often associated with *acidic proteins,* acting as mineralization templates. Specifically, negatively-charged amino acids, particularly glutamate (Glu), are believed to recruit and bind inorganic cations such as calcium thereby initiating the mineralization process. Accordingly, synthetic templates for artificial biomineralization generally utilize surfaces containing acidic amino acids or negatively-charged moieties. The nature and organization of the negative units on the synthetic templates, however, have profound effects upon the mineral assemblies.

Enamel repair and reconstruction is a recent example in which single acidic amino acids have been employed for initiating biomimetic mineralization processes. Enamel, the protective layer of the tooth, has been a particularly difficult challenge for bioengineering, mostly due to its complex needle-like structure comprising tiny crystallites of apatite – a calcium-containing mineral. This configuration underlines the exceptional hardness of enamel, but on the other hand imposes significant hurdles for construction of enamel-mimicking materials in the laboratory. While there have been relatively successful efforts for regenerating enamel through the use of the physiological enamel-template proteins enamelins and ameliogenins as seeding agents, this approach is limited because the supply of such proteins is limited.

An alternative approach introduced by R. Tang and colleagues at Zhejiang University, China, focused on the use of glutamate in conjunction with apatite NPs to generate enamel. The methodology, depicted in Fig. 5.1, is not fully "synthetic" as it relies on existing enamel templates to facilitate assembly of the hierarchical, aligned structure of the apatite crystallites critical for the unusual toughness of the material. The new technique, however, could nicely reproduce the *directionality* of the apatite domains within enamel, a feat which is usually very difficult to achieve in the laboratory, through the inclusion of free glutamate residues in the reaction mixture.

Negatively-charged species other than amino acids have been used as inducers in another physiologically important biomineralization process – **bone formation.** Similar to enamel, bones form through hierarchical mineralization of calcium and phosphate ions forming apatite crystals. Bone morphology and internal organization are intimately connected to the physical properties of *collagen fibers* (see Chapter 4). Specifically, the collagen network hosts numerous microscopic gaps believed to furnish nucleation sites in which inorganic ions bind and crystallize, ultimately yielding the bone framework. While sophisticated techniques have been developed to construct artificial bones through mimicking the collagen-associated mineralization processes, this task is much more complex than initially thought due to the difficulty of mimicking the crystalline inorganic structures within the collagen matrix. As a consequence, alternative approaches have been developed and polymer analogs have been synthesized and used as substitutes for collagen. Such synthetic templates

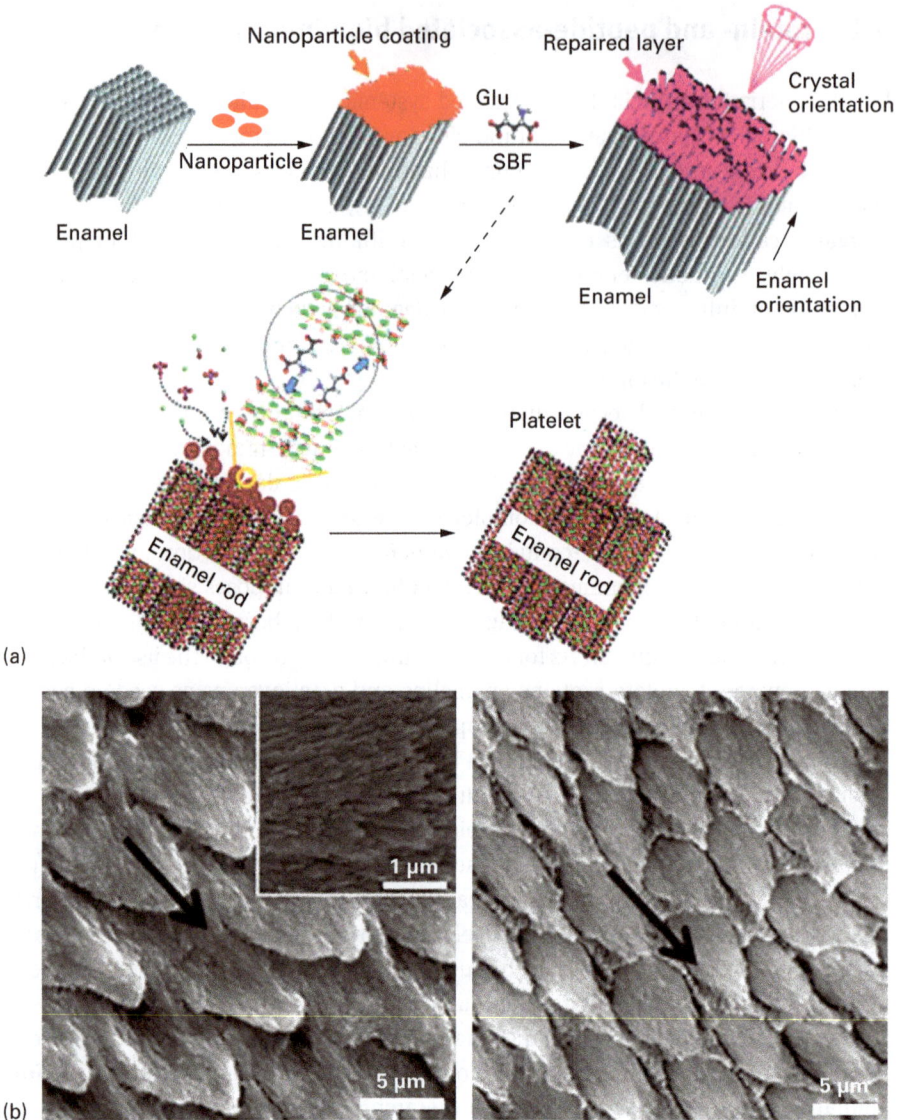

Fig. 5.1: Synthetic enamel through glutamate-induced biomineralization. **(a)** Scheme showing the experimental procedure. A natural enamel template was coated with apatite nanoparticles (NPs). Addition of free glutamic acid to the reaction mixture (simulated body fluid, SBF) resulted in oriented apatite crystal growth upon the NP coating. Regenerated layers of crystalline apatite thus formed through "bridging" by the negatively-charged glutamic acid. **(b)** Scanning electron microscopy (SEM) images showing the similarity in surface morphologies between native and synthetic enamel. **Left:** natural enamel displaying an array of aligned apatite domains; inset shows the individual apatite crystallites. **Right:** enamel produced through the synthetic approach. Reprinted with permission from Li, L. et al., *Adv. Mater.* **2011**, *23*, 4695–4701. Copyright (2011) John Wiley and Sons.

provide greater control of the crystallization dynamics and overall morphology of the artificial bone.

Collagen fibers serve as templates in other biomineralization processes, for example formation of *silica* scaffolds. Biosilica is an important component of skeletons and external protective layers of varied organisms. Unlike inorganic minerals, such as the calcium salts comprising the bone framework, silica is a *polymer* formed through chemical condensation of oxygenated silicon (e.g. silicate) monomers. In many physiological systems, biosilica is formed through the action of soluble (positively-charged) proteins that attract the silicate monomers facilitating their polymerization. In contrast, collagen fibers provide an *insoluble* template for polymerization of the silica framework, occurring through association of the silicate monomers with the collagen surface, and consequent formation of nucleation sites which initiate polymerization of the silica framework.

Modification of the collagen fibers through decoration with positively-charged *polyamine* moieties leads to enhanced attachment of the silicate monomers and substantial biosilica coating both externally on the fiber surface and internally within the porous fiber matrix. The hybrid silica-collagen fibrillar assemblies formed through

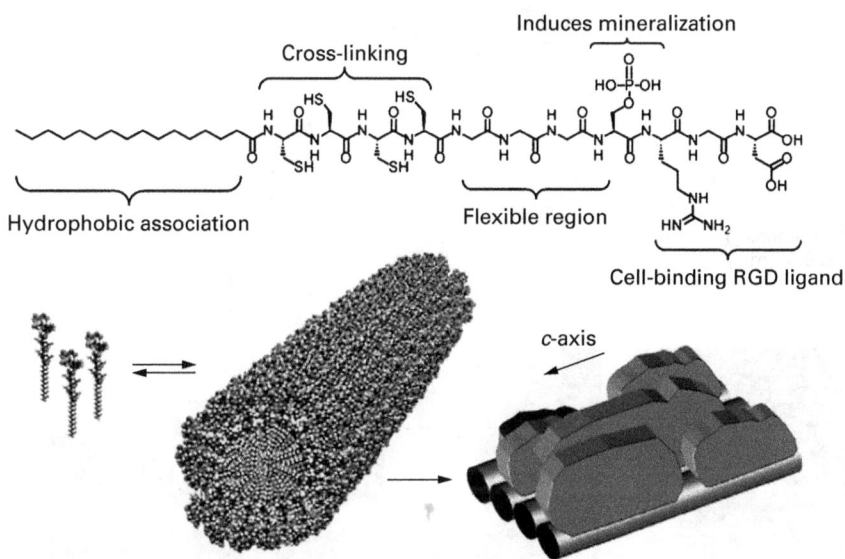

Fig. 5.2: Synthetic peptide-amphiphile fibers mimic bone mineralization. Modular design of the peptide-amphiphile molecule promoted self-assembly into stable cylindrical micelles, mimicking the collagen fiber template for biomineralization in bone. The tubular peptide-amphiphile fibers induced mineralization of hydroxyapatite crystallites, which were aligned along the tubes' long axis, indicated by the arrow (bottom right). Reprinted from Zhang, Z. et al., Applications of functional surfactants, *Curr. Opin. Coll. Interface Sci.* **2002** *7*, 267–275. Copyright (2002), with permission from Elsevier.

this technique not only demonstrate morphological alignment of a non-biological polymer through the use of a biological (collagen) template; silica, in fact, imparts extraordinary mechanical stability to the collagen fibers, pointing to potential utilization of this composite material as bone substitutes and other biological and non-biological applications. Similar to collagen, other fibrous and filamentous proteins, such as silk protein, amyloid peptides, and others have been used as platforms for affecting deposition of inorganic substances.

An example of an artificial, modular peptide-based structure aimed to mimic collagen function and promote artificial bone formation is depicted in Fig. 5.2. In that work, S. Stupp and colleagues at Northwestern University integrated several structural criteria into the design of a self-assembling peptide derivative. Specifically, the synthetic molecule was comprised of several regions, each having a distinct function. *A hydrophobic tail* is a core constituent promoting self-assembly of the molecule into

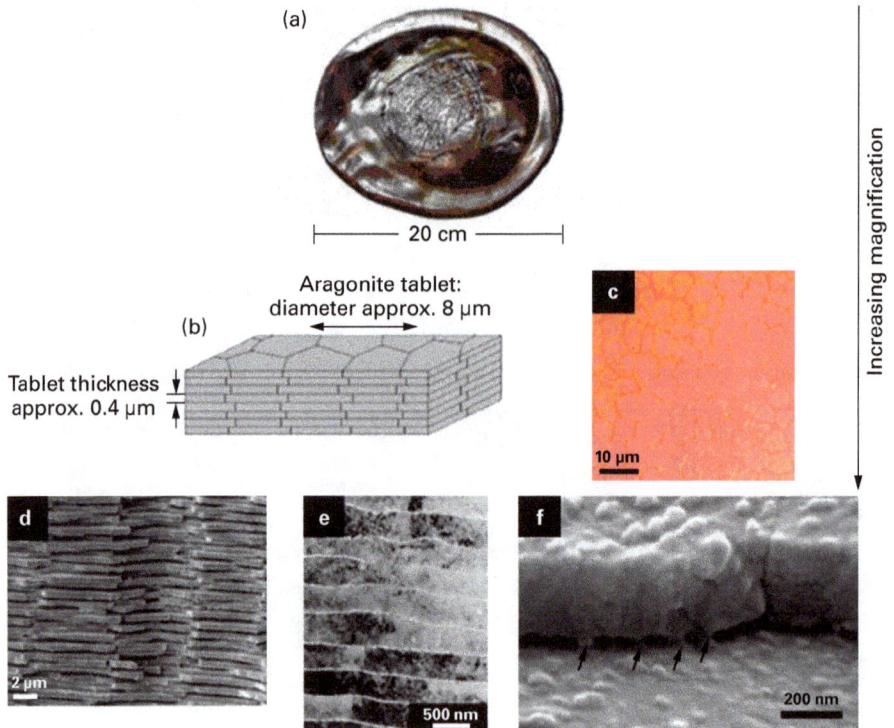

Fig. 5.3: Nacre structure. Nacre ("mother of pearl") morphology in different scales: (a) red abalone shell; (b) schematic of the "bricks and mortar" structure showing the biomolecular constituents interspersed within the layered aragonite crystallites; (c), optical microscopy image of a nacre plate; (d)–(f) SEM images of the oriented, layered Ca(CO$_3$) formations in nacre. The arrows in (f) indicate mineral bridges between individual "tiles". Reprinted from Corni, I. et al., *Bioinspir. Biomim.* **2012** 7, 031001. Copyright (2012) IOP Publishing.

tubular structures in water. Consecutive *cysteine residues* (cross-linking domain) were included to further stabilize the self-assembled fibers through disulfide bond formation upon oxidation of the thiol (C-SH) units. A flexible region connects to a negatively-charged phosphorylated serine residue, which furnishes the site for calcium capture and crystal growth (mineralization inducing domain). The cell-binding motif arginine-glycine-aspartate (RGD) sequence complements the peptide amphiphile structure, facilitating cell adhesion and formation of viable bone tissue. Importantly, the collagen-mimic peptide-amphiphile fibers successfully supported oriented growth of hydroxyapatite crystallites, similar to bone morphology.

While bone is probably the most prominent physiological system targeted in artificial biomineralization applications, there have been attempts to mimic other natural

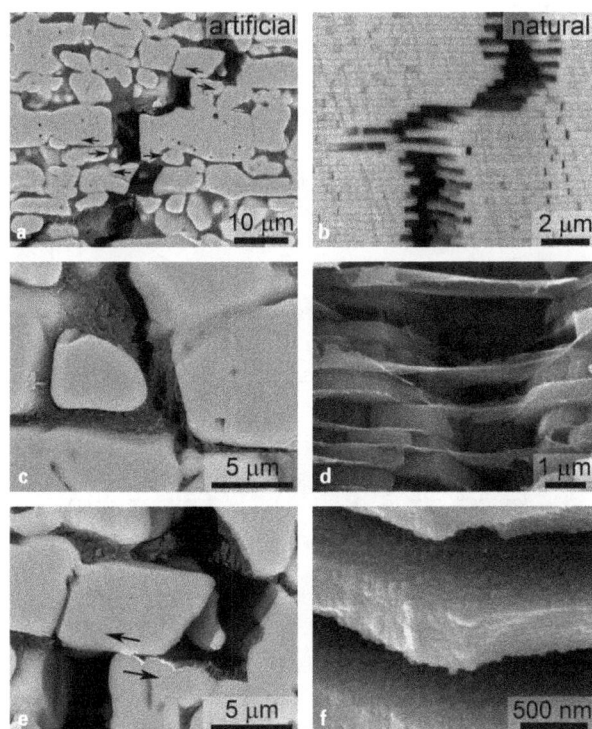

Fig. 5.4: Synthetic nacre. SEM images of artificial nacre comprising Al_2O_3 and PMMA (**left**) and natural nacre (**right**) highlighting the mechanism responsible for nacre toughness attained through crack dissipation and fracture resistance. The images show crack propagation in the two materials at different length scales. (**a**) and (**b**): resisting crack-inducing forces through "pull-out" mechanisms; disintegrated polymer (**c**) and organic phase in natural nacre (**d**) accumulating in interfacial areas and acting as lubricants and "shock absorbers"; (**e**) and (**f**): high surface roughness resulting in enhanced friction resistance between individual particulates. Reprinted from Launey, M.E. et al., Designing highly toughened hybrid composites through natureinspired hierarchical complexity, *Acta Materialia* **2009** *57*, 2919–2932. Copyright (2009), with permission from Elsevier.

biomineralized materials with less of a biomedical relevance. Nacre, or "mother of pearl", exemplifies the challenges to successfully imitate biomineralization, specifically the need to accurately recreate the hierarchical, three-dimensional organization of the inorganic constituents. In nature, mollusks produce nacre through deposition of calcium carbonate (aragonite) crystallites sputtered with organic impurities within a layered biopolymeric template comprising polysaccharides and proteins, in a process which evolved over millions of years. The "brick wall" structure comprising $Ca(CO_3)$ crystals and biopolymer "mortar" (Fig. 5.3) is the main biogenic feature giving nacre its exceptional toughness and optical characteristics. Accordingly, this configuration needs to be reproduced in an artificial nacre system.

Attempts to create synthetic nacre have focused both on mimicking the molecular structures of the aragonite crystallites, as well as generating the hierarchical layered "brick" assembly. R. O. Ritchie and colleagues at the University of California, Berkeley, for example, assembled a hybrid material comprising alumina and poly(methyl methacrylate) (PMMA), an abundantly-used polymer. This material displayed exceptional strength and toughness, ascribed not just to the specific composition, but also associated with the microscopic layered structure of the particulates, which, similar to natural nacre, enabled highly efficient dissipation of cracks and mechanical stress.

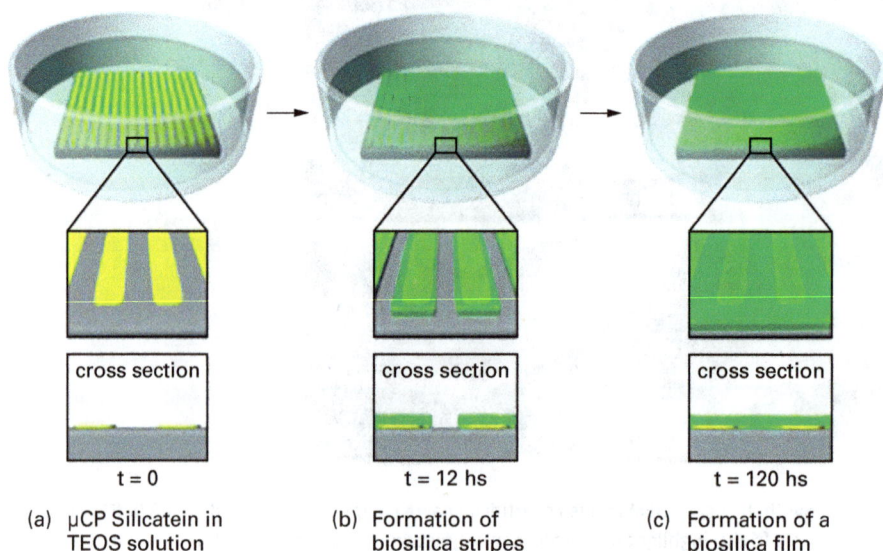

cross section	cross section	cross section
t = 0	t = 12 hs	t = 120 hs
(a) μCP Silicatein in TEOS solution	(b) Formation of biosilica stripes	(c) Formation of a biosilica film

Fig. 5.5: Silica deposition induced by biomolecular surface patterns. Production of thin biosilica layers through polymerization induced by patterned silicatein (produced through a micro-contact printing – μCP – method). The protein stripes catalyze polymerization of a silica layer while incubated in a solution containing the silica precursors. A uniform thin silica layer is progressively deposited upon the patterned surface. Reprinted with permission from Polini, A. et al., *Adv. Mater.* **2011** *23*, 4674–4678. Copyright (2011) John Wiley and Sons.

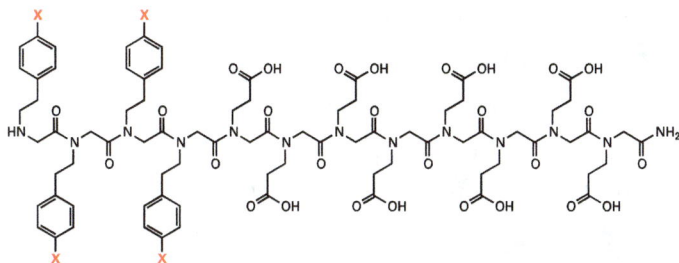

X = **H**, Peptoid-1; = **Cl**, Peptoid-2; = **OMe**, Peptoid-3

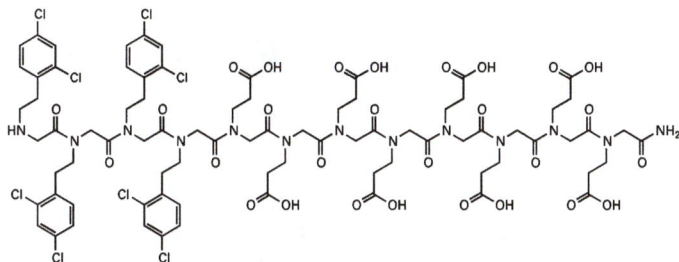

Peptoid-4

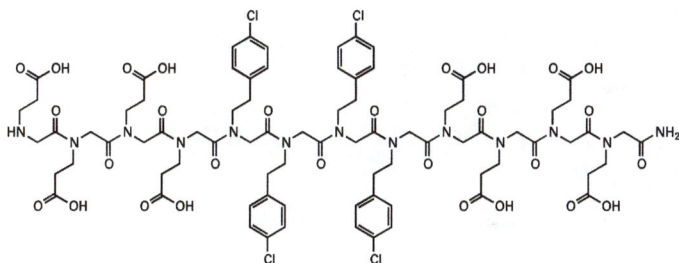

Peptoid-5

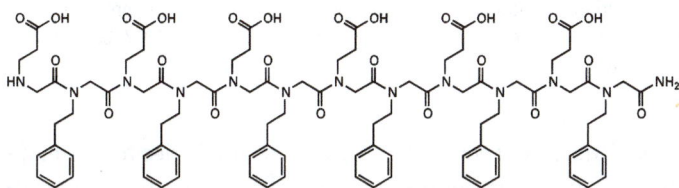

Peptoid-6

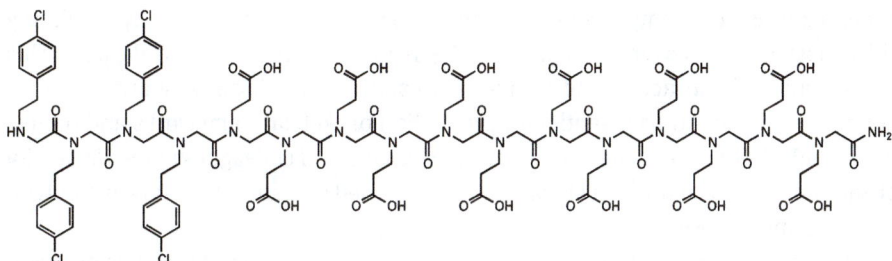

Peptoid-7

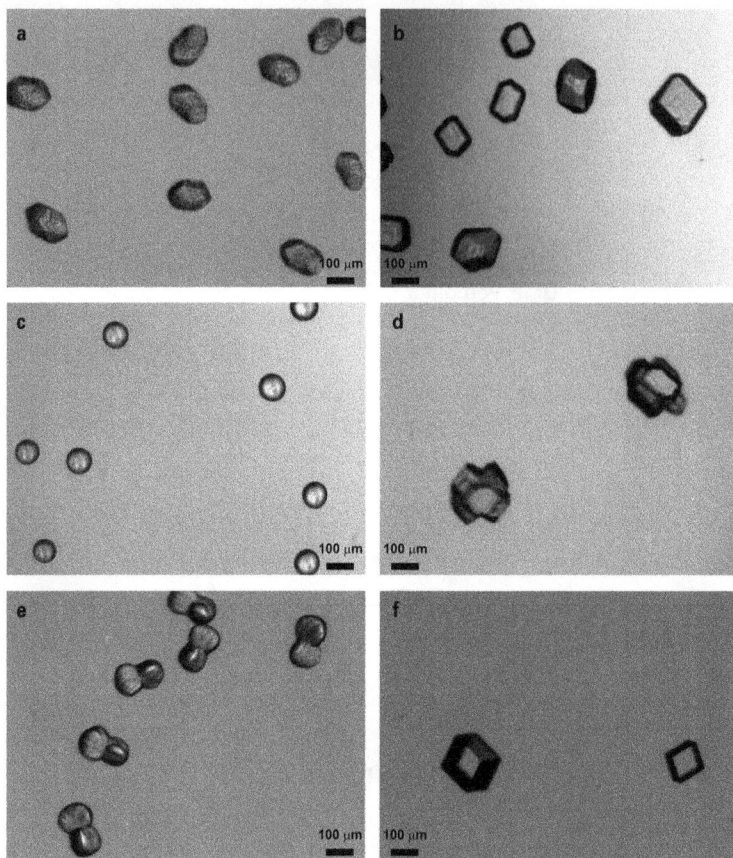

Fig. 5.6: Biomineralization induced by peptoid polymers. Peptoid structures and respective $CaCO_3$ crystal morphologies produced upon incubation of the peptoids with Ca^{2+} and sequestered CO_2. Reprinted with permission from Chen, C.L. et al., *J. Am. Chem. Soc.* **2011** *133*, 5214–5217. Copyright (2011) American Chemical Society.

Biomineralization can be coupled to seemingly unrelated technologies for development of novel applications. *Surface lithography*, for example, can be applied in conjunction with biomineralization reactions for fabricating biologically-induced inorganic patterns. An example is depicted in Fig. 5.5 in which an electrically-insulating silica layer is formed over a lithographically-created surface pattern of *silicatein*. This unique protein is extracted from marine organisms, functioning as a catalyst for silica polymerization and thus essential for production of skeletal frameworks and protective shields. In the system developed by D. Pisignano and colleagues at the University of Salerno, Italy, recombinant silicatein was patterned on a surface, catalyzing polymerization and deposition of biosilica layers.

The silicatein-based biomineralization experiment shown in Fig. 5.5 is noteworthy since the mineralization process was carried out through *enzymatic catalysis*, rather than the more conventional schemes exploiting electrostatic attraction between polypeptides or polymers and inorganic constituents which have been often found lacking in structural control and uniformity. Another advantage of the "biomineral-ization-lithography" approach concerns the use of mild experimental conditions – ambient temperature and physiological pH – designed to retain the functionality of surface-immobilized silicatein. Such conditions are favorable for potential industrial applications of the technique and a more benign environmental impact.

As noted above, a fundamental molecular determinant for biologically-induced inorganic materials is the presence of recognition units and/or domains onto which non-biological substances bind, and which often serve as "nucleation sites" for initi-ation of crystal growth. Indeed, one can use not only strictly *biological* modules (e.g. amino acids or peptides) as crystal growth templates, but also peptide-mimics that might provide molecular facets activating crystal formation. Fig. 5.6, for example, shows different crystalline CaCO$_3$ morphologies induced by *peptoids* – biomimetic peptide-like polymers – in which functional units are substituted on the nitrogen atom within a glycine skeleton. The peptoids examined by J. DeYoreo and colleagues at the Lawrence Berkeley National Laboratory exhibited different positioning of carboxylic acids (providing the negative surface) and aromatic groups (displaying hydrophobic facets), which together promoted distinct crystal morphologies.

Biomimetic protein-initiated biomineralization schemes have grown in scope and sophistication in recent years. Fig. 5.7 illustrates a system designed by S. Heilshorn and colleagues at Stanford University in which protein-based cage-like assemblies were employed as a platform for inorganic crystallization. This study nicely underscores the two essential elements in artificial biomineralization systems. A *template* needs to provide the physical scaffold for nucleation and crystal growth; this role was provided in the experiments summarized in Fig. 5.7 by *clathrin* – a protein forming "cage-like" struc-tures *in vivo* and *in vitro*, employed for transporting molecular cargoes between cells.

The second important constituent in synthetic biomineralization systems is the presence of *nucleation sites* – recognition elements for anchoring the inorganic com-ponents. In the system presented in Fig. 5.7, this function was attained by short syn-thetic peptides which were designed to include sequences binding to specific *epitopes* (short recognition units within the protein sequence) within the clathrin sequence. Importantly, the synthetic peptides also contained modules designed to attract *inor-ganic* ions, essentially acting as nucleation sites.

As a biomimetic mineralization concept the clathrin-based platform offers notable versatility, since in principle diverse peptide sequences can be synthesized and utilized – having different combinations of epitope recognition elements and inorganic targets. The clathrin system exhibits certain limitations, also encountered in other biomineralization strategies that rely on biological species. Specifically, the clathrin-based approach does not offer flexibility in terms of cage *size*, effectively lim-iting crystallite dimensions. In addition, clathrin itself is obtained from tissue sources, potentially limiting large-scale applications.

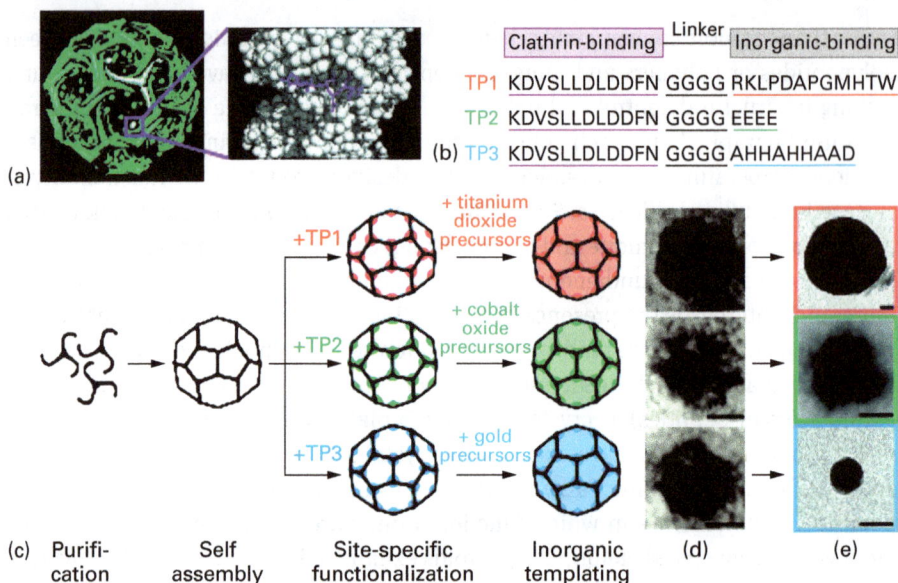

(a)

Clathrin-binding	Linker	Inorganic-binding
TP1 KDVSLLDLDDFN	GGGG	RKLPDAPGMHTW
TP2 KDVSLLDLDDFN	GGGG	EEEE
(b) TP3 KDVSLLDLDDFN	GGGG	AHHAHHAAD

+TP1 + titanium dioxide precursors

+TP2 + cobalt oxide precursors

+TP3 + gold precursors

(c) Purifi- Self   Site-specific   Inorganic   (d)   (e)
    cation   assembly functionalization templating

Fig. 5.7: Biomineralization induced by clathrin nano-cages coupled to short bifunctional peptides. **(a)** Schematic image of the "cage" assembled from the three-arm clathrin monomers. The bifunctional clathrin-binding peptide is shown in purple, attached to the clathrin monomer arm; **(b)** representative bifunctional peptides containing a clathrin recognition element attached through a four-glycine flexible linker to inorganic binding domains. **(c)** Schematic picture of the experimental setup: clathrin cages are assembled from the protein monomers, followed by addition of the bifunctional peptides which are anchored onto the clathrin cage surface. Each peptide-cage construct is incubated with a different inorganic precursor, leading to crystallization of the specific inorganic mineral upon the cage template. **(d)** and **(e)** depict transmission electron microscopy (TEM) images of mineralization stages, confirming the specific deposition of the inorganic substances upon the clathrin cages. Scale bars are 25 nm. Reprinted with permission from Schoen, A.P. et al., *J. Am. Chem. Soc.* **2011** *133*, 18202–18207. Copyright (2011) American Chemical Society.

Cellular membranes have been employed as templates for fabrication of intriguing biomimetic mineralized structures. Specifically, the propensity of lipid molecules to adopt vesicular structures in aqueous solutions has been useful in materials design applications. Indeed, lipid vesicles have been employed both as physical or chemical templates for mineralization processes. Monomeric silica, for example, can be covalently coupled to lipids (or amphiphilic surfactants embedded within lipid bilayers), consequently forming vesicle structures in water. K. Ariga and colleagues at the National Institute of Materials Science, Japan, further showed that polymerization of the silica-lipid self-assembled structures yielded silica constructs which retained the spherical vesicle structures (Fig. 5.8). These silica-coated vesicles, denoted "cerasomes" (e.g. "ceramic liposomes"), retain important characteristics of lipid vesicles, among them an internal volume separated from the external solution with a well-defined barrier. Moreover, the tools developed in decades of vesicle research for controlling size distribution through both intrinsic means, such as lipid chain length, as well as external parameters, such as sonication power and duration, temperature, etc. can be applied for modifying cerasome properties. As generally stable rigid capsules, cerasomes might find diverse uses in materials transport, separation technologies, and possibly drug delivery.

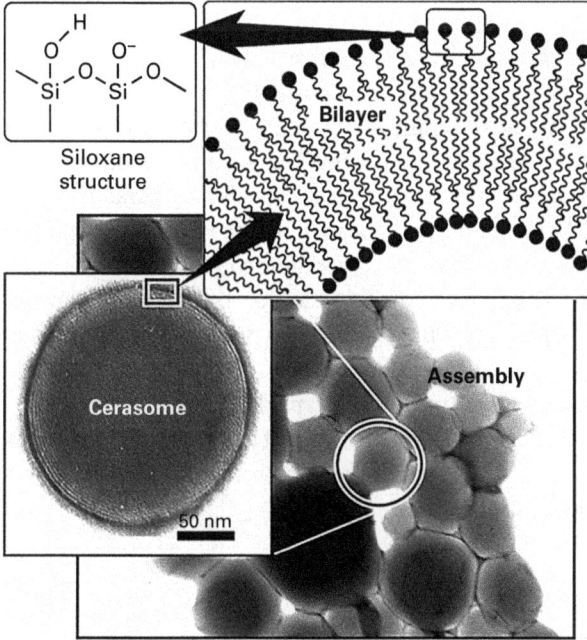

Fig. 5.8: Cerasomes. Synthetic vesicle bilayers comprising silica monomers covalently attached to lipid headgroups. Silica-coated vesicles ("cerasomes") form following polymerization. Reprinted with permission from Ruiz-Hitzky, E. et al., *Adv. Mater.* **2010** *22*, 323–336. Copyright (2009) John Wiley and Sons.

5.2 Organism-templated biomineralization

While biomolecules have been the most widely used platform for biomineralization, either in natural or artificial settings, the use of entire *organisms* (usually *microorganisms*) as biomineralization templates has also attracted interest in this field. Microorganisms can be employed for deposition of inorganic structures either *internally* (i.e. encapsulated by the organism) or *externally* – inorganic coating layers. Nature, in fact, provides ample examples of physiological biomineralization pathways associated with microorganisms. The hugely significant *calcification* processes in which calcium carbonate ($CaCO_3$) precipitation and sedimentation occur in the oceans are mostly mediated by microorganisms such as algae and cyanobacteria. Indeed, bacterially-induced calcification has been proposed as a promising biogenic route for sequestering atmospheric carbon dioxide and for carbon storage – crucial environmental goals in combating global warming.

While numerous microorganism-induced biomineralization phenomena have evolved over eons, *biomimetic* phenomena have also recently thrived. **Virus-encapsulated nanoparticles (VNPs)**, more commonly referred to as **virus-like particles (VLPs)** are interesting biological-inorganic hybrid materials which comprise viral proteins retaining the shape of the virus particles encapsulating *inorganic cores*. VLPs can be also perceived as primitive protein cages encapsulating a cargo that can be delivered to various physiological targets, while maintaining the "stealth" properties of the original virus. Preparation of VLPs usually involves incubation of synthetic nanoparticles or other inorganic substances with viral capsids (e.g. viral *coat proteins*) resulting in embedding the inorganic materials within the virus-like particles.

Importantly, in many instances the sizes of the inorganic cores of VLPs determine the dimensions of the encapsulating capsids. This feature has been recently demonstrated by B. Dragnea and colleagues at Indiana University who demonstrated that gold NPs having distinct diameters gave rise to VLPs comprising different conformations and overall sizes when incorporated within the capsids of *mosaic virus,* a symmetrical icosahedral virus (Fig. 5.9). The dimensionalities of the VLPs, determined by the inorganic cores, further enabled production of ordered particle *lattices* and arrays which might be utilized for optical applications and others.

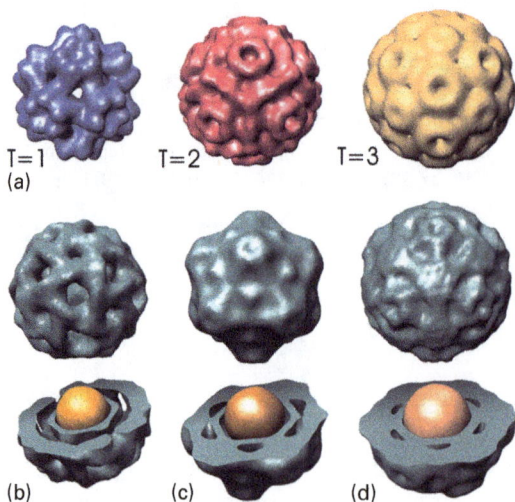

Fig. 5.9: Virus-like particles (VLPs) comprising different sizes of inorganic cores. Electron microscopy-based three-dimensional reconstructed images of *brome mosaic virus* (BMV) capsids encapsulating gold NPs having different sizes. Reprinted with permission from Sun, J. et al., *PNAS* **2007** *104*, 1354–1359. Copyright (2007) National Academy of Sciences, U.S.A.

Fig. 5.9 underscores important advantages for VLPs as advanced materials. First, the encased non-biological core can be manufactured in varied compositions, structures, and sizes, stabilized by the protein coating for long periods in physiological environments. The viral capsids also provide effective means for delivery in biological environments. VLPs are also notable for their *monodispersity* and minimal aggregation (a recurring problem in inorganic NP preparation procedures). Importantly, viral capsid – and thus resultant VLP – properties can be tuned through the powerful arsenal of genetic engineering – providing another lever for control of these biohybrid systems.

VLPs exhibit biomedical applications as vehicles for high-contrast functional imaging. This route usually utilizes magnetic aggregates (usually iron oxide NPs) as the VLP core, making these assemblies useful in magnetic resonance imaging.

In some cases, the morphology and surface characteristics of the native virus are retained, making the system a useful tool for studying viral transport and virus-induced biological processes. In this context, while such applications might be impractical in human studies due to ethical and practical limitations, VLPs could provide useful insight into viral transport in animal models. Similarly VLPs might become a powerful tool for studying viral infections in *plants*, providing a relatively simple means for investigating viral migration in plant compartments.

Whereas VLPs exhibit largely spherical morphologies, template-based mineralization has been also associated with viruses displaying filamentous or rod-shaped structures. Such structures are considered attractive templates precisely because of their *anisotropic* morphologies. *Tobacco mosaic virus (TMV)*, for example, is a highly stable rod-shaped virus (300 nm length, 18 nm external diameter, 4 nm diameter of the inner viral tube) used as a scaffold for assembling inorganic and metal nanotubes. In particular, the charged surfaces of the internal tubing and external surface have been both shown to be effective nucleation sites for initiating crystal growth and binding of metal nanoparticles. Furthermore, amino acids at the TMV surfaces can be chemically modified to bind specific mineral constituents, thereby promoting deposition of targeted mineral structures.

Fig. 5.10 shows representative inorganic structures synthesized by S. Mann and colleagues at the University of Bath, UK using TMV as a mineralization scaffold. The electron microscopy images in Fig. 5.10 highlight the versatility of inorganic morphologies essentially induced by the same parent virus: metallic nanowires, aggregates, and nanotubes. Intriguingly, in some instances *micrometer*-scale TMV-templated tubes were observed (Fig. 5.10c) – which is considerably longer than an individual TMV particle. This result points to the formation of higher order organized structures, such as head-to-tail assembly of the viral particles, which either form prior to the templating and mineralization processes or subsequent to viral coating with the inorganic material. The observation of such hierarchical organizations underscores the potential of viral templates as a useful platform in materials design.

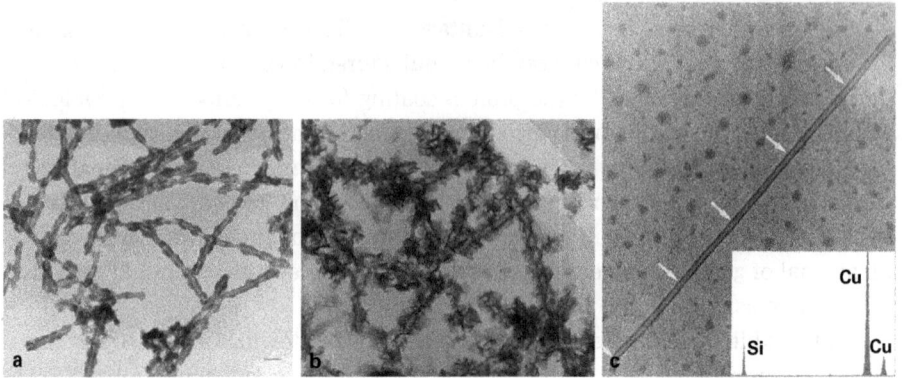

Fig. 5.10: Biomineralization using *tobacco mosaic virus* (TMV) as a template. TEM images depicting inorganic deposits grown on the surface of TMV. **(a)** TMV coated with iron oxide, initial product; **(b)** TMV coated with iron oxide, 6 months incubation. Extensive coating of iron hydride crystals protruding from the TMV rods could be observed. Scale bar in **(a)** and **(b)** is 100 nm. **(c)** TMV coated with silica; the arrows indicate the lengths of individual viral particles, suggesting head-to-tail arrangement of the viral rods. The Si peak is apparent in the energy-dispersive X-ray spectrum on the bottom right, confirming silica deposition. Reprinted with permission from Shenton, W. et al., *Adv. Mater.* **1999** *11*, 253–256. Copyright (1999) John Wiley and Sons.

Another popular filamentous virus used as a vehicle for inorganic templating (as well as many other biological and biotechnological applications) is the **M13 filamentous bacteriophage** (also denoted "M13 phage"). M13 phage is a bacteria-infecting virus, composed of a single-stranded DNA encapsulated by several coat proteins (Fig. 5.11). The phage is linear (or more accurately tubular), approximately 900 nm long and 6.5 nm wide. The viral coat comprises primarily a "major" coat protein (P-8) of which there are 2700 copies, by far the most abundant protein of the phage. Other minor coat proteins (P-3, P-6, P-7, P-9) are expressed at the "head" and "toe" of the phage particles, respectively.

The M13 phage offers a particularly important property that can be utilized in biomineralization applications – the feasibility of displaying peptide *recognition elements* on the viral coat. Essentially, genetic engineering enables peptide display

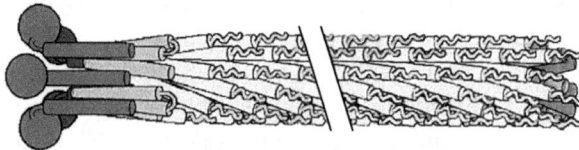

Fig. 5.11: Structure of the M13 filamentous bacteriophage. Reprinted from Rodi, D.J. and Makowski, L., Phage-display technology – finding a needle in a vast molecular haystack, *Curr. Opin. Biotech.* **1999** *10*, 87–93. Copyright (1999), with permission from Elsevier.

through relatively simple modification of the viral coat proteins without adversely affecting phage viability. In practice, the peptides displayed at the tips of the coat proteins could be used for binding and nucleation of different inorganic substances, as depicted for example by A. M. Belcher and colleagues at MIT (Fig. 5.12). The fact that *different* coat proteins are localized in distinct locations within the phage particle allows more complex structural manipulations. For example, the tip proteins of the phage can be modified such that two inorganic substances which exhibit affinities to each other are bound at the two tips of the phage, respectively. Consequently, *ring structures* can be formed through the attraction between the head and toe of adjacent viral particles.

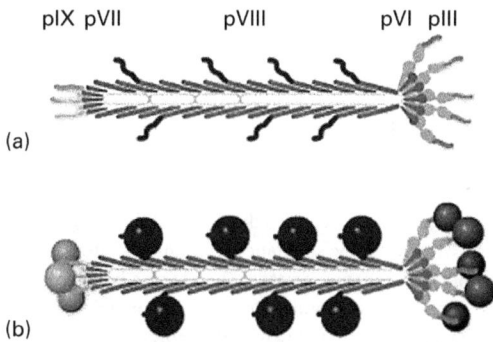

Fig. 5.12: Biomineralization induced through genetically-engineered M13 bacteriophage. The phage genome is modified for display of short peptides at the N-termini of the coat proteins (**top**); the displayed peptides constitute recognition elements for crystal nucleation and inorganic nanoparticles (**bottom**). Reprinted from Flynn, C.E. et al., Viruses as vehicles for growth, organization and assembly of materials, *Acta Materialia*. **2003** *51*, 5867–5880. Copyright (2003), with permission from Elsevier.

A particularly powerful offshoot of the "phage display" technology is the *phage library* concept, in which huge random and semi-random collections of peptides can be displayed on the phage surface through combinatorial engineering of the M13 genes encoding the coat proteins (Fig. 5.13). The modified phages can then be screened against inorganic molecules, colloids, metal nanoparticles, and other entities, extracting the recombinant phage most effective for capturing and promoting the desired mineralized assembly. The DNA sequence of the extracted phages can be isolated and sequenced, and a recombinant phage can be engineered and mass-produced for using as a biomineralization template.

An intriguing use of M13 phage libraries is depicted in Fig. 5.14. The experiment, carried out by A. Belcher and colleagues at the University of Texas comprised two stages. First, an M13 library was created through inserting random sequences into the minor coat protein at the tip of the phage particles. This library was then screened

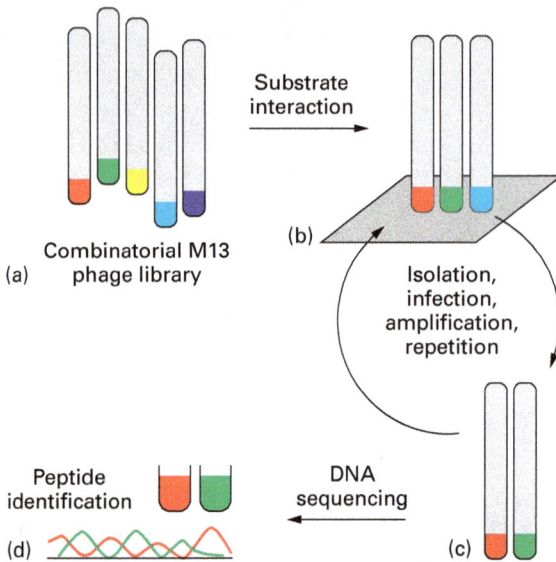

Fig. 5.13: Phage-library technology for identification of effective biomineralization templates. **(a)** A library of recombinant phages is created, comprising different peptides displayed at the phage coat proteins; **(b)** the recombinant phages are reacted with inorganic substances; **(c)** phages interacting with the inorganic materials, or inducing mineralization, are isolated. The process is repeated several times to improve selection; **(d)** the most effective biomineralization-inducing phages are selected and sequenced to identify the displayed peptides interacting with the inorganic substances. Reprinted from Flynn, C.E. et al., Viruses as vehicles for growth, organization and assembly of materials, *Acta Materialia.* **2003** *51*, 5867–5880. Copyright (2003), with permission from Elsevier.

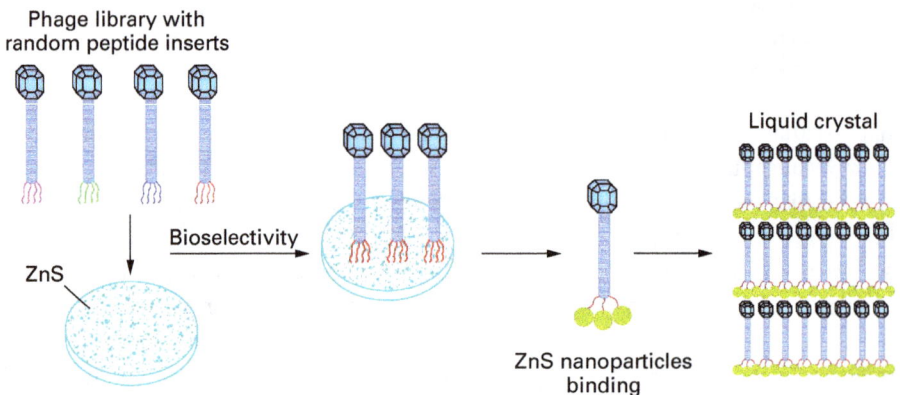

Fig. 5.14: Liquid crystalline quantum dots through M13 library screening. Selection of recombinant phage displaying peptides binding crystalline ZnS. Following amplification, the recombinant phage binds ZnS quantum dots (QDs); concentrated solution of the phage-QD complexes adopts a liquid crystalline organization.

for binding to ZnS crystal surface; the phage displaying the peptide with the highest affinity was isolated and amplified. In the second part of the experiment the researchers exploited another well-known property of filamentous bacteriophages – its *liquid crystalline* organization in concentrated solutions and in thin films, associated with the anisotropic, elongated structure of the phage particle. Indeed, this liquid crystalline ordering of the phage particles can be "tuned" through controlling the particle concentration, ionic strength, or even through externally-applied magnetic fields. Accordingly, the engineered viruses could both capture crystalline ZnS nanoparticles (also known as "quantum dots"), as well as align them in organized liquid crystalline patterns, opening the door to possible optical and spectroscopic applications.

The concept of identifying and selecting effective mineralization-inducing peptides through library screening can be additionally implemented for identifying mutants of known proteins which either accelerate or retard growth of inorganic crystals. This is largely due to the realization that the biological/crystallite interface plays a fundamental role in determining the morphology of mineral crystals. Indeed, in many biomineralization phenomena, for example bone growth, peptide surfaces dictate structure and organization through the combination of electrostatic and hydrophobic interactions with the inorganic ions, crystallites, and colloids that are part of the growing crystal. Phage display libraries have contributed to such research efforts since they enable screening and optimization of a large number of parameters, including type and sequence position of individual amino acids, side-chain orientation, and peptide folds, which intimately affect crystallization.

Phage display technologies have not only aimed to use peptides as vehicles for studying and manipulating biomineralization processes, but also, conversely, to investigate the effects of inorganic surfaces upon bound proteins. This issue is significant in many biotechnological applications in which immobilization of proteins (and/or entire organisms) on inorganic surfaces might affect the biological properties of the bound biomolecules. Here too the phage display technology provides a unique advantage as it allows efficient selection of polypeptides having desired biological properties; in fact, in many cases such library-extracted peptides exhibit superior functions compared to their natural counterparts.

Biomineralization enabled through the recombinant phage display approach can be carried out also in *spherical* viruses. The *T7 bacteriophage* has been a promising candidate for such applications. T7 is an icosahedron-shaped bacteriophage, and in fact well-suited to be used as a template for biomineralization because deposition of inorganic substances can be carried out both externally on the viral coat and *inside* the particle. This is due to the rather uncommon assembly process of the phage; while many viruses first package the genetic component (DNA single strands or double strands) and subsequently build a protein shell around it, the T7 phage first constructs the protein casing and only then condenses the nucleic acids inside it. Consequently, an empty T7 capsid ("ghost virus") can be used as a template for mineralization in its interior space (Fig. 5.15).

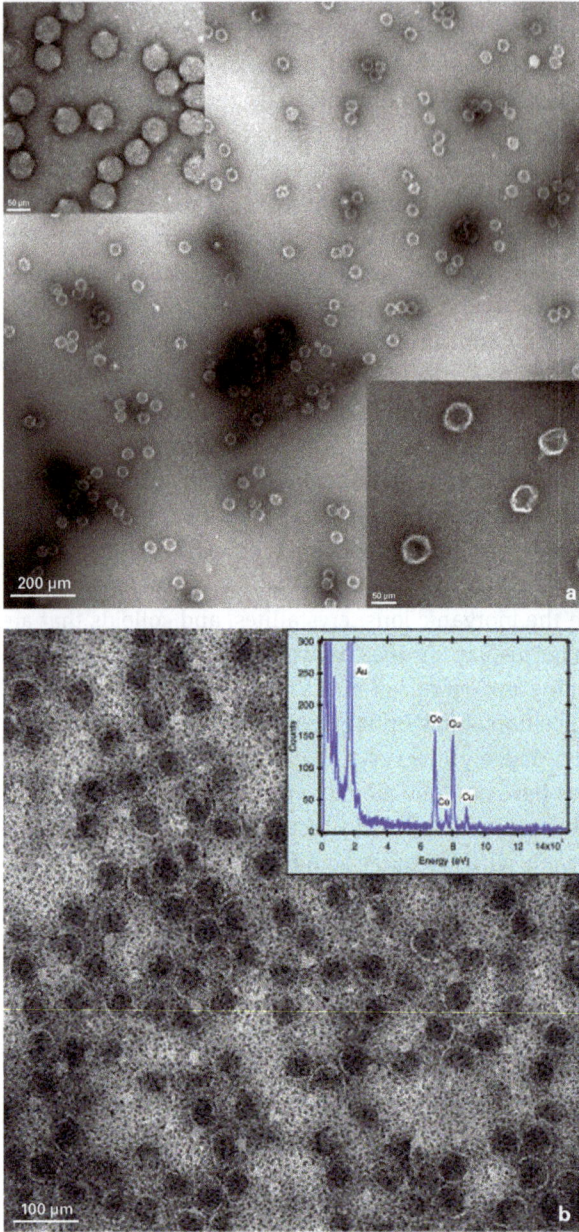

Fig. 5.15: Biomineralization inside T7 "ghost virus". TEM images depicting the biomineralization process. **(a)** "Ghost viruses" comprising the T7 capsids without the encapsulated DNA. Scale bar: 200 nm. Inset in the top left shows the normal T7 particles, inset at bottom right magnification of the empty particles. Scale bars of insets: 50 nm. **(b)** Metallic cobalt grown inside the T7 ghost viruses. Reprinted from Liu, C. et al., Magnetic viruses via nano-capsid templates, *J. Magn. And Magnetic Mater.* **2006** *302*, 47–51. Copyright (2006), with permission from Elsevier.

Microorganisms, particularly bacteria, have been also used as biomineralization templates. Such applications employ both the overall morphologies as physical templates as well as intrinsic biochemical pathways in which inorganic molecules are processed. A case in point concerns sulfur reduction pathways by sulfate-reducing bacteria. Such reactions have been harnessed to precipitate semiconductor nanoparticles including CdS and ZnS on the bacterial cell surface through incubating the bacteria with salts of the divalent metals (such as cadmium).

Similar to the case of proteins and viruses discussed above, bacterial cells have been used as physical *templates* for the creation of inorganic structures. Nucleation seeds for the initiation of biomineralization in such systems are usually achieved via charged amino acids and other functional groups at the cell surface. In many cases, after the biomineralization process, the bacterial cells are removed through annealing (for example through high temperature treatment), leaving behind hollow structures in varied configurations, depending upon the bacterial species encapsulated. Fig. 5.16 depicts highly uniform zinc-oxide hollow spheres produced in the laboratory of D. Zhang at Shanghai Jiaotong Univ., China, using the spherical bacterium *Streptococcus thermophilus* as a template.

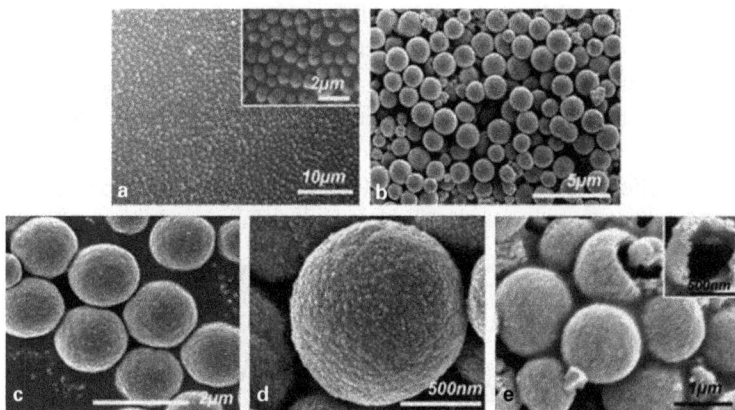

Fig. 5.16: Inorganic hollow spheres produced via bacterial templating. SEM images showing native *Streptococcus thermophilus* bacterial cells **(a)**, and ZnO-coated bacteria **(b)**–**(d)**. **(e)** Hollow ZnO spherical shells after removal of the bacteria cores through calcinations at 600 °C. Reprinted from Zhou, H. et al., Hydrothermal synthesis of ZnO hollow spheres using spherobacterium as biotemplates, *Micro. Meso. Mater.* **2007** *100*, 322–327. Copyright (2007), with permission from Elsevier.

Biomimetic mineralization is not limited to microorganisms such as viruses and bacteria. Other life forms such as plants and algae provide diverse templates and concomitant possibilities for generation of artificial inorganic structures. **Diatoms**, a family of eukaryotic algae, have long attracted considerable attention of chemists, biologists, and nanotechnologists, due to their unique mineralized structures and also as a promising biomimetic platform for new materials. In nature, diatoms are encapsulated within silica coatings called frustules, which come in diverse and visually-attractive species-dependent configurations including cylinders, circles, "donuts", stars and others (Fig. 5.17).

The diverse shell structures of diatoms have been an alluring target for biomimetic practical applications. The interest is due to several important features of diatoms, including the variety of *three-dimensional* morphologies, and particularly the abundance of uniform micro-, and nanopore structures which might provide a platform for constructing organized assemblies of inorganic materials. Indeed, the silica frameworks of diatoms can be used as "stamps" or "replicas" for fabricating arrays of inorganic materials (Fig. 5.18). The replication process carried out by T. Fan and colleagues at Shanghai Jiaotong University, China, shown in Fig. 5.18, employed the silica as a scaffold upon which ZnS nanocrystals aggregate, ultimately yielding crystalline ZnS which precisely traced the contours of the diatom shell. Replicating the silica framework with different inorganic species could be utilized for varied functions, including molecular sieves and filters, drug delivery vehicles, catalytic substrates, sensors, and others. Another notable facet of diatoms is their relatively rapid growth, providing a potentially easy and inexpensive pathway for producing industrial-scale components.

Inorganic replication has been demonstrated in more complex marine animals. **Sea urchins** erect intricate porous calcite skeletons vital for transport of nutrients and refuse substances, as well as for maintaining overall structural stability. Several studies have demonstrated formation of biomimetic materials using the urchin skeleton as a three-dimensional template for deposition of different inorganic substances. F. C. Meldrum and colleagues at Queen Mary and Westfield College, UK, succeeded in coating the inner surface of the urchin scaffold with a thin gold layer (Fig. 5.19); an accurate gold replica

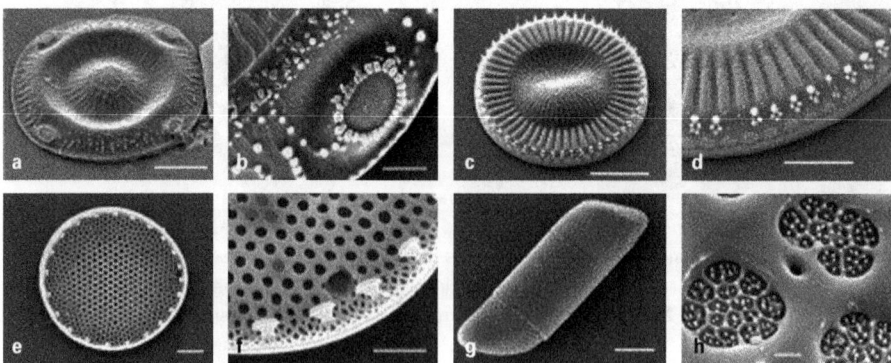

Fig. 5.17: Diatom structures. SEM images of varied diatoms, courtesy of Mary Ann Tiffany. **(a, b)** *Glyphodiscus stellatus*, scale bars are 20 µm **(a)**, close-up in **(b)** 5 µm. **(c, d)** *Cyclotella meneghiniana*, scale bar 20 µm, close-up at 5 µm. **(e, f)** *Roperia tesselata*, scale bar is 10 µm **(e)** and the close up is 5 µm **(f)**. **(g, h)** *Isthmia nervosa*, scale bars are 50 µm and 1 µm, respectively. Reprinted from Drum, R.W. and Gordon, R. Star Trek replicators and diatom nanotechnology, *Trends Biotech.* **2003** *21*, 325–328. Copyright (2003), with permission from Elsevier.

Fig. 5.18: Replicating diatom shell structure with inorganic materials. Electron microscopy images of *Coscinodiscus lineatus* diatom shell after replication with ZnS. The images demonstrate that the shell structure was completely retained, producing an ordered array of crystalline ZnS. **(a–c)**: Top SEM views of the silica-based ZnS replicas in different resolutions; **(d)**: high resolution TEM (HRTEM) demonstrating the crystalline structure of the ZnS replica. Reprinted with permission from Li, X. et al., *Eur. J. Inorg. Chem.* **2009** *11*, 211–215. Copyright (2009) John Wiley and Sons.

was obtained following dissolution of the skeleton template. Such replicas, in which the sub-micrometer pores and channels are preserved, constitute a powerful platform for varied applications such as sensors, chemical filters, catalysts, and others.

The concept of using biological substances as *templates* for assembling inorganic structures is manifested also in **plants**, offering in many instances biomineralization applications on a much larger length scales compared to viruses, bacteria, or diatoms, discussed above. Numerous plants exhibit *fiber* structures that can be coated by metals or inorganic salts. *Cellulose*, in particular, is a polysaccharide scaffold for numerous plant species and often exhibits unique fibrous structures. Cellulose fibers, perhaps the most popularly-known among these is cotton, have been employed as templates for varied inorganic materials – alumina, magnesium oxides, semiconductors, metals, and others.

Plant constituents such as leaves, stems, and bark have been used as mineralization templates. Similar to the systems described above, they combine potentially useful microscopic structures – pores, oriented fibers, etc. – with macroscopic scale organization that can be manipulated for practical applications. In our environmentally-conscious times, plants also offer an abundant supply of the raw material used as templates, chemical treatments that generally require mild and non-toxic conditions, and biodegradable waste (i.e. the annealed scaffolds).

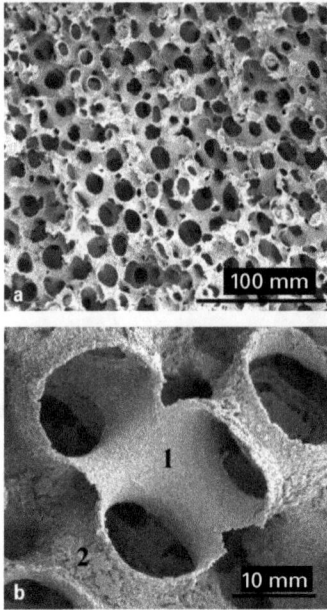

Fig. 5.19: Replication of a sea-urchin skeleton. SEM images depicting gold-coated porous sea-urchin skeleton *after* dissolution of the native calcite template. The replicated gold traces the skeletal microstructure and surface features. Reprinted with permission from Seshadri R. and Meldrum F. C., *Adv. Mater.* **2000** *12*, 1149–1151. Copyright (2000) John Wiley and Sons.

As a source of bio-inspired mineralized structures, *animals* are a class in itself. The enormous diversity of skeletal structures, microscopic and macroscopic morphologies, hierarchical organization, and functionalities found in the animal universe provide ample opportunities for biological-inorganic hybrid designs. For obvious space limitations I will discuss only a couple of examples, numerous other can be found in the scientific literature.

Insects, besides being the most abundant and diverse animal group on the planet, exhibit remarkable structures that can be potentially employed as biomineralization templates. The **chitin** scaffolds of insects are particularly attractive candidates for biomineralization applications. Chitin exhibits stable, rigid, and often highly ordered structures, displaying in many instances interesting optical properties. The beautiful wing colors of many butterfly species, for example, are not related to pigments or other chemical additives, but rather arise from periodical sub-micrometer structures, which result in distinct light refraction patterns.

The unique microstructures of butterfly wings have been employed as "replication templates" for photonic materials. Z. L. Wang and colleagues at Georgia Tech coated the wings of the butterfly *Morpho paleides* with alumina, creating ordered surfaces exhibiting remarkable optical properties (Fig. 5.20). Specifically, the researchers reproduced the periodic surface lamellae, responsible for the striking blue color of the butterfly wing, through atomic layer deposition (ALD) of alumina – a mild procedure for fabrication of thin films through gas-phase reactions. The experiment achieved not only replication of the fine morphological features of the wing surface, but crucially, tuning the *thickness* of the mirror-image alumina coatings produced distinct colors other than the native blue appearance, presumably due to slight changes of the light scattering patterns.

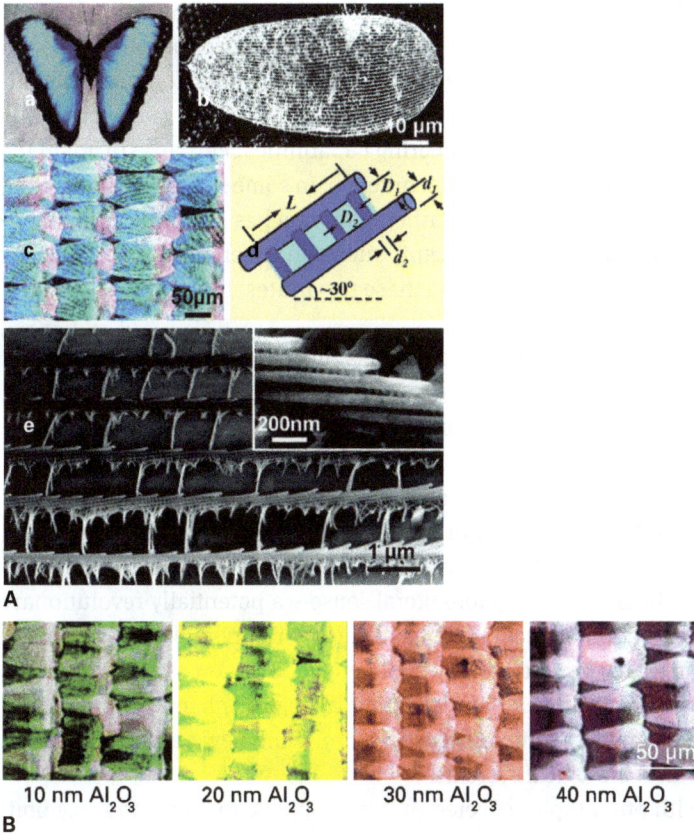

Fig. 5.20: Replicating butterfly-wing microstructures for photonic applications. **A.** Surface morphology of *Morpho peleides* butterfly wing and the relationship with the wing color. **(a)** Butterfly photo; **(b)** SEM image of a single butterfly wing showing the stripe structure; **(c)** an optical microscope image depicting the blue appearance associated with the ordered scales; **(d)** scheme of an individual microstructure within the surface lamellae responsible for the scattered light appearing blue; **(e)** SEM image of the scale surface showing the structure of the lamellae. The inset is a higher resolution SEM image showing the periodic nanostructures. **B.** Optical images of alumina-replicating scales having different layer thickness, demonstrating dramatic color variations. Reprinted with permission from Huang, J. et al., *Nano Lett.* **2006** *6*, 2325–2331. Copyright (2006) American Chemical Society.

6 Synthetic biology

Synthetic biology is a "new" science attracting intense interest which not only testifies to its current achievements, but also points to the potential to infuse new ideas, concepts, and experimental schemes at the interface between biology, chemistry, and materials science. A notable caveat, however, is that the *definition* of synthetic biology is still somewhat in flux. A literal interpretation of this new discipline invokes expectations of the construction of biological systems in which cells and even whole organisms are built from completely artificial components, essentially from the bottom up. Naturally, however, there are doubts whether such a feat can ever be achieved, considering the enormous complexity and sophistication of living systems. Some recent developments might bring us closer to this goal, mostly involving advanced bioengineering capabilities. Perhaps surprisingly, despite the advances of current synthetic techniques, in some senses it is becoming apparent that traditional "bottom-up" synthetic approaches might be fundamentally inferior to a radically different concept, successfully employed for eons by nature: evolution. This chapter will discuss these two routes, their differences, and intersections.

Current ongoing "synthetic biology" research is inherently interdisciplinary and encompasses biology, chemistry, bioengineering, computer science, and related fields. The overarching goals of these research efforts is to produce new functional biomolecules, genetic traits, cell components, and potentially new organisms based on "first principles", i.e. not relying on nature as a source but using a synthetic drawing board instead. While activities in this field overlap to a certain degree with more established disciplines such as genetic engineering and protein design, many scientists treat synthetic biology in a more literal sense – a potentially revolutionary concept for construction of biological systems using laboratory tools. A claim being occasionally made is that synthetic biology could engender a new understanding of living systems and the molecular factors affecting biological phenomena and processes. This knowledge might be implemented for designing new entities that might even outperform natural biological systems.

As is the case with other topics covered in this book, due to scope and size limitations this chapter cannot cover all facets of synthetic biology, which is a huge and highly-active field. Specifically, I cannot discuss all cutting-edge research in chemical biology, genetic engineering, and protein chemistry which can be broadly included in synthetic biology. Nor will I thoroughly discuss the enormous body of research on structural and functional modifications of proteins aiming to improve properties such as stability, activity, etc. The emphasis here will be on synthetic biology concepts and systems pursuing "biomimicry", mostly in a *functional* sense (rather than simply mimicking structural aspects of biological entities). Thematically, this chapter is divided according to molecular hierarchy: from *molecules* (engineered proteins, oli-

https://doi.org/10.1515/9783110709490-006

gonucleotides), to larger *macromolecular assemblies* (artificial chromosomes), and *entire microorganisms* (artificial viruses and bacteria). Newly-minted biomimetic *concepts* (such as artificial genetics) will be presented as well.

6.1 Synthetic proteins

Proteins – the building blocks and "molecular workhorses" of life – come in numerous structures and exhibit diverse functions, so it might appear that nature has already spanned the available structural pool and exhausted all possible combinations. Surprisingly, however, it appears that human ingenuity continuously expands the protein universe, assembling new amino acid combinations that present new structures and surprising new functionalities.

The fundamental factors underlying protein structures and functions are embedded in the sequence of the amino acids comprising the protein. However, the relationships between distinct domains within the protein as a whole are what determine the crucial structural features – the secondary conformation, and even more importantly – tertiary and quaternary structures (Fig. 6.1). The repertoire employed for construction of protein structures is not infinite – the most abundant building blocks are the *α-helix, β-sheet*, and *turns* and those structural units are organized in different abundance and varied ways in relation to one another.

Protein engineering is considered one of the bedrocks of synthetic biology. Even though synthetic biology as a distinct discipline is still in its infancy, protein engineering has been practiced for decades, mostly aimed at improving industrial applicability of proteins, enzymes in particular. Modern protein engineering methodologies have centered on improving desired protein functions through changing the protein amino acid sequence by genetic alterations, chemical modifications, and combinations of the two. Engineered proteins have been also a staple tool of modern cell biology research. *Green fluorescent protein* (GFP), for example, is an indispensable protein employed for microscopic imaging of biological processes, structures, and entire cells and organisms (Fig. 6.2). This small protein, originally isolated from a jellyfish species, is intensely fluorescent in green color as its name implies. The broad utilization of GFP in biology arises from insertion of the gene encoding the protein into gene sequences corresponding to other proteins, or genes belonging to different organisms. The fluorescence of GFP can be subsequently utilized for *imaging* applications following expression of the engineered proteins containing the GFP sequence.

Protein engineering has also made strides into the materials world, contributing useful properties of proteins to the design of novel materials. An example is *stimuli-responsive materials*. These materials respond to external stimuli, such as temperature, light, electric fields or chemical substances, through macroscopic, often dramatic changes in size, shape, solubility, and others parameters. While the protein

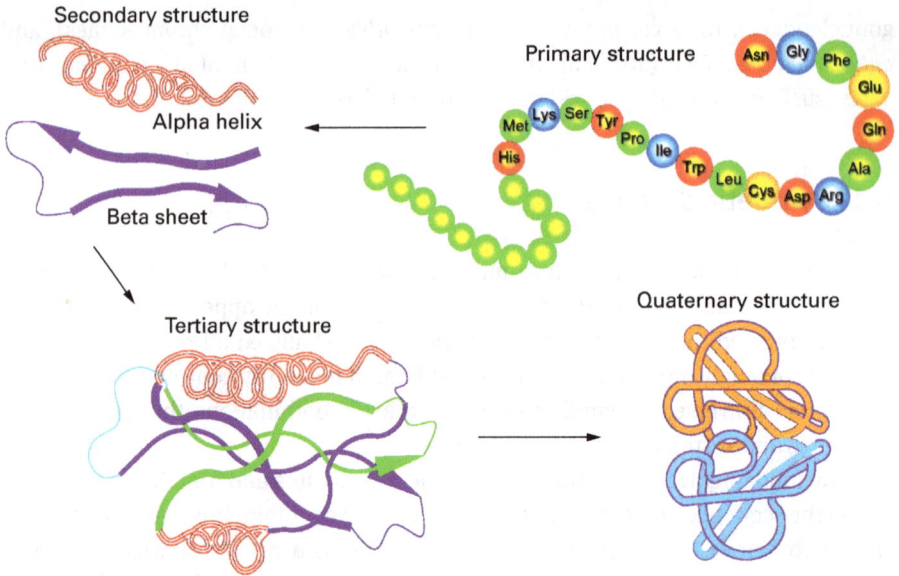

Fig. 6.1: Protein structural elements. *Primary structure* consists of the amino acid sequence, which in turn determines the *secondary structure* (alpha-helix, beta-sheet, turns, and others). The secondary structures further fold into the protein *tertiary structure*. Some proteins form *quaternary structures* comprising several protein subunits.

Fig. 6.2: Fluorescence microscopy imaging using green fluorescent protein (GFP) expression. A subset of cells isolated from the brains of adult rats with ubiquitous expression of GFP. Image courtesy of Professor Ronald Kalil, University of Wisconsin-Madison.

world provides many examples of stimuli-responsive materials (e.g. pH-dependent protein conformations, muscle contraction affected by protein action), until quite recently synthetic materials exhibiting such properties were almost wholly non-biological, comprising polymers, hydrogels, and other inorganic substances. Integrating engineered proteins in materials design, however, has upended the sole reliance on synthetic chemistry in such applications.

Indeed, strategies for identifying and engineering functional protein domains in new stimuli-responsive materials have achieved some notable success. *Elastic materials*, for example, have utilized structural motifs from well-known protein families such as *elastins*, of which the primary biological role is to withstand external pressures and deformations and maintain cell shape and morphology. *Gelatin* has been also used in advanced materials as a *temperature-sensitive* component, exploiting the phase transition temperature of the protein which changes between solution and gel phases. Engineered proteins that respond to *several* stimuli (such as pH and temperature, or two temperatures) have been also utilized by engineering "chimera proteins" which fuse distinct peptide domains from different proteins into a single composite sequence.

Mutagenesis is a generic term for a technology in which individual amino acids are substituted within the protein sequence, resulting in modification of protein functions. Isolating mutated proteins having better traits and improved functionalities are perhaps as old as civilization itself – domestication of animals and agricultural crops can be included in this definition. The sophistication of techniques developed to induce and identify mutated proteins has increased in parallel with our understanding and control of genetic reactions and gene pathways. Many chemical and physical factors have been introduced in order to modify the genetic code, and resultant protein mutants. The limitations of "mutagen-based" protein manipulation approaches are primarily the lack of control of the exact *site* of mutation within the DNA sequence, consequently requiring development of functional screening assays and protein isolation procedures for identifying the exact mutation products and extraction of the proteins having the desired (altered) functionalities. These requirements have for many years hindered the development of protein engineering as a rigorous and widely-practiced scientific and applied tool.

The introduction of **site-directed mutagenesis**, aided by development of the powerful polymerase chain reaction (PCR) technology which enabled DNA amplification, had a tremendous impact in putting protein engineering at the center stage of modern molecular biology. Many variations, protocols, and technical tools for carrying out site-directed mutagenesis have been developed over the years. The basic tenet of the technique is the substitution of native oligonucleotide sequences encoding for specific amino acids within a protein with synthetically-produced DNA strands – resulting in replacement of amino acids (single or multiple) within desired, rather than random, sites at the protein.

The first rudimentary site-directed mutagenesis methodology was demonstrated in the early 1980s. The experiment utilized substitution of a single nucleotide base within a short DNA sequence complementing a native sequence, producing a single base-pair mismatch (Fig. 6.3); binding of the mutated oligonucleotide to the native sequence still occurred despite the single base-pair mismatch because of the complementarity of the adjacent base-pairs. The complete DNA strand containing the substituted nucleotide was subsequently amplified and translated through conventional protein expression procedure in bacteria, producing the mutated protein in large quantities.

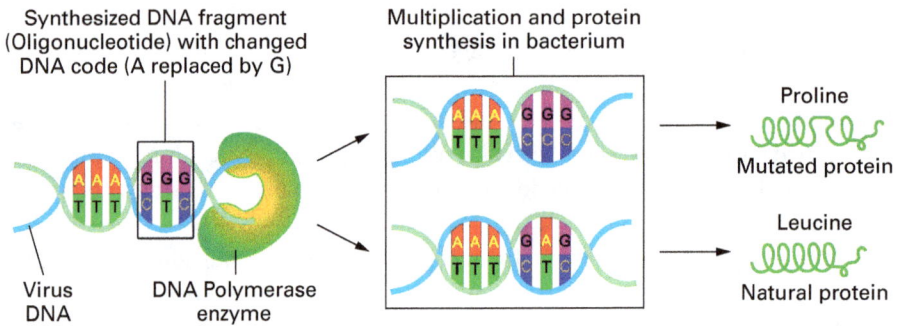

Fig. 6.3: Site-directed mutagenesis. Oligonucleotide sequence corresponding to the desired protein is inserted into a single-strand viral genome. A single nucleotide mismatch is then introduced in a *synthetic* complementary DNA strand (A replaced by G in the scheme shown) which is annealed onto the viral strand. The DNA is amplified and expressed in bacteria, yielding the native protein and mutant protein in which a single amino acid was changed.

A more versatile and technically efficient site-directed mutagenesis approach involves *PCR* and is depicted in Fig. 6.4. This technique and its numerous variants have gained broad popularity since it overcomes many of the technical and practical limitations encountered in the original mutagenesis schemes, such as multiple steps, lower yield of mutated DNA and respective proteins, and the need for isolating *mutated* proteins from the *native* ones. PCR uses two synthetic oligonucleotides (shown in light blue in Fig. 6.4) as *primers* to amplify a nucleotide sequence of interest – encoding for the sequence containing the amino acid or acids to be substituted. The two central primers, however (b and c in Fig. 6.4), contain a complementary "foreign" oligonucleotide sequence designed to insert an additional amino acid into the sequence. Amplification of the oligonucleotide target is carried out through consecutive steps of double-strand denaturation, annealing, and strand extension by a DNA polymerase enzyme. The amplified mutated DNA sequence is subsequently reintroduced into the original plasmid through a cut-and-paste procedure (for example by using restriction enzymes), and the modified DNA is employed for producing the desired protein mutant in large quantities.

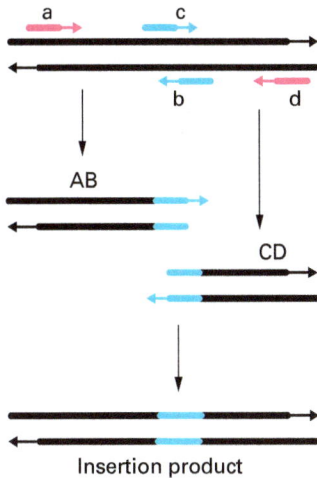

Fig. 6.4: Site directed mutagenesis through polymerase chain reaction (PCR). See text for details.

Site-directed mutagenesis has opened new horizons in protein engineering. Most initial (and many current) applications targeted enhanced enzymatic activities. This goal has been made possible through systematic modifications through sequence changes in different enzyme domains, including the active sites, catalytic domains, and cofactor-binding regions. Transformations of enzymes effected through such routes could be significant; enzymes have been made to recognize and act on non-native substrates, in some cases substrates that are totally different to the native ones. This industrially important field is also referred to as "metabolic engineering" since the parameters screened in such studies are the yields and purity of the enzyme metabolic products.

A powerful approach for modifying enzyme activities through protein engineering has been the alteration of **regulation pathways**. Metabolic processes in biological systems are highly regulated as organisms need to respond and adapt to varied environmental cues such as temperature, salinity, light, etc. Regulation of enzyme (and other proteins) activities is carried out in many cases on a *genetic* level via interactions of specialized proteins with regions in the DNA strands. Accordingly, such *transcriptional regulatory proteins* modulate gene expression in response to external stimuli. For most enzymes, for example, such regulatory proteins (i.e. promoters, repressors) can adjust the level of enzymatic activity depending on the level of activators or inhibitors present in the reaction mixtures. Protein engineering methods thus tune regulation pathways through modification of specific genes, altering proteins involved in the regulation processes themselves, and controlling the types and concentrations of molecules interfering with the action of the regulatory proteins.

Genetic regulatory mechanisms constitute a core technology in the burgeoning field of *genetically modified organisms* (GM, or GMO) – also referred to as "transgene technologies". In this huge and commercially-active field, useful genes are identified in one organism (plant, bacterium, algae, mammals . . .), and implanted inside the genome of a *different* organism, with the goal of imbuing the latter organism with desired, non-native properties and functions. Most current GMO applications operate through an "inducible expression system" concept, depicted schematically in Fig. 6.5. The thrust of this approach is to not just insert the desired foreign gene (or genes) into the genome of the target organism, but to *regulate* the expression of the implanted gene. This task is achieved by incorporation of protein activation or repression elements that are directly coupled to the expressed transplanted gene. The actual control mechanism is usually based upon *inducing* the action of the regulatory element by small molecules which are either naturally present in the environment of the GMO, or externally added (known as *chemical* induction).

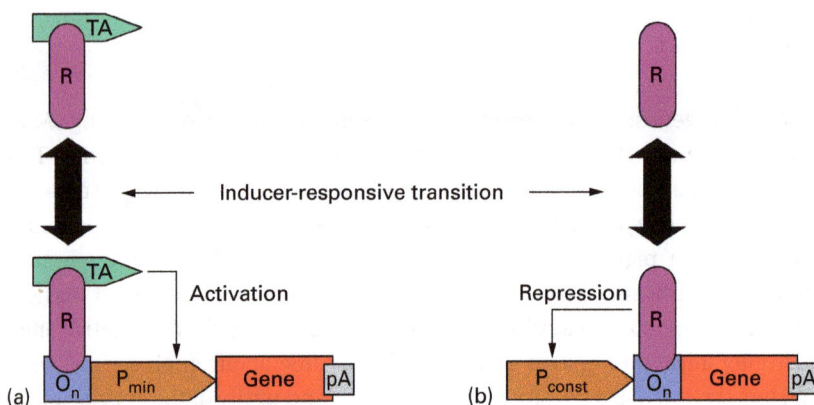

Fig. 6.5: Genetic engineering through induced expression. Principles of induced regulation strategies. **(a)** Gene *activation*. The bacterial regulatory protein **(R)** is coupled to a transcription activator domain **(TA)**. Binding of the TA-R complex to the cognate operator domain in the DNA **(O)** induces expression of the target gene through activation of a promoter **(P)** by the TA. **(b)** Gene *repression*. Binding of R onto the operator O induces steric inhibition of the constitutive promoter (P_{const}). *Control* over both activation and repression pathways is accomplished through *small molecule inducers* which modulate DNA binding or dissociation of the regulatory protein R.

An interesting interface between metabolic engineering and synthetic biology has been the use of protein motifs as structural *scaffolds* facilitating metabolic processes. This experimental approach has been implemented, for example, to improve the yields of enzyme "chains" in which the product of one enzymatic reaction acts as a substrate or ligand for another enzyme. Such chains are usually constructed in bacterial expression systems, and imbalances in the metabolic products of the enzyme members in the chain might be toxic to the host cell. An elegant solution to this problem has been to embed in the bacterial cell a specific protein scaffold containing appropriate binding sites for immobilization of the individual enzymes in the "correct" metabolic order and in close proximity, thereby maintaining reactivity of each enzymatic product with the subsequent enzyme in the chain (for which it is the *substrate*).

Numerous sophisticated protein engineering schemes have been developed. M. W. Chang and colleagues at Nanyang Technical University, Singapore, have recently introduced an innovative approach for engineered "microbe-killing" bacteria. In this technique, *E. coli* bacteria were engineered to produce proteins that recognize substances secreted by *Pseudomonas aeruginosa* – a well-known pathogenic bacterial strain. The complexes formed through this binding subsequently activate two additional genes engineered into the *E. coli* genome: one gene produces a toxin which kills *P. aeruginosa*, while the other activated gene produces a protein that essentially lyses the *E. coli* cells – thereby releasing the toxin and killing the *P. aeruginosa* in the surrounding area.

While protein engineers have mostly tinkered with substitutions of amino acids selected from the pool of twenty natural residues comprising human proteins, intriguing technologies have been introduced to incorporate **unnatural amino acids (uAAs)** into protein sequences. In seminal studies, P. Shultz and others have shown that the genetic machinery can be engineered (or "tricked") into inserting uAAs into protein sequences through a conceptually-simple manipulation (Fig. 6.6). Essentially, each of the natural 20 amino acids is encoded by three sequential nucleotides – three-letter "codons" containing three of the four bases A, T ,C, and G. The codons are embedded within the DNA sequence, transcribed to the messenger RNA (mRNA), which in turn complements the transfer RNA (tRNA) molecules which actually carry the amino acids (each amino acid is in fact encoded by several codons, denoted "degenerate codons"). The complementary binding between the mRNA codon and tRNA "anticodon" results in assembling the amino acids in the correct

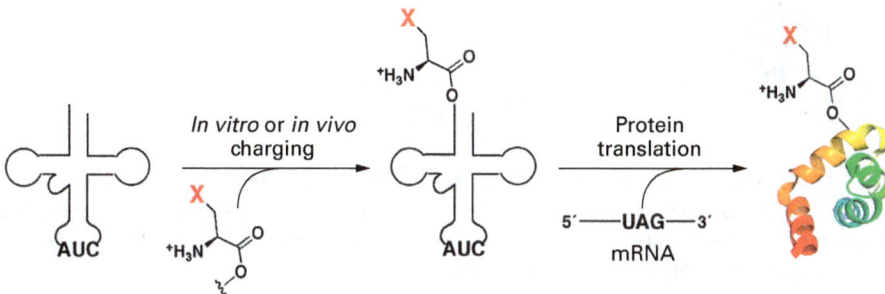

Fig. 6.6: Insertion of unnatural amino acids through nonsense codon translation. tRNA containing a nonsense anticodon (AUC) is loaded with an unnatural amino acid (either synthetically or through biochemical means). The protein translation machinery matches the modified tRNA to the nonsense codon within the mRNA (following insertion of the codon at the desired site in the gene of interest), consequently using the uAA-loaded tRNA to incorporate the unnatural amino acid into the protein at a specific site. Figure reproduced from F1000 Biology Reports [http://www.f1000.com/reports/b/1/88/]

order within the protein sequence. In addition to codons that encode amino acids, there are in fact three codons denoted "nonsense" or "stop" codons, which result in terminating protein translation through recruitment of a specific "suppressor tRNA" in the ribosome. The key feature lying behind uAA incorporation in proteins is "hijacking" the nonsense codons to encode for new residues. This can be accomplished through chemical or enzymatic attachment of the unnatural residue to the suppressor tRNA displaying the *stop anticodon*. In parallel, the nonsense *codon* is inserted through means of genetic engineering into a desired position in the gene encoding the protein of interest (in which the uAA will be incorporated). Accordingly, when the mRNA gets translated in the ribosome, instead of terminating the protein assembly upon translating the stop codon, the modified tRNA actually adds the uAA to the protein sequence.

Since the early demonstration of uAA incorporation into proteins, the technology has expanded considerably. Varied uAAs have been inserted, for example residues in which the functional groups were modified in order to examine steric or electrostatic effects in protein folding, biomolecular interactions, enzymatic activity, and other functions. More sophisticated applications of the uAA technology concern the inclusion of uAAs that further exhibit functional "handles" (Fig. 6.7). This strategy, also referred to as "bio-orthogonal protein labeling", relies upon inclusion in the protein (through the expressed uAA) of molecular units that can be further conjugated through chemical means with specific targets – fluorescent markers, bio-active species, inorganic materials, and others. This generic experimental scheme considerably increases the possible applications of proteins engineered by this manner.

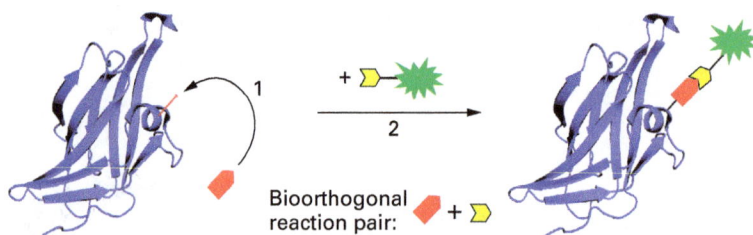

Fig. 6.7: Bio-orthogonal protein labeling. Exogenous molecular unit (red) is incorporated into the protein through the uAA technology; a complementary functional group (green-yellow) is conjugated through chemical means. Reprinted with permission from Hao, Z. et al., *Acc. Chem. Res.* **2011** 44, 742–751. Copyright (2011) American Chemical Society.

uAAs are, in fact, one facet of the efforts to expand the "genetic lexicon". In this context, methods have been developed to not just identify new uAAs, but also introduce into the genetic machinery unnatural *base-pairs* (in addition to the A-T and G-C pairs), which would integrate with the transcription and translation mechanisms and exhibit high fidelity. Experiments have also demonstrated new schemes for encoding amino acids via *four letter* codons (rather than three), consequently significantly expanding the number of available codons (and thus the possibility to introduce new amino acids encoded by these codons).

Significantly, while uAA technology has been initially validated in cell-free systems, i.e. "test-tube mixing" of the ingredients of the protein biosynthesis machinery, later developments have enabled insertion of uAAs into proteins in actual cells. *In vivo* application of uAA protein engineering poses technical challenges since genetic modification and protein translation need to be carried out independently within the cell in its complex environment. Furthermore, the process should proceed without adversely affecting cell viability on the one hand, and on the other hand generate sufficient quantity of the modified proteins. From a methodological standpoint, introduction of a generic biomimetic platform for *in vivo* engineering of proteins containing uAAs needs to adhere to several conditions. First, new three-base codons unique to desired uAAs need to be added to the DNA in a controlled manner. In addition, synthetic machinery for chemical coupling of the uAAs to tRNA has to be developed. The requirements here are stringent: the suppressor tRNA selected can be functionalized *only* with an unnatural amino acid, not one of the *natural* 20 amino acids. Similarly, uAAs should be coupled *only* to the tRNA of choice, not to other suppressor tRNA molecules present in the cell. These requirements have been addressed through enabling the coupling of uAAs to suppressor tRNA via intracellular biochemical routes. Alternative synthetic schemes in which uAA-tRNA complexes were assembled and then micro-injected into the cell have had limited success.

The enormous promise and practical potential of protein engineering cannot be exaggerated. The technology has made possible the creation of a wide range of "designer proteins" performing desired tasks, borrowing beneficial properties and discarding weaknesses and "Achilles' heels". Using protein engineering one can envisage the invention of novel enzymes, hormones, ion channels, signaling molecules, and other critical components of life machinery. Such biomolecules might have improved functional capabilities, physical stabilities, and superior biological properties.

6.2 Directed evolution

Richard Feynman declared more than 50 years ago that "What I cannot create I don't understand." Indeed, the nascent field of "directed evolution" (DE) has led, through creation of new enzymes and proteins, to dramatic expansion of our understanding of evolutionary concepts, processes, and progression. DE can be perceived as "making Nature work fast-forward," through inducing rapid evolutionary processes in a laboratory setting, rather than in a cell (or organism) in its natural environment where natural selection is left to work its way over millions of years.

In a sense, DE research touches on the fundamental question of how to produce a new protein with defined functions and properties (particularly *catalytic* properties in the context of enzymes). This challenge has been traditionally addressed through an "engineering" approach (generally pursued in the realms of bio-organic chemistry and chemical biology), which starts with a desired function and from there aims to design the molecular structure that would facilitate this function. The "evolutionist" approach, on the other hand, works in a trial-and-error fashion, much in the same way that nature has worked for eons. In this methodology, random mutations in the genetic code produce different protein structures (and functions), occasionally leading to "better" traits or functions which increase the chances of survival of the organism. DE as a platform for protein design aims to pursue the evolutionary concept to its extreme – creating artificial systems in which evolution of a protein can be carried out on a significantly rapid timescale. Importantly, the mutations occurring in the genetic code in DE experiments can be random (just like natural evolution), but the "selection rules" (i.e. identifying the "superior" functions) are predefined and depend upon the desired functional target (i.e. more stable enzyme, enhanced ligand binding, etc).

DE has been predominantly used for designing and improving performance of *enzyme* molecules. This is mostly due to the importance of identifying enzymes with superior catalytic properties in biological systems, and practical and commercial applications. In addition, *screening* for enzymatic activities is well-established with diverse assays available for evaluating functionality and catalytic action. Indeed, since its inception about two decades ago, DE has been successfully used for both designing new enzymes, as well as providing insight into fundamental issues such as the presence of conserved structural scaffolds in enzyme families and the relationship between different enzyme properties.

Fig. 6.8 compares the strategies and steps of directed evolution vs. conventional rational design for enhancing the activities of proteins and enzymes. *Rational protein design* is usually based upon (usually known) protein structure and functional information gained through substitution of single amino acids within the protein sequence (i.e. site-directed mutagenesis studies). The protein variant (having the predetermined amino acid sequence designed to achieve the desired functions) is then expressed in a bacterial cell, purified, and analyzed. In particular, assays are applied to evaluate whether the protein adheres to the desired functionalities. This effort is reproduced until the most optimal protein functions sought are attained. Rational design is effective when structure-function relationships are firmly established for the protein under study, but nevertheless the technique is relatively slow and labor-intensive, since optimization and analysis are implemented individually for each protein synthesized.

DE, in comparison, starts with constructing mutant *gene libraries* by random modulation of the genetic code, rather than modifying the sequence of an individual protein in a direct manner. Protein libraries are subsequently produced through bacterial expression and screened (in conventional multiwell plate settings) through application of a range of selection parameters, such as pH sensitivity, stability in organic solvents, temperature sensitivity, and others. This protein characterization

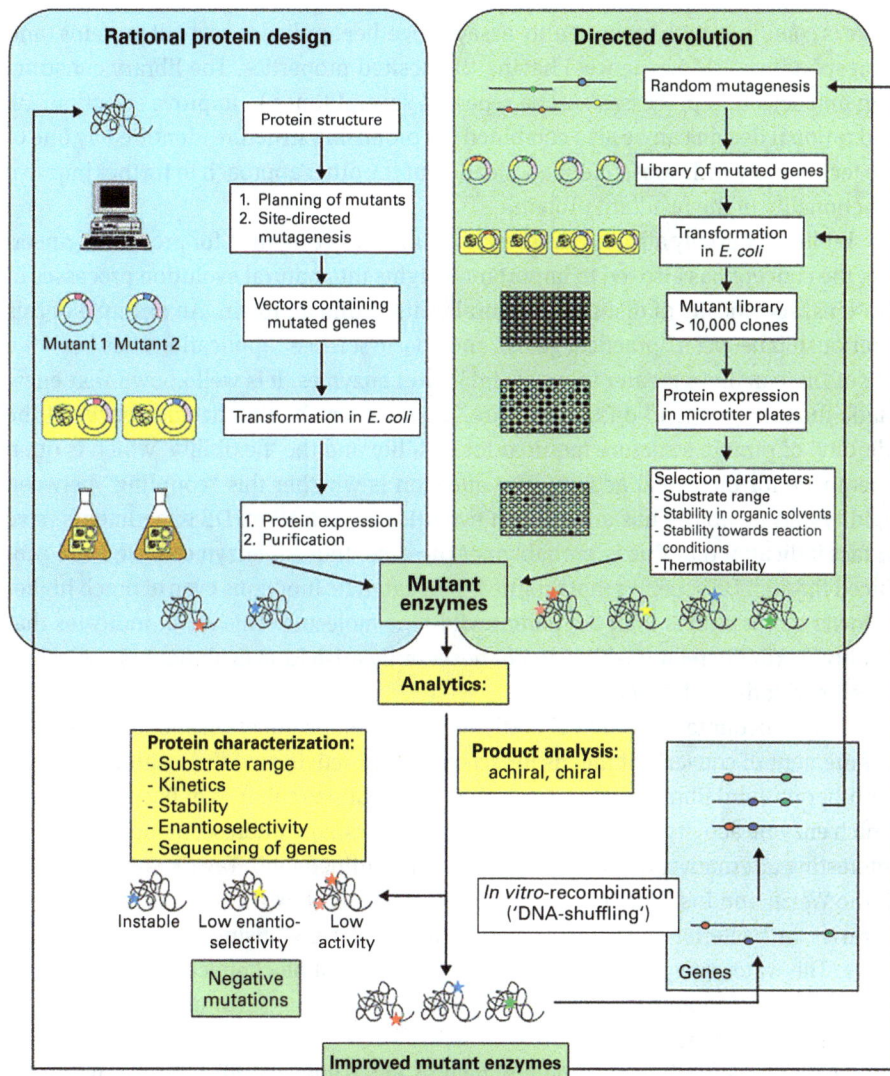

Fig. 6.8: Enzyme optimization through rational protein design approach vs. directed evolution. *Rational protein design* is based upon pre-selection of protein mutants having known structures and functions. The mutants are prepared through site-directed mutagenesis, expressed in bacteria, and analyzed for desired properties. *Directed evolution* utilizes libraries of mutant genes created by random mutagenesis. The genes are then expressed and the resultant protein libraries are screened in microtiter plates using a variety of selection assays. Both approaches can be repeated until proteins (enzymes) with optimal, desired properties are generated. Reprinted from Bornscheuer U.T. and Pohl M., Improved biocatalysts by directed evolution and rational protein design, *Curr. Opin. Chem. Biol.* **2001** *5*, 137–143. Copyright (2001), with permission from Elsevier.

stage is combined with functionality assays, together applied to sort out proteins (and corresponding DNA sequences) having the desired properties. The library construction and selection processes can be repeated several times to improve selection. DE and rational design can be also combined – a promising structure identified by one of the techniques, with subsequent application of the other approach to further improve functionality of the resultant protein.

While DE initially attracted interest as an innovative method for protein engineering, the concept has also led to important insights into natural evolution processes of proteins, specifically of distinct structural units within a protein. An example of this relationship between "practical goals" and "basic science" applications has been the use of DE to achieve greater thermal stability of enzymes. It is well known that enzymatic functions depend on temperature, arising from the fine balance between the 'rigidity' of protein structure required for stability and the 'flexibility' which is often a requisite for biological activity. The question is whether this "coupling" between rigidity and flexibility has arisen from evolutionary selection. DE experiments have in fact indicated that this is probably *not* the case. Indeed, enzymes have been produced through DE schemes that retained their catalytic functions even at much higher or lower temperatures than the parent wild-type molecule. This result indicates that rigidity/flexibility relationships do not necessarily arise from evolutionary "selection pressures" induced by temperature.

An oft-encountered technical bottleneck in development of DE applications is the requirement of convenient means for *screening* the activities of the resultant enzyme (or other protein) libraries. Most screening assays utilize colonies of bacterial cells in which enzyme activity yields detectable products (usually fluorescent molecules). An interesting alternative, depicted in Fig. 6.9, was outlined by D. Tawfik and colleagues at the Weizmann Institute of Sciences, Israel. Instead of expressing the mutant gene libraries in bacterial cells, they created "artificial cells" using water-in-oil emulsions. The water compartments facilitate occurrence of biochemical processes, such as genetic transcription and protein expression, while the surrounding oil prevents leakage of the reaction products.

The purpose of the experiment shown in Fig. 6.9 was to identify, through a DE approach, inhibitors to a *DNA nuclease* enzyme (an enzyme which clips a DNA at specific locations). The artificial cell-like compartments contained a cell-free extract containing the DNA transcription apparatus, as well as an *inactive* nuclease. Also embedded was a gene library which was initially allowed to produce proteins in the enclosed space. A crucial regulatory mechanism that was built into this cell-like system was the ability to *reactivate* the nuclease enzyme through delivery and binding of nickel or cobalt ions via addition of "nano-droplets" to the oil emulsion. Accordingly, the actual selection (i.e. evolution) process could be triggered by the metal ions; since the metal-activated nuclease cleaves the DNA (thus preventing protein expression), only genes producing *nuclease inhibitors* were expressed inside the compartment. Moreover, the most effective nuclease inhibitors encoded in the gene library yielded the most pro-

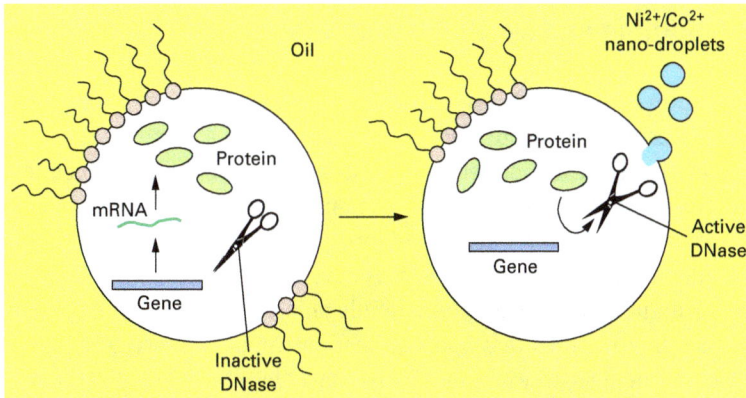

Fig. 6.9: Directed evolution carried out in artificial cell systems. An experimental scheme designed to identify, through DE, inhibitors to the *DNase* enzyme. Library of genes is added to a translation extract isolated from cells, and the mixture is encapsulated in water-in-oil (w/o) droplets. Inactive DNase is also incorporated within the aqueous compartments. The genes are allowed to transcribe and translate, and the DNase is subsequently activated through insertion of nickel or cobalt ions into the droplet. Active DNase digests and destroys the enclosed gene, *except* for genes encoding a DNase *inhibitor* which are subsequently isolated and sequenced. Reprinted from Bernath, K. et al., Directed evolution of protein inhibitors of DNA-nucleases by in vitro compartmentalization (IVC) and nano-droplet delivery, *J. Mol. Biol.* **2005** *345*, 1015–1026. Copyright (2005), with permission from Elsevier.

nounced enhancement of the respective genes in the compartment. These genes could be sequenced through conventional PCR, overall underscoring the capability of this elegant cell-free approach to apply and identify directed evolution products.

Numerous DE experiments have been described in the literature and detailed analyses of the different applications are beyond the scope of this book. While the significant majority of DE applications have focused upon enzymes and other proteins, the technique has been applied also for optimization of structural parameters of other biomolecules, such as glycoconjugates. Like many novel scientific concepts, however, DE has its limitations and inherent weaknesses. While the technique in its early days gave rise to the hope that one might be able to completely supplant natural evolution, thus enable the discovery of completely new proteins having previously unknown properties, later studies indicated that this is not the case. In fact, as the selection methodologies have been developed and varied DE platforms have been tried, it dawned on the DE community that there should be a "starting point" of protein structure and functionality in order for the experiments to succeed. In other words, it is likely impossible to "start from scratch" and evolve a protein property that was not there to begin with. This realization might have fundamental significance beyond the technical limitations of DE, since it might signify that there is, in fact, a *finite* pool of protein functionalities that nature can select from, and in that sense evolution is not totally random.

6.3 Protein modeling and synthetic proteins

Recent years have witnessed an explosive growth in the attempts to both identify specific amino acids and sequences responsible for distinct structural features of protein, and use this information to design proteins "from scratch" (or from "first principles" in the structural biology lingo). Such "*de novo*" protein design has had some remarkable successes and researchers have achieved impressive feats in creation of highly complex and functional macromolecules. Proteins have been fabricated in which the shapes, sizes, and structural properties were carefully tailored and calibrated. In some instances, even protein structures that had not been previously observed in nature were created. Fig. 6.10 depicts an example of a *de novo* "four-bundle helix" protein designed by M. Hecht and colleagues at Princeton University. The rather complex protein conformation in Fig. 6.10, essentially predicted from the primary amino acid sequence, attests to the significant progress and sophistication of the field.

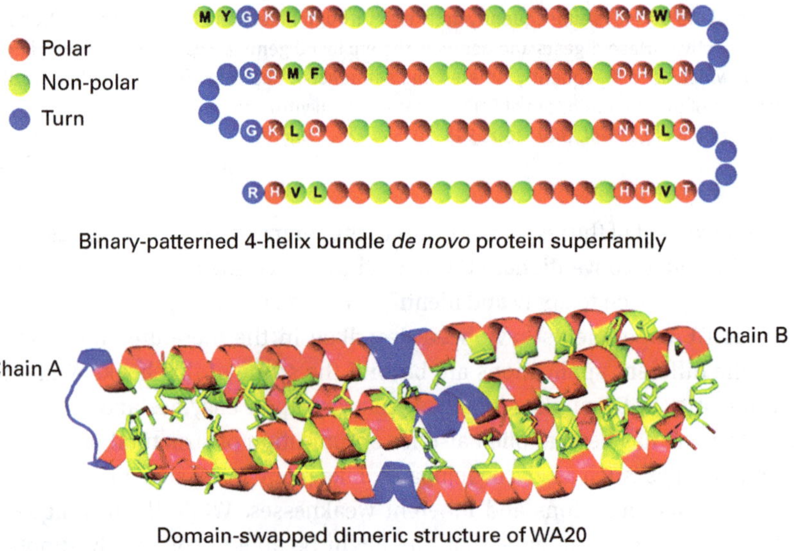

Binary-patterned 4-helix bundle *de novo* protein superfamily

Domain-swapped dimeric structure of WA20

Fig. 6.10: Synthetic *de novo* protein design producing a four-bundle helix protein. The primary amino acid sequence (top) highlights the positions of the different residues – polar, non-polar, and turn residues – which affect folding of the helixes and their orientation in respect to one another. Reprinted with permission from Daniels, D.S. et al., *J. Am. Chem. Soc.* **2007** *129*, 1532–1533. Copyright (2007) American Chemical Society.

Molecular modeling and a variety of computational methodologies have been developed over the years to predict protein structures (and use such predictions for protein design). Predicting and calculating the folded conformation of a protein can follow two main routes. One approach is to essentially take into account *all factors* affecting protein folding, starting from protein biosynthesis in the ribosome, through post-translation modifications (such as chemical modifications of side-chains, glycol-conjugation, etc.), binding of ligands and metal cofactors, interactions with protein structure modulators such as chaperons, and others. Carrying out such a comprehensive analysis involving so many parameters is currently unrealistic.

The second avenue towards protein structure prediction is to just use the amino acid sequence of a protein as the input for algorithms which could predict a three dimensional structure. This approach is perhaps more limited in scope, but it can be broadly implemented in practice and has achieved remarkable success in many instances. Indeed, research in this field, which combines modeling, bioinformatics, physics, and theoretical biology, has made significant advances in recent years, partly aided by the vast expansion and availability of computer power. This progress has led to proliferation of studies predicting not only *secondary* structural elements of proteins, but also *tertiary* and even *quaternary* organization.

Varied algorithms have been developed and applied in computational protein structure studies, some of which are based solely upon physical principles (i.e. the use of "force fields" – energy calculations related to the forces between atoms and molecules comprising the protein), while others are referred to as "knowledge-based" – generally utilizing prior information on known protein structures. An ultimate goal of computational protein design has been the development of "free models", capable of predicting protein structures just from their amino acid sequences, without any prior knowledge of the structures of the proteins. Free model predictions have been demonstrated for small proteins, however they were shown to be extremely challenging for polypeptides with more than 150 amino acids.

A popular and likely the most widely-used method for protein structure prediction and design is denoted "template-based modeling" (TBM). TBM uses "templates" – known structures already determined for protein sequences that are similar or related to the target macromolecule. A TBM procedure starts with a "homology search", in which close sequences are identified in protein structure databases. This step is combined with "alignment" of the target sequence – the goal being to achieve the most optimal overlap between the sequences. Both steps are naturally carried out with the aid of computer algorithms. Once the best template is determined (based upon database screening and sequence alignment, a procedure referred to in the literature as "tethering"), a structural frame is constructed using the template as a starting point, further taking into account the non-aligned regions of the protein, side-chain orientation, solvent effects, and other external parameters.

Numerous "tethering algorithms" have been developed since inception of TBM, aiming to efficiently mine the Protein Data Bank (PDB) – the central depository for protein structures – and identify optimal templates as starting points for structure prediction. In many cases, however, insufficient sequence homologies are located in the PDB. In these situations, protein design needs to be carried out through "free model" approaches as described above (i.e. the *de novo* approach). Even though such methodologies do not rely on extensive structural templating the way TBM does, *de novo* protein design does take into account prior knowledge concerning local structural constraints, evolutionary considerations determining structural motifs, and similar parameters which overall enhance the predictive capacity of these computational approaches.

It is generally accepted that the higher the accuracy required in a protein structure prediction, the more rigorously the prediction procedure should be applied. Thus, free model schemes usually produce generic protein families and protein folds, overall topology of a protein, and domain orientations. TBM modeling, in comparison, is capable of producing more detailed descriptions of protein features, for example ligand docking sites in protein receptors and substrate binding domains in enzymes. Indeed, in situations where the sequence homologies between the analyzed sequences and the templates are high, one can even distinguish between protein binding of structurally-similar small molecules.

While TBM and free model methods have been applied mostly to predict protein structures, molecular modeling can aid the quest for improved protein *functions*. A generic depiction of molecular modeling applied towards engineering new and improved proteins, especially enzymes, is presented in Fig. 6.11. The starting point is a known protein or enzyme, for which the basic overall structure and substrate binding or catalytic sites have been determined. To improve protein functionality, one can proceed along three main avenues (usually a combination of the three is pursued): 1. *Domain modification* (e.g. binding site of a ligand, catalytic region of an enzyme), which is usually carried out through substitution of single (or several)

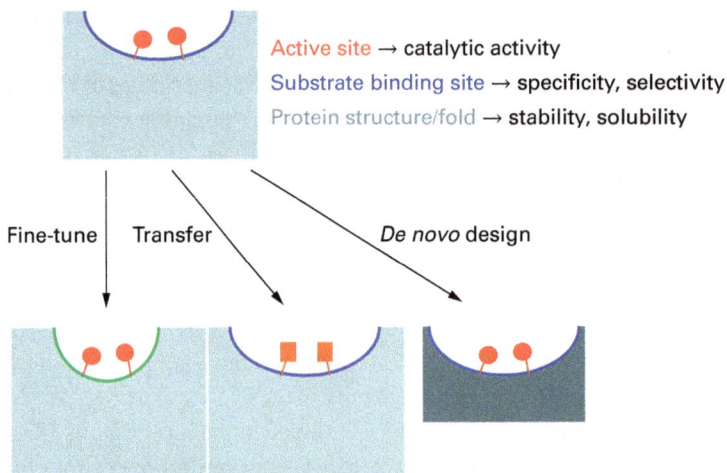

Active site → catalytic activity
Substrate binding site → specificity, selectivity
Protein structure/fold → stability, solubility

Fine-tune | Transfer | *De novo* design

Fig. 6.11: Molecular modeling for improved protein functions. Molecular modeling strategies aimed at improving overall protein functions: *fine-tuning* of the substrate or ligand binding sites, *transfer* of an active site from a different protein, *de novo design* of the protein scaffold supporting the active site. Reprinted from Pleis, J., Protein design in metabolic engineering and synthetic biology, *Curr. Opin. Biotech.* **2011** *22* 611–617. Copyright (2011), with permission from Elsevier.

amino acids in the active domain; 2. *Entire domain replacement* – essentially "transferring" functional domains from related proteins or enzymes, thereby altering the biological activity of the engineered protein; and 3. *Scaffold modification* – applying methodologies such as *de novo* protein design to change or optimize the structural features of the protein "scaffold" supporting the active sites.

The advance of protein modeling and *de novo* techniques for protein design raises a fundamental question which essentially defines synthetic biology – can a *completely new* protein sequence be synthesized and functionally substitute an *existing, natural* protein in living systems? This question can be defined in somewhat different terms: almost all applications and advances in synthetic biology have their roots or starting materials in natural components, i.e. genes, proteins, and regulatory elements such as RNA, which are modified in order to accomplish specific structural or functional goals. However, is there indeed something "unique" in the existing pool of genes and protein sequences identified in living systems? Or can one create functional proteins – through library screening, directed evolution, modeling, or all approaches combined – comprising amino acid sequences that correspond to completely new protein structures and functions?

Experiments aiming to address these intriguing questions were reported. M. Hecht and colleagues at Princeton University, for example, designed an elegant system to test the above concept (Fig. 6.12). The researchers created a peptide library in which *de novo*-designed peptides folded into a known structure (a "four-helix bundle"), albeit *without* a known function. The artificial peptides were then screened for their ability to "rescue" bacteria (i.e. enable bacterial growth) in which different genes, vital for growth, had been deleted. The remarkable observation in this study and similar other investigations was that in some cases synthetic peptides could indeed adopt the "missing" functions of the bacterial cells, functions that in the native bacteria were fulfilled by proteins having *completely different* sequences compared to the synthetic molecules. Particularly striking was the fact that the artificially-constructed peptides retained *life-supporting* roles – i.e. crucial functions that were essential for bacterial viability.

Research such as depicted in Fig. 6.12 has broad implications to the advancement and future of synthetic biology. Indeed, if our synthetic toolbox is *not limited* to shuffling and creating new combinations within *existing* genes and proteins, then it should be possible in principle to identify additional folds and functions associated with new protein sequences. Combined with the introduction of unnatural amino acids, discussed above, one can create, in principle, whole new families of biomacromolecules with potentially novel properties and biological roles. In a broader context, such studies might pave the way for attaining the goal of constructing complete novel synthetic genomes.

An active research area in which protein design and molecular modeling have made considerable contributions concerns **protein misfolding** diseases such as Alzheimer's disease, Parkinson's disease, prion-related diseases. These devastating and

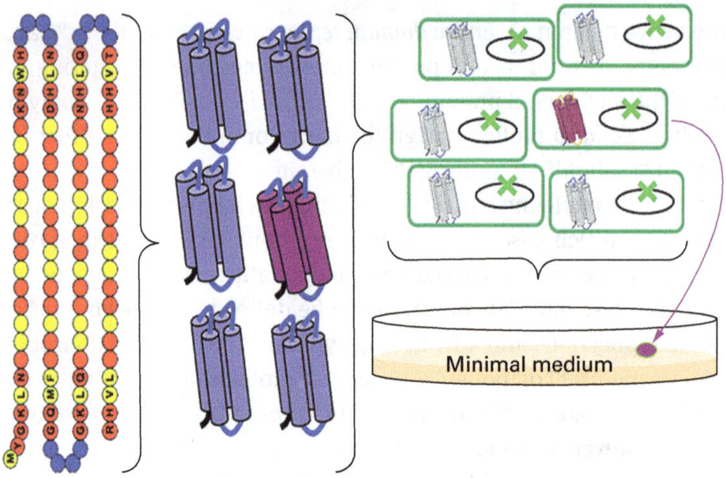

Fig. 6.12: Novel peptides attain distinct biological functions. Library of *de novo* peptides yields four-bundle helix structures. A member of the library (schematically depicted in a purple color) is capable of "rescuing" bacteria for which vital genes have been deleted, preventing their natural growth. Reprinted from Fisher, M.A. et al., *PLoS ONE* **2011** *6*, e15364. Copyright (2011) Fisher et al.

non-curable pathological conditions are associated with, and possibly caused by, aggregation of peptides in neuronal tissues – both inside and outside of cells. These peptides – *amyloid-beta* and *tau* in Alzheimer's disease or *alpha-synuclein* in case of Parkinson's disease – exhibit propensity for adopting beta-sheet conformations resulting in formation of insoluble fibrils. The peptide aggregates constitute the main ingredients in the highly entangled insoluble plaques which render (or are associated with) cell death in such diseases.

Two main fundamental questions arise on the possible causes of amyloido-genic diseases, which might also point to potential therapeutic remedies. The first mystery is rather physiological: virtually all known amyloidogenic peptides associated with diseases are in fact present in "normal" cells and tissues, without undergoing uncontrolled aggregation which is the hallmark of disease progression. The second question focuses on the specific factors inherent to the peptides themselves (sequence, structural features) which precipitate their self-assembly into insoluble aggregates and plaque formation. Many research groups have addressed these questions through careful design of short peptides which exhibit distinct assembly and aggregation profiles.

Artificial peptides can address important questions pertaining to fibrillation and aggregation phenomena. Specifically – whether and what are the pivotal amino acids inducing and shaping aggregation, and what are their positions within the peptide sequence? Are there distinct sub-sequences responsible for peptide self-assembly and fibrillation? Are there *minimal* peptide lengths that are required for peptide aggregation? Are there specific amino acid sequences that induce aggregation through interactions with *other* biological species, such as cellular membranes? Representative experiments designed to address some of these points are summarized below.

One of the most fundamental questions in this field is whether there are critical amino acids affecting peptide aggregation. The "usual suspects" have been hydrophobic amino acids such as alanine, leucine, phenylalanine, tyrosine – since these residues are expected to reduce the propensity of the peptide to form stable soluble conformations, and conversely promote formation of aggregates in aqueous solutions through hydrophobic interactions. Some models, however, point to *ionic binding* between alternating positively-charged residues (such as lysine) and negative amino acids (glutamate) as a significant driving force for fibril assembly. Experiments carried out in recent years using synthetic model peptides suggest that amino acids displaying aromatic side-chains, primarily phenylalanine, are the primary culprits in peptide aggregation. The underlying mechanism of peptide aggregation according to these theories is the "π-stacking" phenomenon, in which parallel orientation of adjacent phenyl rings results in a low-energy conformation which aids self-assembly of elongated fibrils.

It is notable that many of the models outlined above have been backed by experiments in which synthetic peptides having predesigned sequences were shown to undergo aggregation and fibril formation, often exhibiting distinct morphologies. Fig. 6.13 presents an example of the profound effects of peptide composition upon fibril morphology. In that work, E. Gazit and colleagues at Tel Aviv University, Israel, discovered that

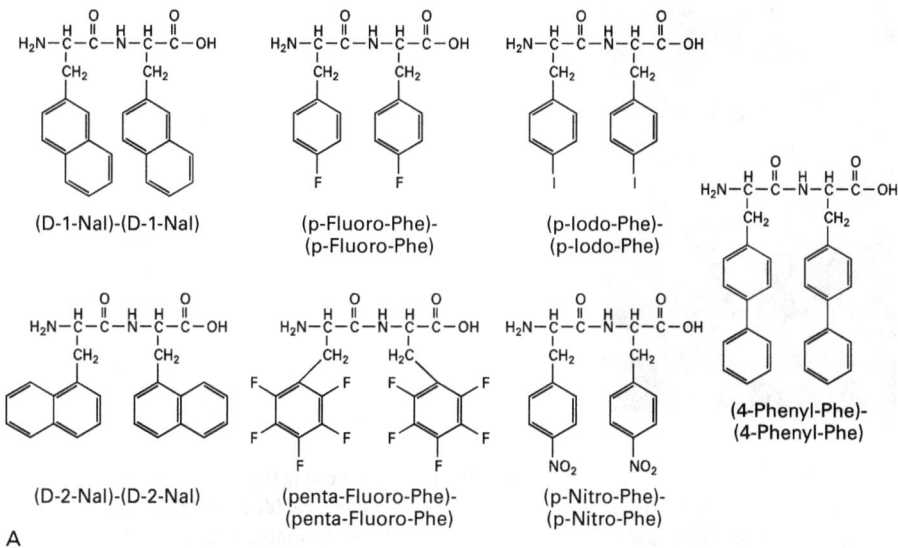

Fig. 6.13: Distinct morphologies formed by dipeptides comprising phenylalanine derivatives. **A.** Molecular structures of the dipeptides.

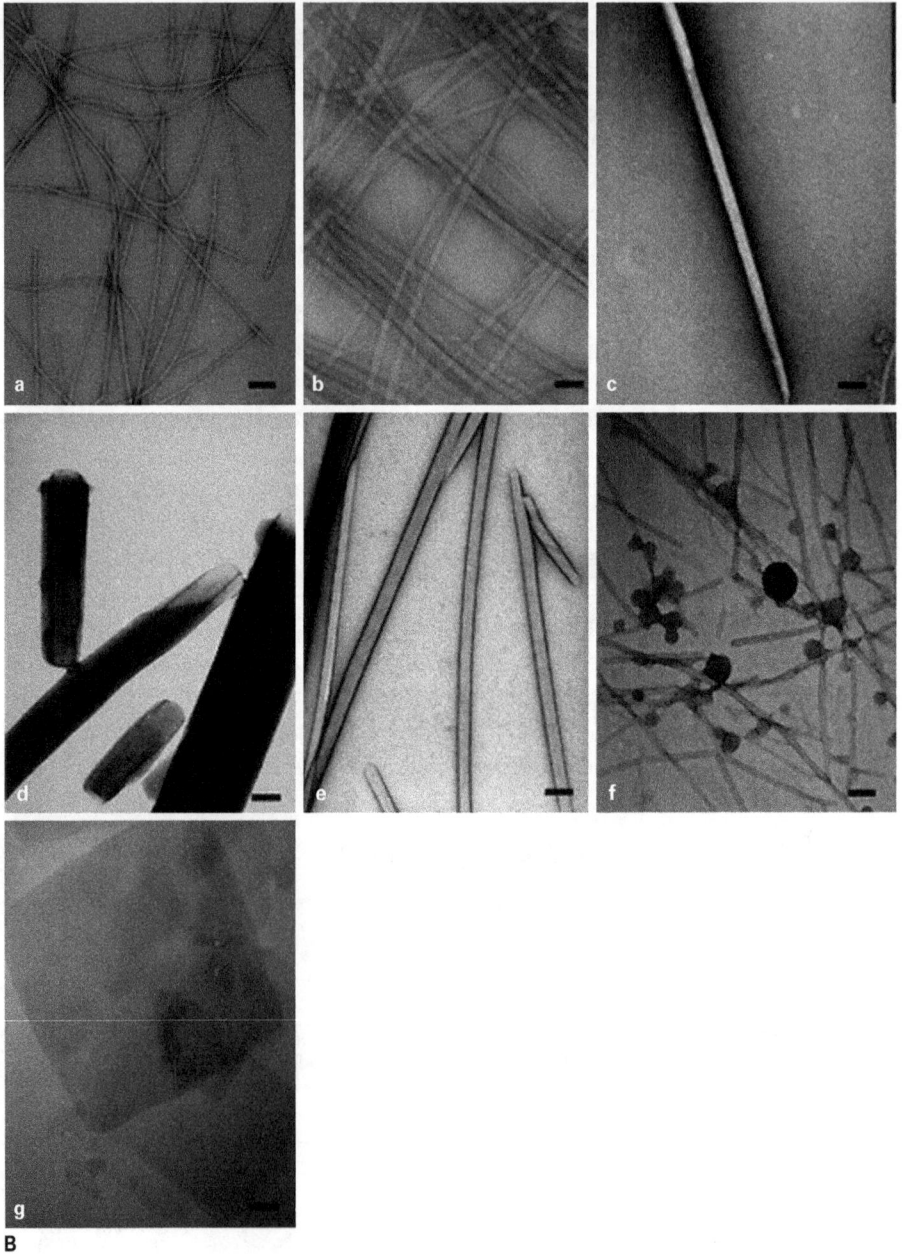

B

Fig. 6.13: B. Transmission electron microscopy (TEM) images showing the morphologies of the dipeptide fibrils. **(a)** di-D-1-Nal; **(b)** di-D-2-Nal; **(c)** di-para-fluoro-phe; **(d)** di-pentafluoro-phe; **(e)** di-para-iodo-phe; **(f)** di-para-nitro-phe; **(g)** di-4-phenyl-phe. Reprinted from Reches M. and Gazit, E Designed aromatic homo-dipeptides: formation of ordered nanostructuresand potential nanotechnological applications, *Phys. Biol* **2006** *3*, S10–S19. Copyright (2006) IOP Publishing.

dipeptides comprising phenylalanine derivatives gave rise to tubular-fibril structures that were clearly distinct and dependent upon the phenylalanine derivative employed.

The realization that fibril morphology can be intimately controlled through the amino acid sequence has attracted further interest in the context of materials applications, possibly using peptide fibrils as material *templates* or components. The key challenges in such applications is both the design of specific amino acid sequences that would produce fibers with the desired physical and mechanical properties, as well as develop methodologies to organize the synthesized peptide fibers in a manner that would endow functionality on a macroscopic scale.

6.4 Engineered oligonucleotides and synthetic genes

Beside the extensive work devoted to engineering proteins through manipulations of the amino acids sequences, there have also been significant efforts directed at modulating the **genetic framework** of cells to achieve new and optimized information processing and regulation pathways. DNA manipulation has naturally been at the forefront of this field (as well as other disciplines such as *DNA nanotechnology*, discussed in Chapter 9). However the other crucial and highly diverse component of genetic circuitry – the RNA molecule – has also attracted significant interest in this context in recent years.

Indeed, if one defines the core objective of synthetic biology as the design of biological systems for which structure and function can be synthetically programmed, RNA biology might provide even more diverse toolbox than DNA. RNA molecules, responsible for transcription of the genetic code and regulating the protein translation machinery, display diverse structural and functional features. RNA occupies a prominent place in modern chemical biology, and numerous applications have been described in the literature and pursued in industrial R&D settings.

Sensitive *RNA thermometers* can be derived from natural regulatory elements of RNA, as depicted in Fig. 6.14. Essentially, functional RNA units such as "RNA switches" (or "riboswitches"), that are responsible for activating or deactivating protein translation in the ribosome, are largely held together through hybridization interactions (e.g. hydrogen bonds between complementary base-pairs). Such interactions are highly sensitive even to slight changes in temperature. Accordingly, in the low-temperature state the RNA module is properly folded, bound to the ribosome, and represses protein translation by preventing docking of the transfer RNA (tRNA). However, even a slight temperature increase results in destabilization of the RNA conformation through "fraying" of the hybridized RNA structure, activating the translation machinery.

RNA aptamers are among the most well-known RNA-based sensing elements, similar in prominence to their *DNA* aptamer counterparts. They are usually synthesized in libraries, with the optimal compounds identified via selection against desired ligands serving as binding targets. This "evolutionary pressure" yields RNA fragments that exhibit very high affinities and specificities to the particular ligands. Selected

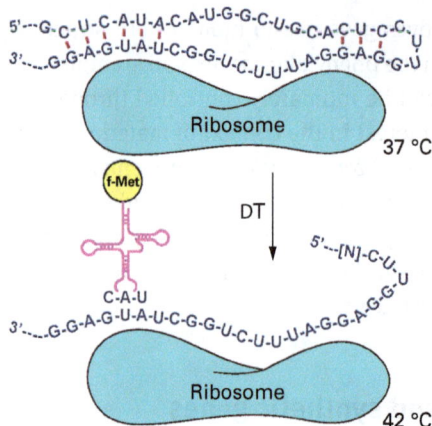

Fig. 6.14: An "RNA thermometer". A regulatory RNA element (e.g. "riboswitch") undergoes temperature-induced conformation change which enables protein translation, thus providing the means of detecting increased temperature. At *low temperature* the riboswitch is folded while embedded upon the ribosome surface. In the *high-temperature* state, the hybridization between the RNA nucleotides is destroyed, the RNA strand is opened and made available to binding the complementary tRNA units and consequent protein translation.

high-affinity aptamers could be employed as recognition elements in various biological and chemical constructs, such as pathogen sensors, diagnostics vehicles, drug carriers, inhibitors, and others. In particular, since the binding between aptamers and their targets is often exceptionally strong, structural characterization of small molecule-RNA aptamer complexes could aid the attempts for making RNA a potential therapeutic target. Indeed, the ubiquitous involvement of RNA molecules in gene expression, transport, catalysis, and other cellular processes renders aptamer screening a potentially important tool in drug development. RNA aptamers have been also proposed as components in innovative therapeutic approaches, potentially substituting conventional biomolecules such as antibodies, which often exhibit less efficient binding to their respective biological targets (Fig. 6.15).

While oligonucleotides (RNA and DNA) have mimicked proteins in varied applications, their properties as *information bearing* molecules probably offer the most tantalizing possibilities in synthetic biology. However, while scientists have tinkered with proteins for decades, introducing and exploiting novel ways to transmit genetic information is more demanding in a *synthetic* context. This is due in large part to the fact that effective, high-fidelity information transfer is conceptually more challenging compared to modulating structural features of a protein. Indeed, altering the chemistry of the genetic code needs to enable efficient storage of the genetic information (via base-pair complementarity), propagate and translate this information, as well as enable *evolutionary* routes.

The four DNA and RNA base-pairs do quite a good job as genetic building blocks, but scientists continuously try to augment these two fundamental components of the

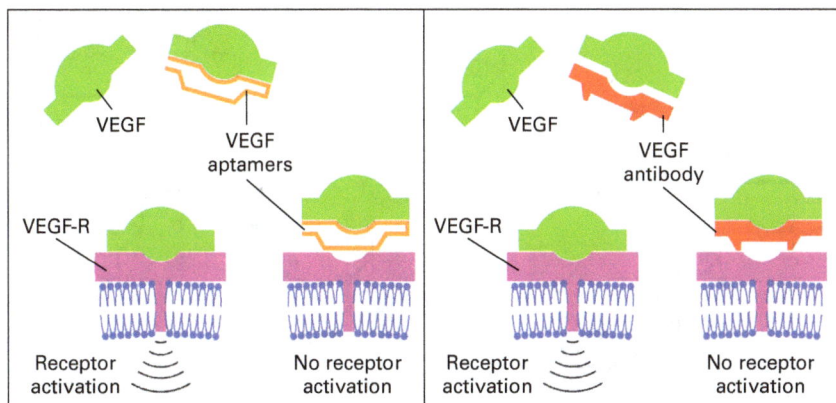

Fig. 6.15: RNA aptamers as substitutes for antibodies. Aptamers (left panel) can be designed to function in a similar way as antibodies (right panel) – binding the target molecules (e.g. ligands or other bio-activators) thereby preventing their interactions with their cognate receptors.

biological information transfer mechanism. Initial steps in this direction have been the technologies developed by P. Schultz in the 1980s for incorporation of unnatural amino acids into proteins through modifying transfer RNA molecules (see previous sub-section). This technique, however, still harnesses the same DNA (and RNA)-based toolbox for protein design. A truly "artificial genetics" platform would substitute (or mimic) the DNA or RNA units with completely different building blocks that would be capable of encoding and translating information. Recent exciting synthetic work carried out by P. Holliger and colleagues at Cambridge University has been a significant step towards this goal. The scientists synthesized a biomimetic sugar backbone that was different than the ribose and deoxyribose frameworks in RNA and DNA (Fig. 6.16), but could effectively substitute the natural sugars, exhibiting the hallmark complementarity of the natural nucleotides.

The importance of the work depicted in Fig. 6.16 is not just the introduction of synthetic genetic building blocks (denoted *xeno-nucleic acids – XNA),* but the demonstration of actual hereditary and translation processes, similar to those demonstrated with natural nucleotides. Specifically, the researchers succeeded to modify (mutate) *polymerases,* the enzymes responsible for assembly of the complementary oligonucleotide strains in replication processes, and these modified enzymes could translate XNA sequences to DNA and the other way around. Remarkably, experiments showed a rudimentary evolution occurring following implementation of external selection pressure, resulting, for example, in a XNA sequence which "evolved" to bind to specific protein and RNA targets.

The significance of such "biomimetic genetics" studies focusing on chemical systems which enable transfer of information between generations is the demonstration that heredity and evolution – two of the most fundamental concepts in biology, are not unique to DNA and RNA. Whether other molecules within the vast universe of polymer

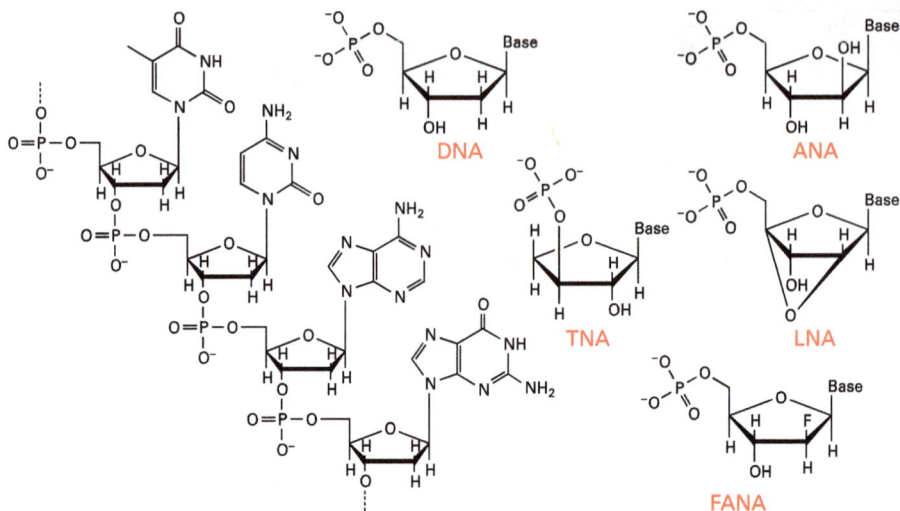

Fig. 6.16: Biomimetic nucleotides. Synthetic "xeno-nucleic acids" used as substitutes for DNA and RNA bases. Different sugar backbones enable base-pairing with DNA nucleotides or other XNAs. Shown are the native deoxyribose DNA unit and a fragment of single strand DNA (left), and several XNAs: arabinonucleic acids (ANA), 2'-O,4'-C-methylene-β-D-ribonucleic acid [e.g. locked nucleic acid (LNA)], α-L-threofuranosyl nucleic acid (TNA), 2'-fluoro-arabinonucleic acids (FANA).

chemistry could effectively substitute these oligonucleotides and form a basis for new information storage and propagation capabilities is indeed one of the most exciting questions in synthetic biology and related research into "artificial life" (see Chapter 7).

While considerable efforts have been invested in engineering proteins through manipulations and substitutions of single genes, numerous cellular processes do not just rely on the activities of single proteins, but rather operate through *networks* of proteins, genes, and other biomolecules. Such networks function through the *combined* action of all constituents, which usually communicate with one another and where sophisticated regulation pathways control the expression and activity of the individual molecules. Indeed, recent efforts in synthetic biology aim to reconstruct entire networks rather than modify individual gene or protein elements. *Synthetic networks* could open new avenues for exerting external control over biological processes, introduce new cellular processes, and contribute to better understanding of naturally occurring signaling and regulation networks.

6.5 Mimicking functional biomolecules

The biological universe consists of molecules other than oligonucleotides and proteins, including polysaccharides, hormones, and other molecules. Indeed, research efforts have centered on the introduction of new classes of artificial synthetic molecules that might play roles in biological processes. Activities in this field generally aim to generate synthetic molecules that on the one hand mimic structural characteristics of natural molecular species, but on the other hand exhibit additional superior functionalities. A common strategy in this field has been the use of a "shotgun approach" in which a large number of molecules are produced, usually through combinatorial library platforms, and subsequently screened for their biological properties. Similar to the directed evolution schemes discussed above, the use of random combinatorial libraries (or more likely "semi-random" libraries in which some predetermined structural constraints are imposed) combined with development of selection functional assays, constitutes a powerful instrument for identification of new synthetic compounds with improved biological properties.

Peptoids are a case study of synthetic biomimetic molecules exhibiting useful biochemical functionalities whose properties can be finely tuned. As schematically depicted in Fig. 6.17, peptoids differ from natural peptides by a key structural feature: in contrast to peptides in which the side-chains are attached to the backbone alpha-carbons, in peptoids the functional residues are bonded instead to the *nitrogen*. As a result, the amide hydrogen, which plays a central role in shaping secondary structures of natural proteins, is absent in peptoids, giving rise to distinct structural space that is *not* based upon hydrogen bonding. Indeed, peptoids have been found to exhibit high structural stability in comparison with polypeptides – maintaining their structural features in varied organic solvents and at high temperatures in which natural peptides usually unfold and lose functionality.

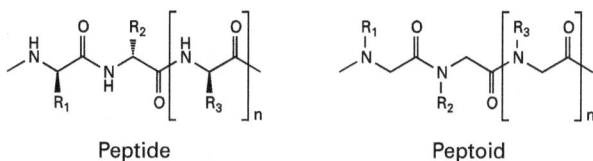

Peptide   Peptoid

Fig. 6.17: Peptoid vs. peptide structures. The functional units in peptoids are covalently attached to the backbone *nitrogen* instead of *carbon* as is the case in peptides.

Another notable consequence of the substituting of the amide protons in peptoids is their resistance to cleavage and degradation by protease enzymes; this has been, in fact, the primary motivation for developing this class of molecules in the first place. Also attractive is the relative ease of peptoid synthesis. The polymerization reactions which were developed for coupling the individual peptoid units generally have higher yield than the comparable peptide synthesis procedures. Furthermore, nitrogen-substituted monomers (i.e. the amine monomers in the peptoid chains) constitute a highly diverse source of building blocks displaying diverse side-chain structures and compositions.

Peptoids have been investigated in varied biochemical and biomedical contexts. *Combinatorial peptoid libraries* have attracted interest as a route for creating very large pools of compounds for functional screening. Peptoids are particularly amenable to library screening due to their modular architecture, the diversity of the N-substituted side-groups, and readily available synthetic routes. Peptoid libraries produced many bioactive molecules, including antimicrobial agents, and ligands for varied biological targets particularly protein receptors (both *agonists* – which trigger receptor response, and *antagonists* – blocking receptor activity upon binding).

Fig. 6.18 depicts an example of a peptoid combinatorial library and screening of the members of the library for protein binding (specifically – binding to *phosphoproteins*, for which the generation of high-affinity antibodies has proven difficult). The library, designed in the laboratory of T. Kodadek at the Scripps Research Institute, is based upon a common "one bead one compound" approach, in which each bead displays only a single (peptoid) compound. As shown in Fig. 6.18, the synthesized peptoids comprised four repeats, thus using combinations of the 14 amine monomeric building blocks in each synthesis step yielded a theoretical diversity of 14^4 (38,416) different compounds.

How to accurately and efficiently identify the most active member within a library is a recurring challenge for taking full advantage of the molecular diversity in combinatorial peptoid libraries (or bioactive compound libraries in general). Indeed, development of effective *screening assays* is an integral part of compound library applications. In the experiment depicted in Fig. 6.18, functional screening of the compounds was carried out directly on the bead surface through monitoring protein binding; exposing the bead pool to a fluorescent target protein enabled a simple visual identification of the binding event, selection of the bead, and straightforward analysis of the bead-displayed peptide.

While the peptoid systems discussed above have been designed to build and improve upon naturally occurring counterparts, other biomimetic "supramolecular" systems were introduced to carry out important biological tasks. *Artificial chaperones* are a case in point. Chaperones are cellular proteins which play a crucial role in generating the "physiologically correct" three-dimensional (3D) conformations (folding) of proteins. This function is generally carried out through binding of the chaperone to an initially unfolded or partially folded target protein, assisting its folding towards the functional 3D structure. Chaperone-assisted protein folding pathways are complex and significant efforts have been invested in development of artificial systems which may induce folding of disordered proteins. Indeed, attaining this goal may have significant health benefits since varied diseases are associated with improperly folded proteins, including well-known incurable neurodegenerative diseases such as Alzheimer's disease.

Fig. 6.19 outlines protein reorganization pathways, illustrated by L. Shi and colleagues at Nankai University, China, specifically the core function of an artificial chaperone in regulating protein-folding processes. In general, protein folding is a dynamic process usually involving an intermediate, partially folded, state (Fig. 6.19, middle). In some instances, a protein does not reach the native, functional conformation but

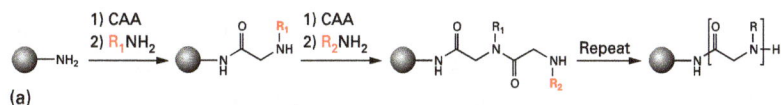

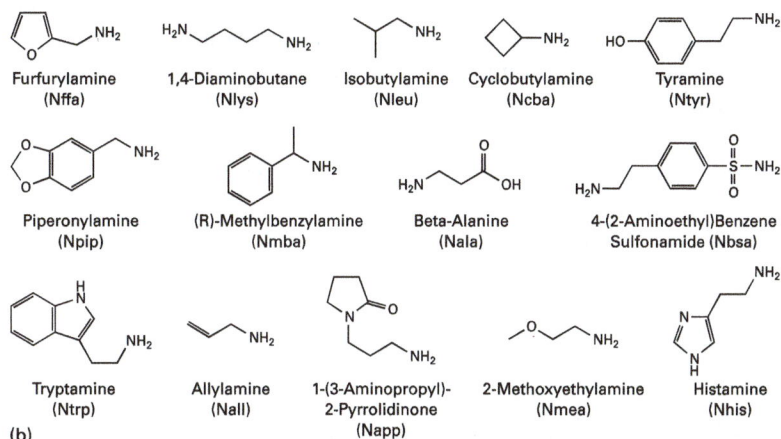

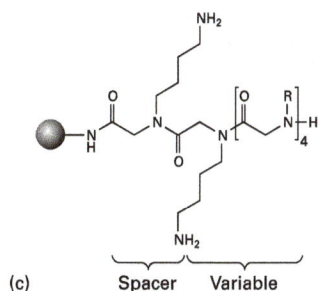

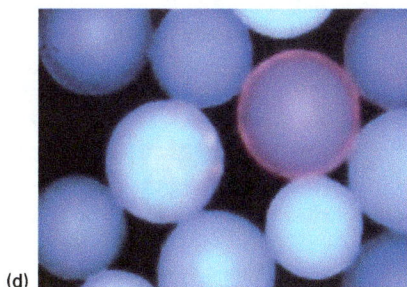

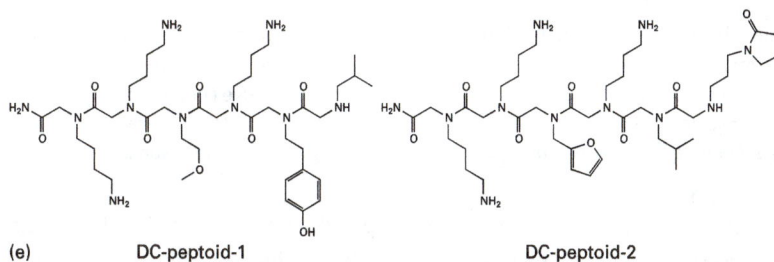

Fig. 6.18: Peptoid library synthesis and screening. **(a)** Synthesis scheme depicting the placement of two consecutive residues, R1 and R2, selected from the library of 14 functional units shown in **(b)** (CAA corresponds to activated ester of chloroacetic acid). The step is repeated four times yielding the general peptoid structure shown in **(c)**. **(d)** Fluorescence microscopy image showing a *single* bead displaying a red halo due to binding of a peptoid to a phosphorescent protein. **(e)** Structures of two peptoids (DC-peptoid 1 and DC-peptoid 2) exhibiting high-affinity to the phosphoprotein phospho-PDID. The two peptoids were selected from the peptoid library. Reprinted from Cai, D. et al., Peptoid ligands that bind selectively to phosphoproteins, *Bioorg. Med. Chem. Lett.* **2011** *21*, 4960–4964. Copyright (2011), with permission from Elsevier.

rather forms aggregates (Fig. 6.19, top) which might exhibit adverse physiological properties. As indicated in Fig. 6.19, an artificial chaperone exhibits several roles. Specifically, the synthetic chaperone captures unfolded (or partially folded) protein molecules, thereby preventing their undergoing misfolding pathways which might lead to aggregation. In parallel, functional units at the chaperone surface promote proper folding of anchored protein, ultimately releasing the molecule in a biologically native conformation.

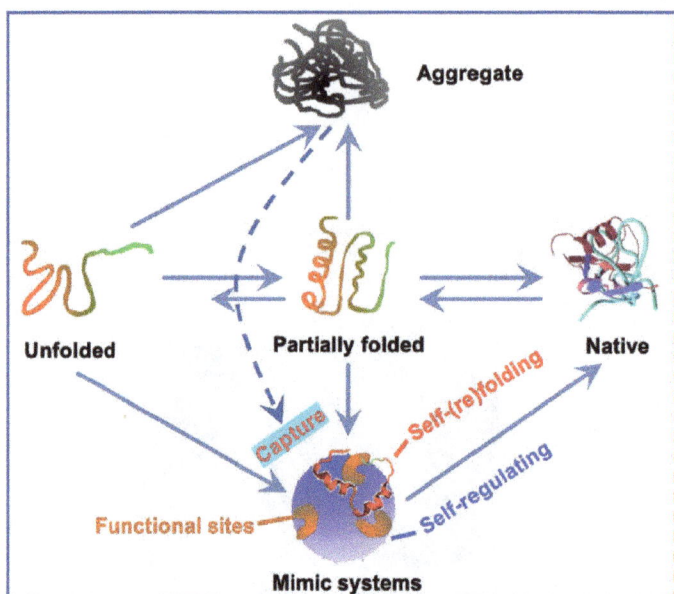

Fig. 6.19: Protein folding aided by artificial chaperones. The scheme shows the folding/misfolding pathways of a protein and the roles of a chaperone mimic in both preventing misfolding and aggregate formation as well as inducing folding of the protein into the native conformation. Reprinted with permission from Ma et al., Mimicking Molecular Chaperones to Regulate Protein Folding, *Advanced Materials*. 2020, *32*, 1805945, Copyright (2020) John Wiley and Sons.

Varied chemical assemblies have been developed displaying artificial chaperone functions. Fig. 6.20 presents an example of spherical polymeric micelles, referred to as "nanochaperones", designed to transform a denatured protein into its bioactive folded state. The thrust of the experimental system presented in Fig. 6.20, developed in the L. Shi laboratory, is the formation of highly branched surface organization within the nanochaperones, which, importantly, display abundant hydrophobic domains (represented as the green cylinders in Fig. 6.20). The hydrophobic areas have a crucial role since they allow binding of the denatured protein through its hydrophobic segments (which generally maintain folded conformation of the protein when solubilized in aqueous environments).

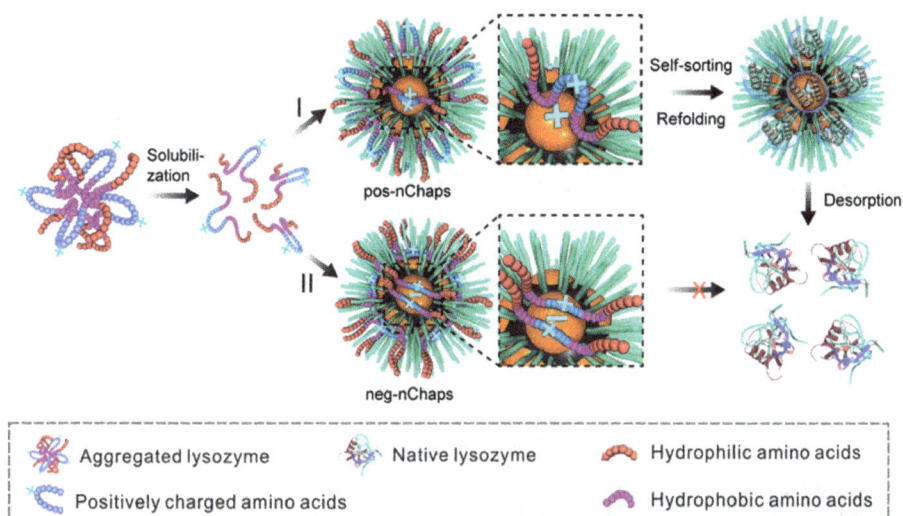

Fig. 6.20: Protein refolding induced by charged polymeric micelles. Binding and refolding of denatured lysozyme through binding to charged polymer micelles. Protein refolding and release in the native conformation occurs only when the polymer displays positive charge (top), while irreversible binding between the protein and polymer is observed in the case of a negative polymer surface (bottom). Reprinted with permission from Ma et al., Synthetic Nanochaperones Facilitate Refolding of Denatured Proteins, *ACS Nano.* 2017, 11, 10549–10557, Copyright (2017) American Chemical Society.

Fig. 6.20 depicts two alternative pathways for the denatured protein (lysozyme, a well-known immune protein, was selected by the researchers as a model) depending upon the surface charge of the nanochaperones. Lysozyme is intrinsically positive, however its binding to the nanochaperones occurs through affinity between the hydrophobic protein domains and the hydrophobic polymeric pockets upon the polymer micelles' surface. Importantly, match or mismatch between the protein and nanochaperone charges constitute an important factor for both refolding of the denatured protein as well as subsequent release of the folded molecule. Indeed, the combination of electrostatic repulsion between the positive microdomains of the lysozyme/nanochaperone, and affinity between the hydrophobic regions endows the protein with sufficient flexibility essential for its refolding. In contrast, when mismatch occurs between the positive protein and negative nanochaperone, the electrostatic binding between them prevents the refolding process. Importantly, refolding of the nanochaperone-bound lysozyme into its native conformation results in surface exposure of the positive microdomains at the protein surface, consequently weakening the affinity between the folded protein and the polymeric chaperone, leading to release of the natively folded protein into the solution.

Echoing the system above, in many instances, control of the refolding/release processes has been accomplished through intimate tuning of the interactions between the protein and artificial chaperone matrix. Such tuning has been carried out also through application of external stimuli. Fig. 6.21 depicts a "soft tubing" polymer system capable of capturing a denatured protein, inducing refolding while bound within the tubes, and eventually releasing the properly folded protein to the solution. Notably, these processes can be induced through modulating the solution acidity. The elegant experiment, which was carried out by N. Kameta and colleagues at the National Institute of Advanced Industrial Science and Technology (AIST), Japan, utilized an amine-displaying polymer as the nanotubes' building block. Amine units undergo protonation (e.g. proton uptake, generating positive charge) and deprotonation (proton release) depending upon the *acidity* of the surrounding solution. Accordingly, the extent of (positive) surface charge can be readily tuned through modulating the solution pH.

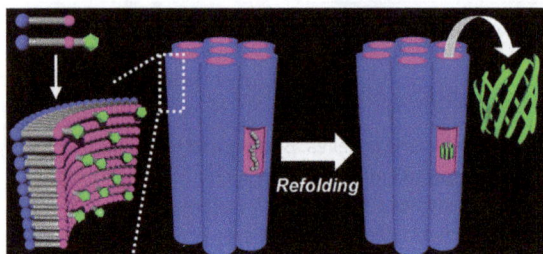

Fig. 6.21: Polymer nanotube artificial chaperones. The inner surface of the nanotubes displays positive molecular amine units (green spheres) which attract the soluble (negatively charged) protein through electrostatic affinity. The captured protein undergoes refolding inside the nanotubes' channels. Ultimately, the folded protein is released through reducing the acidity of the solution, giving rise to negative charge upon the nanotubes' surfaces and consequent electrostatic repulsion with the embedded protein. Reprinted with permission from Kameta et al., Soft Nanotube Hydrogels Functioning As Artificial Chaperones, *ACS Nano*. 2012, 6, 5249–5258, Copyright (2012) American Chemical Society.

Specifically, the researchers utilized a well-known protein – the *green fluorescent protein (GFP)* which emits intense green-blue fluorescence thus allowing visual monitoring of the condition and physical location of the protein throughout the process. As illustrated in Fig. 6.21, the polymeric nanotubes display amine units upon the internal channel surface (green spheres). In the initial experimental conditions, the solution is kept acidic thereby making the amine residues positively charged and accordingly attracting and encapsulating the negatively charged GFP molecules. Incubation of the GFP/nanotubes for long time periods resulted in slow refolding of the protein inside the channel (Fig. 6.21). Importantly, however, reducing the solution acidity (pH increase) induced release of the refolded protein from the nanotubes. This process is due to deprotonation of the amines at the higher pH, resulting in greater repulsion between the negative GFP and nanotube surface.

While the design of synthetic, artificial chaperones has flourished, endowing possible new tools to understand and control protein folding, there are obvious challenges in this field related to the intrinsic complexity of protein structures and intricate folding pathways. Indeed, most reported systems have manifested simple "two-state" scenarios – a denatured protein directly transforming into a folded state. While accomplishing such transformation is remarkable, it is important to note that in the actual cellular milieu proteins exist in a variety of states in the conformational space – partially folded, existing in transition states. Native chaperones are primed to handle such protein mixtures, often maintaining a delicate balancing act among different protein conformations required for proper functionality of cellular processes. It should be also noted that most studies thus far have utilized model protein systems, which are easy to work with and their folding states can be readily monitored; most physiological proteins, however, are difficult to readily control via simple chemical means.

Artificial enzymes or "enzyme-mimics" constitute a broad class of molecules and molecular assemblies designed to mimic the catalytic properties of naturally occurring enzymes, while overcoming their limitations in term of stability, temperature sensitivity, recyclability, and overall performance. Artificial enzymes have been shown to accelerate both biological as well as chemical reactions. As such, enzyme-mimics have attracted interest in industrial applications as cost-effective vehicles designed to provide higher efficiency in the case of different chemical processes.

Fig. 6.22 depicts an elegant synthetic chemical system exhibiting useful biocatalytic properties. The "supercages" depicted in Fig. 6.21 were fabricated by W. Tan and colleagues at the University of Florida from copper hydroxide "nanoribbons" that were microscopically aligned to form microscale supercages. Notably, the supercages feature important aspects which usually characterize enzymes – chemically reactive surface domains which anchor the target substrates (in the case here, the nanoribbons displaying polar hydroxide units and copper ions), and constrained "pockets" within the cage in which molecular transformation of the substrate occurs. Interestingly, the researchers have demonstrated that the copper hydroxide supercages were capable of catalyzing biologically important oxidation reactions.

Nanozymes, defined as nanomaterials with enzyme-like characteristics are considered among the promising "new frontiers" in biomimetic enzyme research. Nanozymes constitute carefully constructed nanoscale materials designed to exhibit superior catalytic properties compared to natural enzymes. Nanozymes are perceived as capable of overcoming some of the main limitations of both natural and artificial enzymes through harnessing the advantages of the "nanomaterial universe" – tailored structural and functional properties, reproducibility, large-scale production and synthesis from inexpensive and environmentally benign building blocks.

Fig. 6.23 depicts a representative nanozyme system employed in anticancer therapy. The spherical nanozymes are designed to catalyze production of oxygen molecules, converted to *reactive oxygen species (ROS)* which locally oxidize and destroy cancer cells. As such, a sufficient supply of O_2 molecules is needed at the site of the

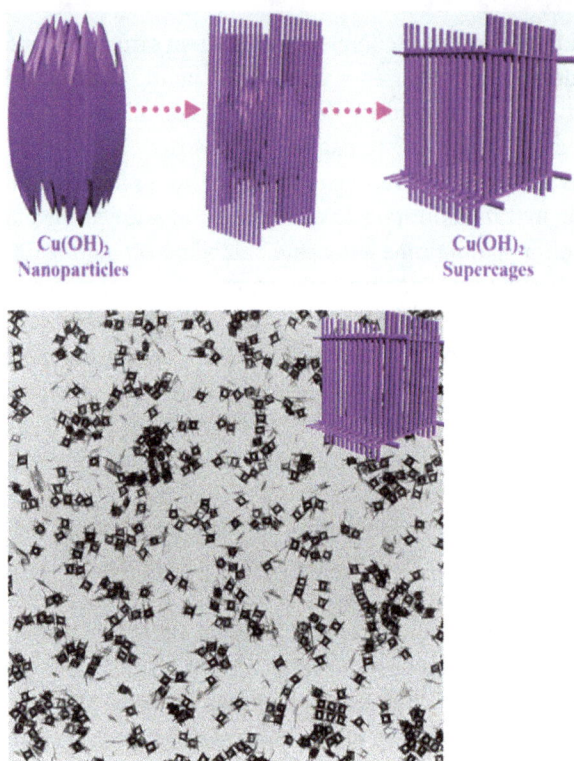

Fig. 6.22: Self-assembled supercages as artificial enzymes. **Top:** formation of the supercages from Cu(OH)$_2$ nanoribbons transforming into the nanocage faces. **Bottom:** transmission electron microscopy images of the supercages; sides of the nanocubes are approximately 150 nm. Reprinted with permission from Cai et al., Single Nanoparticle to 3D Supercage: Framing for an Artificial Enzyme System, *J. Am. Chem. Soc.* 2015, *137*, 13957–13963, Copyright (2015) American Chemical Society.

tumor, and the artificial enzyme system illustrated in Fig. 6.23 aims to accomplish this goal through decomposition of hydrogen peroxide (H$_2$O$_2$). The nanozymes, designed by T. Hyeon and colleagues at Seoul National University, Republic of Korea, comprised of porous silica nanospheres anchoring manganese ferric oxide NPs (black specks in the electron microscopy image in Fig. 6.23b).

The manganese ferric oxide NPs play the key role in accelerating O$_2$ generation as metal oxide NPs exhibit in many instances chemical catalysis properties. In parallel, the silica nanospheres provide a biocompatible framework for surface display of the NPs thereby facilitating their catalytic properties. The cartoon in Fig. 6.23a underscores the importance of the nanoscale dimensionalities of the silica/manganese ferric oxide nanozyme as the composite particles can transport and migrate in the blood capillaries, penetrating through the capillary walls to the tumor, wherein ROS production is carried out through the catalytic action of the particles.

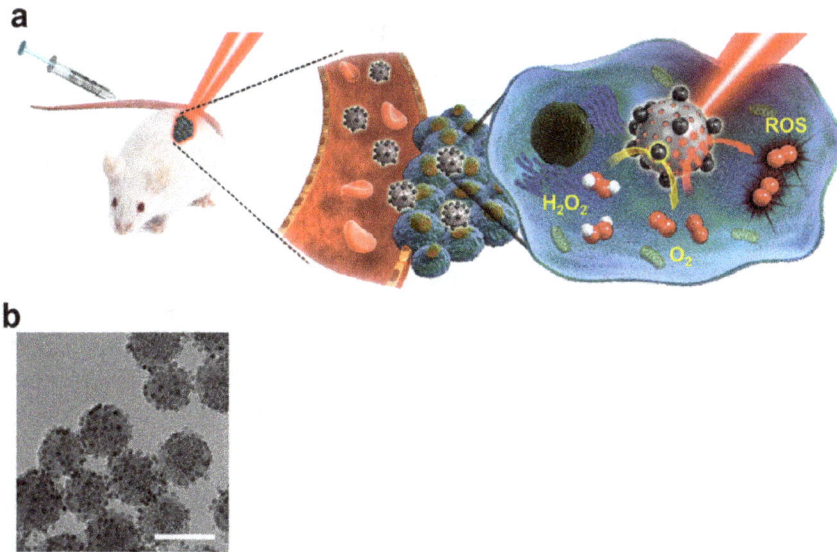

Fig. 6.23: Nanozymes for anti-cancer activity. **(a)** Schematic depiction of the nanozyme action. The nanozymes are injected into the bloodstream. They subsequently penetrate through the leaky walls of the blood vessel into a tumor. In the tumor, the nanozymes catalyze production of both O_2 and reactive oxygen species (ROS) which are toxic to the tumor cells. **(b)** Transmission electron microscopy (TEM) image of the nanozymes, comprising silica nanospheres displaying manganese-ferric-oxide nanoparticles (black dots). Scale bar corresponds to 60 nm. Adapted with permission from Kim et al., Continuous O2-Evolving MnFe2O4 Nanoparticle-Anchored Mesoporous Silica Nanoparticles for Efficient Photodynamic Therapy in Hypoxic Cancer, *J. Am. Chem. Soc.* 2017, *139*, 10992–10995, Copyright (2017) American Chemical Society.

6.6 Artificial chromosomes

While protein engineering is now routinely applied through genetic alterations and related chemical or biochemical schemes as discussed above, a much more ambitious goal in synthetic biology is to create wholly *synthetic chromosomes*. Such an accomplishment would enable the production of not one protein at a time, but rather a multitude of proteins through embedding the synthetic chromosomes in the target microorganism. In a more fundamental sense, constructing a synthetic chromosome would bring science closer to what is perhaps the ultimate goal of synthetic biology – the creation of a functioning "synthetic" organism. Artificial chromosomes possess other practical attractive features. Desired proteins can be positioned in the chromosome with fewer constraints compared to site directed mutagenesis, for example. Also, synthetic chromosomes allow straightforward encoding of *unnatural amino acids* (see above) in almost any desired location within a protein. Similarly, enzymatic action sites, peptide recognition elements, and other functional modules can be placed throughout the synthetic genome, producing a pool of carefully-determined protein functionalities, regulation and metabolic pathways.

Bacterial artificial chromosomes (BACs) have been an important thread in the quest for creating artificial human chromosomes, even though these constructs do not truly represent mammalian chromosomes. BACs, depicted schematically in Fig. 6.24, have been highly successful as vehicles for cloning foreign genes, particularly mammalian genes, in bacterial expression systems. The technique is based upon inserting the DNA sequence of interest into a bacterial "cloning vector" – i.e. bacterial plasmid that can be cleaved in specific locations by restriction enzymes. The resultant recombinant-DNA BAC is then embedded within a bacterial expression system (usually *E. coli*); transformed bacterial cells expressing the BACs can be subsequently identified and isolated via conventional antibiotic selection. BACs are amenable to *library screening*; combinatorial libraries of genes can be incorporated within cloning vectors, and the selection procedure enables identification and sequencing of the genes responsible for bacterial survival.

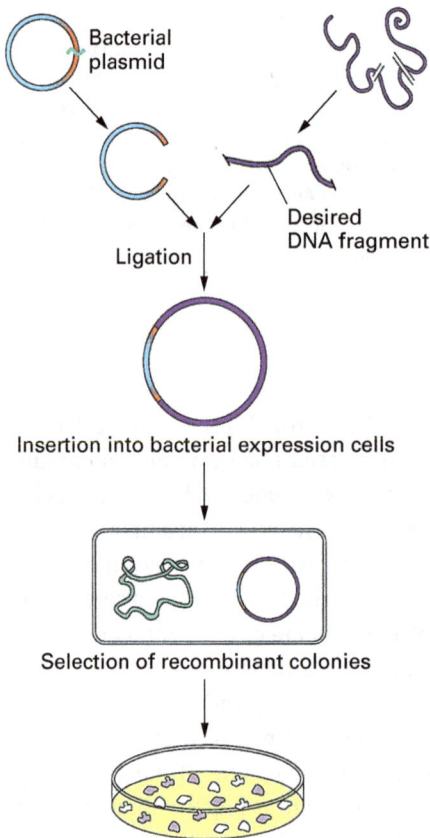

Fig. 6.24: Bacterial artificial chromosomes (BACs). BACs constitute bacterial cloning vectors (plasmids) containing foreign DNA. The target DNA strand is inserted into the cloning vector through the action of cleavage enzymes (which clip the bacterial plasmid) followed by DNA ligation. Subsequent incorporation of the recombinant BAC into bacterial cells is carried out through electroporation. Bacterial cells expressing the BACs are isolated through antibiotic selection via inclusion of antibiotic-resistance genes in the cloning vector plasmid.

The advantages bestowed by artificial chromosomes carry tradeoffs. While modifying the DNA encoding single proteins in a way that would not adversely affect the overall genomic functionalities (e.g. site-directed mutagenesis) can be usually accomplished, designing a complete artificial chromosome is more likely to interfere with the viability of the host organism. In particular, synthetic chromosomes would have to perform the complex tasks of division and segregation within the two daughter cells. In this context, one has to not only identify and reconstruct the proper DNA sequence within the chromosome, but also ascertain that the DNA strands are suitably localized within the chromosome through cell division. In addition, synthesis of a complete, sizeable genetic sequence embedded in an artificial chromosome is labor-intensive, even when using automated DNA synthesizers "around the clock".

Despite the significant difficulties, artificial chromosomes have been synthesized and successfully introduced into varied microorganisms; from the workhorse of biotechnology – the *E. coli* bacterium, all the way to the more complex eukaryotic cell – the *S. cerevisiae* yeast. Even though the task of synthesizing an effective artificial chromosomal genome is formidable, synthetic chromosomes reported thus far have been surprisingly resilient and well-accommodated by their host cells. In particular, artificial chromosomes were shown to undergo smooth division following their incorporation into host cells, and perhaps even more striking – minimal adverse phenotypic effects were apparent in most of these engineered cells.

The transition from constructing synthetic chromosomes of a simple eukaryotic cell (yeast) to artificial *human* chromosomes is extremely challenging, however. Human chromosomes are significantly larger than yeast chromosomes (two orders of magnitude greater DNA content) and the biological processes they undergo throughout the cell lifecycle are much more complex. *Cell division*, in particular, dictates multistage coordinated processes which ultimately lead to chromosome replication and partition between the two separate cells. Artificial human chromosomes have been assembled, albeit usually at smaller sizes (and supposedly greater stability) than the naturally-occurring ones. Such "micro"-chromosomes were shown to undergo mitotic transformations during hundreds of cell generations without losing their structural and functional integrity.

Artificial human chromosomes could have concrete therapeutic potential, as they might help circumvent some of the intrinsic hurdles facing *gene therapy* (see discussion of gene therapy techniques in Chapter 9). Specifically, delivery of genes to the desired targets within natural chromosomes has emerged as extremely difficult and the main impediment to realizing the promise of gene therapy. One way to overcome this barrier would be to create entirely synthetic chromosomes containing the "correct" genes, obviating the need to deliver and substitute discreet DNA fragments into existing chromosomes.

A related, highly active area in which synthetic human chromosomes could have a major impact is in the field of *stem cell* research. Reprogramming cells that are already differentiated (i.e. "somatic" cells having specific defined physiological functions) back into stem cells that could then differentiate upon external stimuli (denoted "induced pluripotent stem cells", or iPS cells) is among the most promising albeit challenging goals towards development of stem cell therapy. Current efforts aim to reprogram somatic cells through delivery of specific genes into the chromosomes. The problem with most current strategies involving gene delivery is that these approaches introduce significant risks of mutagenesis (and its adverse consequences). Furthermore, a crucial requirement for implementing somatic cell manipulations designed to produce iPS cells is that reprogramming should not involve genetic changes that might adversely affect cell properties. In this context, human artificial chromosomes exhibit notable advantages. The synthetic chromosomes can carry practically unlimited genetic material designed to activate cell reprogramming, without relying on complex and often unreliable gene-cargo vectors. Equally important, synthetic chromosomes essentially constitute an "autonomous" system that participates in cell division and related processes without interfering with the biology of the other (native) chromosomes.

6.7 Engineered microorganisms

Creation of wholly synthetic microorganisms, such as viruses and bacteria, is considered among the "holy grails" of synthetic biology. **Viruses** are known as disease causing agents, occasionally lethal, and something to avoid at all costs. Among biomedical scientists, however, viruses offer possible therapeutic routes to many diseases – as vehicles for transporting genetic materials into cells. In particular, viruses are critical participants in "gene therapy" – trying to address the root causes of various diseases through "reprogramming" the genetic codes of defective cells (see more detailed discussion in Chapter 9). Despite this potential, modification of native viruses for gene therapy applications is fraught with technical and ethical problems. Indeed, *synthetic viruses* might offer an alternative attractive platform.

Viruses possess quite simple structures, which make the creation of synthetic viral particles feasible. Specifically, viruses are composed of oligonucleotides (DNA or RNA) – encoding the genetic information – and a relatively small number of essential proteins, such as protective coat proteins and proteins required to hatch and insert onto the surface of the target cell. Fig. 6.25 shows an example of an artificial virus architecture which incorporates basic structural elements designed to facilitate seamless integration of the viral DNA into the gene translation machinery of the host organism.

An important and useful property of viruses in the context of synthetic biology is their relative "promiscuity" – the fact that often extensive genetic alterations do not adversely affect viral viability and functions. Accordingly, inclusion of relatively sizeable exogenous proteins within viral particles has been demonstrated. Fig. 6.26,

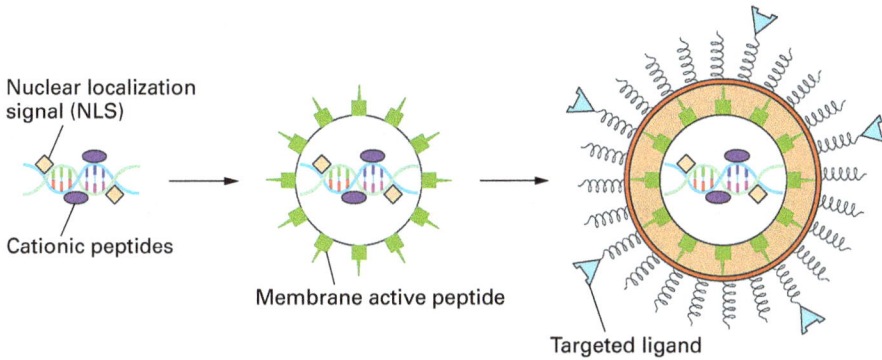

Fig. 6.25: Artificial virus. Schematic design of an artificial virus, comprising a "nucleus" containing the *oligonucleotides* packaged with *cationic peptides* (condensing and stabilizing the anionic DNA) and *nuclear localization signal peptides* (for transporting the DNA of the artificial virus into the nucleus of the target organism). The viral DNA is further coated with *membrane-active peptides* assigned with disrupting the host membrane and enabling insertion of the viral genetic material. An additional "protective shell" comprising synthetic or biological molecules can be included, serving as a scaffold for recognition units (i.e. ligands) latching onto cell-spxecific receptors on the cell surface (Fig. inspired by Mastrobattista et al., *Nature Reviews Drug Discovery* 5, 115–121).

for example, depicts a fluorescence microscopy image of a cell infected with a recombinant virus in which one of its capsid proteins was engineered to incorporate green fluorescent protein (GFP). Even though GFP, with its 238 amino acids, is a relatively sizeable protein in relation to the endogenous viral proteins, infectivity of the virus was not diminished.

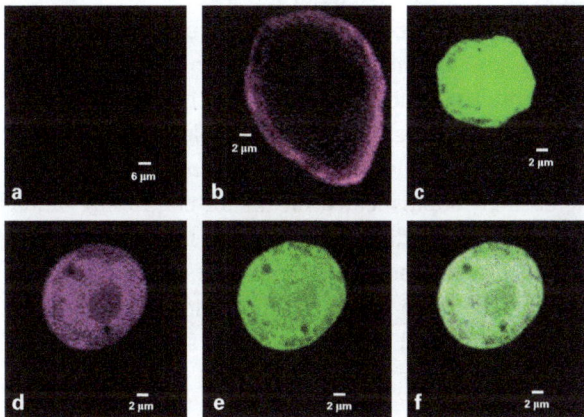

Fig. 6.26: Recombinant virus containing GFP retains its infectivity. Confocal fluorescence microscopy images of *Sf9* cells infected with recombinant GFP baculovirus vector. **(a)** mock infection (no viral particles added); **(b)** infection with GFP-viral particles. Reprinted from Gilbert L. et al., Assembly of fluorescent chimeric virus-like particles of canine parvovirus in insect cells, *Biochem. Biophys. Res. Commun.* **2004** *313*, 878–887. Copyright (2004), with permission from Elsevier.

Construction of **synthetic bacteria** is conceptually and practically more difficult than viruses, which have much smaller, easier to manipulate genomes. Scientists, however, have continuously expanded the technological frontiers in this field. Researchers led by J. C. Venter, in particular, have demonstrated several important feats. First, the entire genome of a *mycoplasma*, the smallest known bacterial cell, was synthesized and assembled from individual nucleotide building blocks. Next, the chemically-synthesized bacterial genome was amplified (i.e. producing a large copy number) in an external expression system – this was done in the Venter laboratory using a yeast host system (*Saccharomyces cerevisiae*), found to be an efficient, high-fidelity expression vector for the sizeable synthetic genome. Finally, and probably the most difficult task, the synthetic genome was incorporated into a *non-native* cell, resulting in a transplant microorganism whose properties (i.e. "phenotype") were determined *solely* by the new, implanted genome.

This remarkable achievement exemplifies the promise, limitations, and ethical constraints of "whole organism synthesis". For one thing, the above experiment did not produce a true "synthetic bacterial cell" – since the technique relied on an existing bacterial cell framework into which the synthetic genome was embedded. Moreover, successful amplification of the synthetic bacterial genome depended upon the effectiveness of existing expression systems, which might fail in situations where very long DNA sequences are used. Nevertheless, the biotechnological and therapeutic potential of the technology are enormous; since the introduced genome is essentially synthetic, it can include any number of genetic substitutions, deletions, or additions, which might, in principle, create "new" organisms with new and tunable traits. Significant ethical issues need to be considered as well, since the technology can be applied for constructing pathogenic, invasive, and resistant bacterial strains.

Synthetic biology has made huge strides towards developing and implementing a comprehensive "toolbox" for creating and modifying artificial biological systems. Indeed, the fundamental molecular units at the core of synthetic biology research – new proteins, genes and genetic circuits, novel classes of biomolecules – could all be employed in "bottom-up" construction of new biological materials and systems. These components provide increasingly sophisticated means for controlling fundamental properties in biological systems, including structural transformations, cooperative processes, information transmission, molecular motion, and others. As such, modern synthetic biology will likely continue to be an active and exciting field in which physics, chemistry, bioengineering, and related molecular sciences converge, breaking new paths in biomimetic applications.

7 Artificial cells

Creation of artificial cells touches on one of the most profound questions in biology: whether life – or a *living* system – can be created artificially from basic molecular building blocks. Such a goal, in fact, goes against a fundamental observation – that all cells originate from pre-existing cells. Even in the laboratory, no living cells have been assembled yet just from a combination of molecular units, even mixtures that attempt to closely mimic "biological" compositions. Perhaps not surprisingly, building artificial cells (and thus creating "life"), is both extremely difficult technically, as well as a field fraught with ethical issues.

Dramatic contours of synthetic cell systems have come to light at the Craig Venter Institute in Rockville (USA) where bacterial cells containing wholly synthetic genetic codes have been created (see Chapter 6.7). These artificial bacteria were claimed to be fully "computer-designed" and represent the considerable overlap between "synthetic biology" and artificial cell research. Indeed, synthetic bacteria could have uses as the future "factories of life" – programmed, in principle, to produce proteins, drugs, therapeutic compounds, and other biological molecules. While these systems do not represent completely artificial cells (as they rely on existing bacterial scaffolds) they underscore both the achievements made and potential of this field.

While the molecular organization and functions of cells are interwoven and complex, it is not inconceivable to synthesize an artificial cell from its basic elements. The ingredients required in an artificial cell are recognized: a membrane layer that separates the cell interior from the outside world and creates a distinct entity (e.g. compartment) within which the individual cellular processes occur. The artificial cell should also include the basic machinery for carrying out chemical processes that would transform nutrients into constituents required by the cell for functioning. This molecular machinery should be able to harness *energy* for initiating, and terminating, these chemical reactions. The most critical and perhaps most difficult property to mimic in an artificial cell is the *replication capability* – the fundamental characteristic of life itself.

Echoing the diverse literature on the subject, the discussion below uses various terms aimed to characterize the synthetic assemblies designed to mimic cells. "Artificial cells", "protocells", "synthetic (or semi-synthetic) cells" all have some distinct nuances depending upon the properties of the system (and the laboratory conducting the experiments). While the ultimate goal of a functional synthetic cell is universally accepted, the experimental and conceptual efforts to reach this target are generally divided into two main routes. One approach aims to first create the physical entities facilitating cell-like behavior, generally a membrane-encapsulated compartmentalized assembly, endowing it with the chemical means for spontaneous biochemical processes and replication machineries. An alternative strategy has been directed towards first developing artificial systems for transmission of biological information,

https://doi.org/10.1515/9783110709490-007

i.e. *cell-free* DNA and RNA systems, subsequently incorporating the information-bearing molecules or networks within cell-like assemblies. While research in this field has often integrated both approaches, the emphases in each route are distinct: one stresses the *composition* as the primary vehicle for achieving a functional cell, the other focuses on *sequential information flow* as the core feature which needs to be reproduced in an artificial cell.

This chapter first discusses *biomimetic membranes*, a huge research field by itself. Some aspects of membrane research are only peripheral to artificial cells, nevertheless the prominent role of the cellular membrane means that synthetic cells constructed in the laboratory would have to mimic membrane functions to be truly viable. A subsequent discussion focuses on efforts to develop cell-like compartments containing biological molecules capable of undergoing spontaneous, self-controlled molecular processes. Designing systems enabling *replication* and related processes of intergeneration information flow are also presented.

7.1 Biomimetic membranes

The *cellular membrane* is one of the most abundant structures in the biological world, functioning as the cell protective shield, and plays prominent roles in intracellular and extracellular communication, molecular transport phenomena, and other processes. Artificial membranes, or "model membranes", have attracted interest as vehicles for diverse applications ranging from drug carriers and therapeutics to nanotechnology and scaffolds for new materials. Model membranes exhibit both advantages and disadvantages. On the one hand, in light of the complexity and variability of real physiological membranes, model membranes allow one to focus on the roles and contributions of specific molecules. On the other hand, over-simplification could overlook the interplay among the different constituents contributing to membrane properties.

The membrane barrier in *artificial cells* has two primary roles: *compartmentalization* and *selective permeation*. Specifically, a membrane has to restrict the specific intracellular constituents into a defined space, allowing chemical reactions to occur between molecules. A membrane should also allow communication and molecular transport of nutrients, metabolites, and signaling elements between the cell interior and the outside environment. *Regulation* of the two-way molecular flow across the membrane is indeed among the most intricate membrane phenomena observed in cells. While biomimetic membrane designs address some of these requirements, model membranes are still far from precisely imitating the properties and functions of physiological membranes. As discussed below, biomimetic membranes have mainly played prominent roles as platforms for studying membrane events and as vehicles for biotechnology and pharmaceutics applications.

Artificial membranes mostly comprise *phospholipids* – the basic building blocks of real cellular membranes. Phospholipid-based model membranes usually constitute spherical unilamellar bilayer structures, denoted *vesicles* or *liposomes* (Fig. 7.1). The basic vesicle architecture can be modulated in terms of size (small unilamellar vesicles, large vesicles, giant vesicles). From a molecular perspective, model membranes can mimic physiological scenarios in which the lipid bilayer hosts a broad range of molecules embedded within the lipid scaffold: proteins, sugars, glycoconjugates, and others.

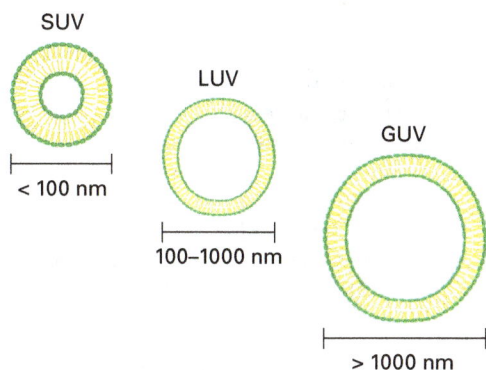

SUV

LUV

GUV

< 100 nm

100–1000 nm

> 1000 nm

Fig. 7.1: Lipid vesicles. Common bilayer membrane models in different diameter ranges: *small* unilamellar vesicle (SUV); *large* unilamellar vesicle (LUV); *giant* unilamellar vesicle (GUV).

Numerous procedures have been developed to assemble vesicles. These processes rely on the fundamental tendency of lipids (and related amphiphilic molecules) to minimize exposure of their hydrophobic moieties to the water environment, enhancing interactions of the *polar* components with the aqueous solution. In many instances, the energy needed to overcome the barrier to the self-assembly of lipids into vesicles in aqueous solutions is provided by ultrasound waves – i.e. *sonication*. This process promotes dissolution and dispersion of the amphiphilic molecules, which subsequently assemble into layered structures (*mono*layers or usually *bi*layers in case of phospholipids) encapsulating internal aqueous volume.

Giant vesicles (GVs, also referred to as giant *unilamellar* vesicles, GUVs) exhibiting diameters in the micron to tens-of-micron range, have been a popular experimental tool in artificial cell studies, mostly due to their size regime which is similar to actual cells. Many experimental schemes have been developed for the construction of GVs. Fig. 7.2, for example, shows an elegant experimental procedure for GV formation through a "microfluidic jetting" technique. The method, developed by D. A. Fletcher and colleagues at the University of California, Berkeley, provides means for incorporating specific molecules both into the lipid bilayer coat as well as into the aqueous volume enclosed inside the GV. As outlined in Fig. 7.2a, small vesicles merging into oil layers separating two aqueous solutions were employed to deliver lipid (and amphiphilic) molecules into a planar lipid bilayer which subsequently forms the GV encasing bilayer. The planar bilayer is subsequently deformed through "jetting" by

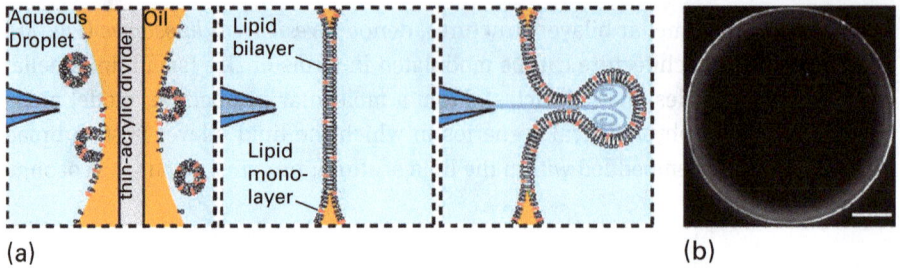

(a)   (b)

Fig. 7.2: Giant vesicles formed through microfluidic "jetting". **(a)** Oil layers are deposited on both sides of a thin acrylic sheath, separating two aqueous solutions. **Left:** Small vesicles comprising pre-selected lipids or amphiphilic molecules fuse into the oil phase; **middle:** a planar lipid bilayer forms following removal of the acrylic divider and slow exclusion of the oil. The bilayer comprised of the amphiphilic molecules delivered by the fused vesicles; **right:** a microfluidic needle produces a fluid jet, resulting in bilayer deformation and formation of a GUV. **(b)** Confocal fluorescence microscopy image showing a GUV formed through the jetting method incorporating a fluorescence amphiphile within the bilayer. The scale bar corresponds to 1 μm. Reprinted with permission from Richmond, D.L. et al., *PNAS* **2011** *108*, 9431–9436. Copyright (2011) National Academy of Sciences, U.S.A.

the microneedle, ultimately resulting in a GUV formation; the internal content of the vesicles is thus determined by composition of the injected fluid.

The system depicted in Fig. 7.2 represents common methodologies in which membrane vesicles are produced through *mechanical* processes. In some cases, however, scientists have taken advantage of *physiological* cell processes for assembling lipid vesicles. Vesicles, in fact, are generated in important biological processes, including neuronal cell communication (in which vesicles migrate in the synapses separating two neurons), internalization of substances from the cell environment, and others. In most of these phenomena, new vesicles "bud" from parent membrane bilayers, usually through complex protein-promoted processes. These budding processes have both inspired and exploited biomimetic generation of vesicular particles. Such "biologically induced" vesicle formation processes generally build upon known intracellular pathways, particularly enzyme-catalyzed *endocytosis* (cell uptake of external objects) and *exocytosis* (release from the cell into the external space). Fig. 7.3 shows a route for cell-harnessed vesicle production recently reported by D. Wang and colleagues at the Max Planck Institute, Potsdam, Germany. The approach involves three main steps: First, the desired vesicle-encapsulated cargo (drug molecules, nanoparticles, other agents) is inserted into the cell through conventional endocytic uptake; second, enzymes known to catalyze vesicle production are added to the cell or expressed inside it, encapsulating the cargo within the vesicular assemblies; and third – the vesicles formed inside the cells are released, possibly through induction of pores or channels in the external plasma membrane of the cell.

Biomimetic vesicles that *do not* contain phospholipids as primary components have been created. Vesicles of *high density lipoprotein (HDL)*, popularly known as "good cholesterol", are involved in cholesterol (and other lipids) transport and metab-

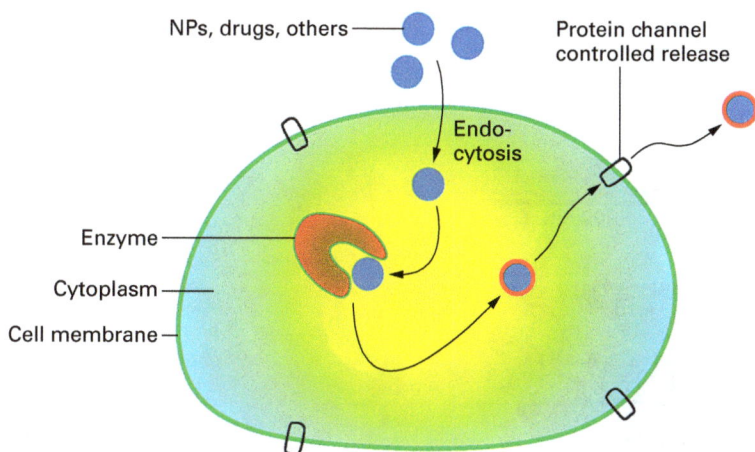

Fig. 7.3: Vesicles assembled through a biogenic process. Harnessing intracellular enzyme catalysis to assemble vesicles and embed molecular cargoes inside them. The vesicle particles are subsequently released from the cell.

olism. These natural "biological nanoparticles" comprise lipids and embedded proteins (belonging to the *apolipoprotein* family) responsible for capturing cholesterol or other lipid targets, and for stabilizing the overall particle structure. The compact organization of HDL particles is their defining feature, facilitating seamless mobility within the bloodstream and effective sequesteration of cholesterol from artery walls to clearance locations in the body (liver).

Reconstituted HDL (rHDL) particles (Fig. 7.4) have been touted both as possible replacements for defective or disease-impacted natural HDL, and also as a possible platform for vaccine and drug delivery. rHDL particles can be produced through self-assembly of the lipid and protein constituents. The rather straightforward preparation procedures enable further incorporation into the biomimetic particles of a broad range of hydrophobic guest molecules, which could then be transported in the bloodstream. rHDL particles have been shown, in particular, to package *membrane-associated proteins* without significantly diminishing their functions; this is not trivial since membrane proteins are known to be highly sensitive and their reconstitution in artificial environments is difficult.

The accessible surface of rHDL allows presentation of molecular recognition elements for diverse applications. The displayed molecular units might function as "triggers" of the immune system, and in such applications the rHDL particle essentially mimics an "adjuvant", e.g. the molecular scaffold facilitating effective presentation of the immunogen and ensuing activation of an immune response. In a conceptually similar scheme, A. N. Parikh and colleagues at the University of California, Davis, have shown that exogenous molecules displayed on the rHDL surface could act as "baits" for pathogens (Fig. 7.5). Such "decoy rHDL" particles could be part of therapeutic strate-

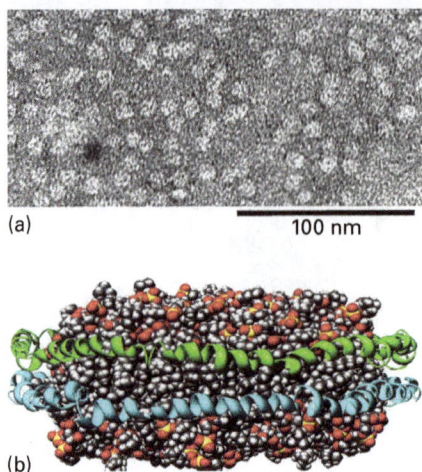

(a)

100 nm

(b)

Fig. 7.4: Reconstituted high density lipoprotein (rHDL) particles. **(a)** Transmission electron microscopy (TEM) image of negatively-stained rHDL particles indicating their relatively uniform dispersion having ~10 nm diameters; **(b)** scheme of an rHDL particle, illustrating the lipid bilayer (backbones in grey, headgroups in red and yellow) and apolipoprotein (apo-A) protein dimer (green and blue helixes) spanning the bilayer. Reprinted with permission from Whorton, M.R. et al., *PNAS* **2007** *104*, 7682–7687. Copyright (2007) National Academy of Sciences, U.S.A.

gies in which the engineered particles are injected into the bloodstream, subsequently reducing or even eliminating pathogenic targets through binding and sequestration.

While biomimetic membrane designs have generally employed molecules and molecular assemblies encountered in *human* physiology, such as phospholipids (natural or synthetic) and lipoproteins, other organisms have also inspired research and development in this field. *Archaea* are single-cell microorganisms found in diverse ecological habitats and niches. Their survival in extreme environments – particularly high temperature and high salinity conditions, has attracted significant interest, and scientists have investigated the underlying mechanisms responsible for archaea viability in harsh conditions. Indeed, the *archaeal membranes*, particularly the *molecular compositions* of the membrane, are believed to significantly contribute to archaeal stability and resilience in extreme conditions.

Archaeal membrane lipids exhibit important differences compared to lipids extracted from bacteria or eukaryotes, enticing researchers to use those lipids in artificial vesicle constructs. Unlike phospholipids which contain *ester bonds* between the headgroups and the alkyl chain, archaeal lipids exhibit *ether bonds* between the alkyl chain and the glycerol unit (Fig. 7.6A). The ether bond bestows resistance to hydrolysis and degradation, and also contributes to the stability of archaea in conditions of high salinity and acidity. Another notable distinction of archaeal lipids is the branched (i.e. "isoprenoid") nature of their hydrophobic acyl chains (compared to the unidirectional, unbranched configuration of the acyl chains in phospholipids). This property underlies the resilience of the archaeal membrane shell.

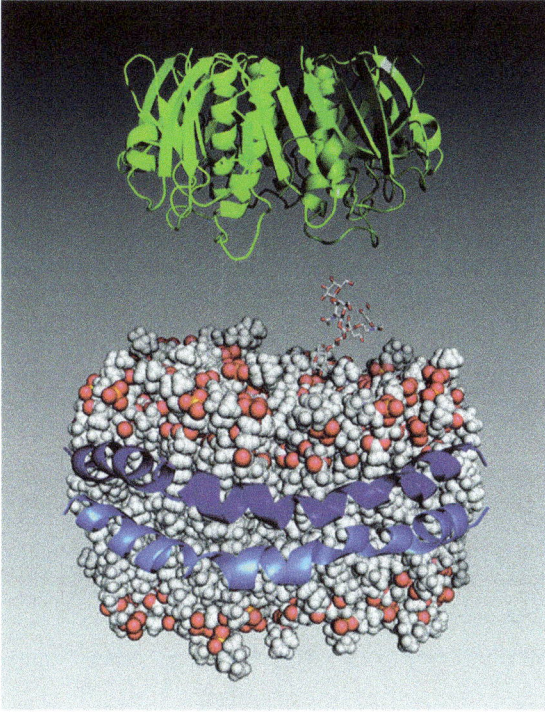

Fig. 7.5: rHDL "decoy". The rHDL particle incorporates *GM1* – a glycoconjugate receptor for cholera toxin (green). The displayed molecule can act as "bait" for the toxin.

A third fundamental property encountered in archaeal membranes and generally absent in bacteria or eukaryotes is the formation of membrane *monolayers* comprising archaeal lipids in which the polar headgroups are *covalently linked* through the acyl chains (Fig. 7.6B). The lipid monolayers result in higher membrane stability on the membrane compared to a lipid *bilayer* – which can be more easily disrupted due to the separation between the internal and external sheaths. Moreover, the monolayer organization combined with the branched lipid structures make archaeal membranes less permeable to ions and other small molecules – an important feature in saline environments.

Overall, purified archaeal lipids or their synthetic counterparts have been used for construction of biomimetic vesicles, potentially used as drug delivery vehicles and other applications. Archaeal liposomes, denoted also as "archaeosomes", are less prone to rapid enzymatic degradation in the bloodstream since many lipid-digesting enzymes only hydrolyze *ester bonds* of phospholipids. Archaeasomes should be stable for long time periods, and can encapsulate useful biological and therapeutic cargoes.

Efforts to mimic the cellular membrane have expanded the arsenal of bilayer constituents beyond phospholipids (or lipids in general), to include also *non-biological* molecules. Some polymers, in particular, adopt structures that closely mimic lipid

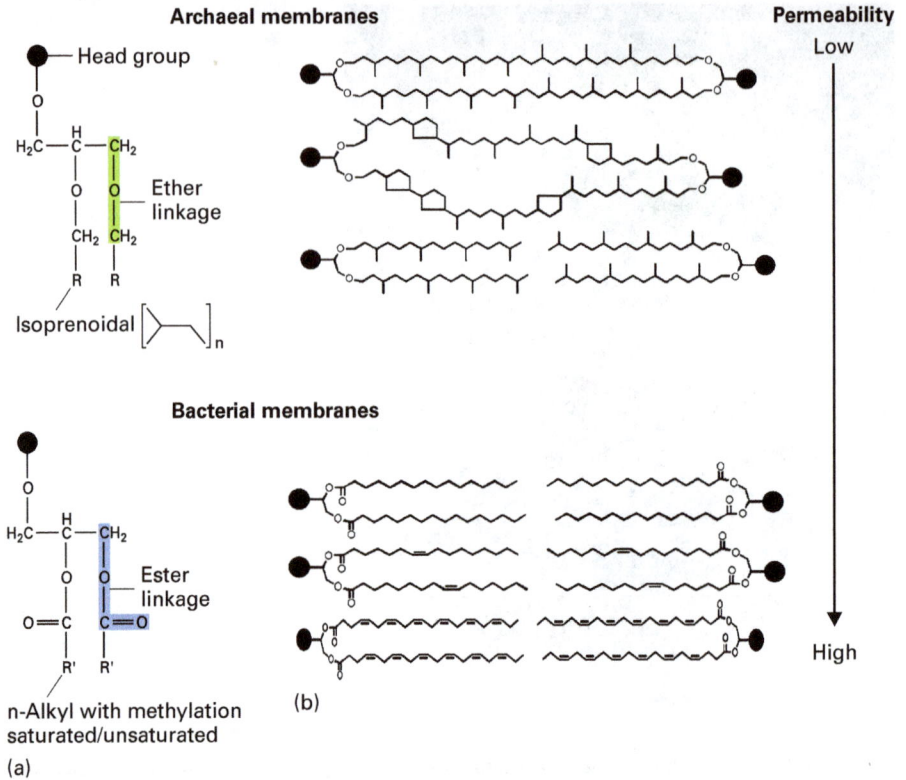

Fig. 7.6: Archaeal lipids vs. bacterial lipids. Differences in structure and organization between archaeal and bacterial (also eukaryotic) lipids. **(a)** Archaeal lipid structure highlighting the *ether bond* and the branched *isoprenoidal* acyl chains. In comparison, bacterial lipids contain *ester bonds* and *unbranched* acyl chains. **(b)** Lipid organization. Archaeal lipids often form *monolayers* (in which the headgroups are displayed on both sides of the monolayer) in comparison with *bilayers* which are the almost universal membrane organization in bacterial and eukaryotic cells.

vesicles in terms of size, curvature, fluidity, and incorporation of guest species. The key structural feature of such polymeric units is the *amphiphilic* nature of the molecule, i.e. a hydrophilic (polar) component (usually the headgroup) and hydrophobic (a-polar) part. *Block copolymers* are particularly amenable for generating vesicular architectures. These polymers comprise distinct "blocks" covalently linked to each other; accordingly, blocks can be designed such that they display different hydrophilic or hydrophobic properties which dictate vesicle self-assembly. Fig. 7.7 presents examples of *diblock* copolymers forming bilayer vesicles, designed by D. E. Discher and colleagues at the University of Pennsylvania. Interestingly, diblock copolymers were synthesized forming "ideal" (symmetrical) bilayers or "interdigitated" bilayer (i.e. comprising interspersed chains), allowing investigating the effects of interdigitation upon membrane properties.

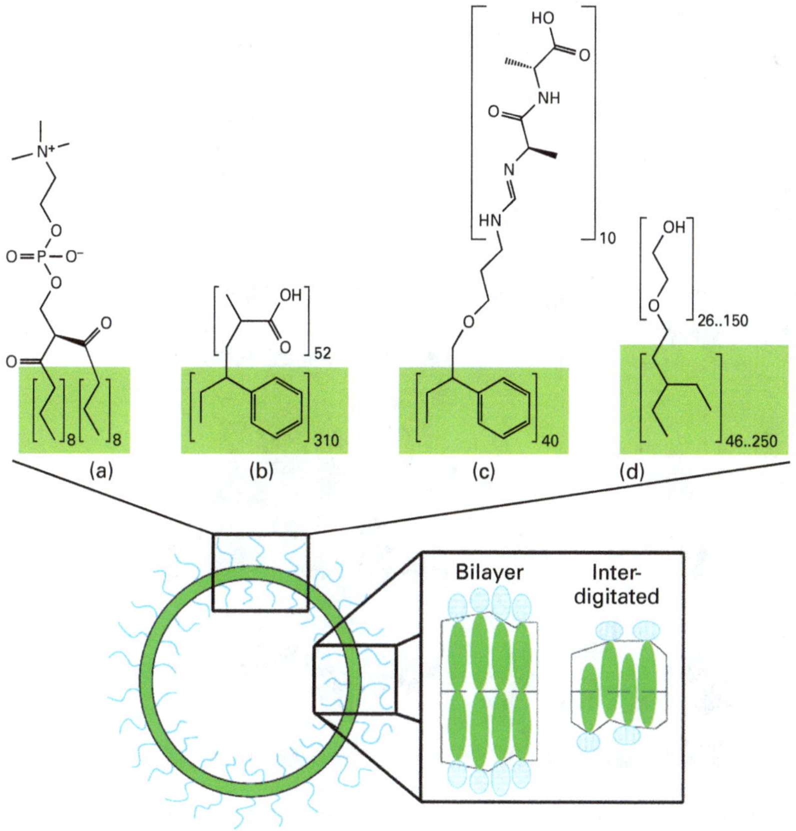

Fig. 7.7: Vesicles comprising synthetic diblock copolymers. Similar to the abundant membrane lipid phosphatydilcholine **(a)**, diblock copolymers **(b)–(d)** contain hydrophilic headgroups and hydrophobic side-chains resulting in the formation of stable vesicles in aqueous solutions. Depending on the side-chain properties, the bilayers formed can be symmetrical or interdigitated.

Unilamellar vesicles (i.e. single-layered coated vesicles) assembled from block copolymers are sometime referred to as "polymersomes". The "tuning capacity" inherent in the polymersomes' chemical design has made them attractive candidates for practical applications such as drug delivery. Remarkably, polymersomes were shown to undergo wholly "biological" processes such as membrane fusion. The microscopy experiment depicted in Fig. 7.8, carried out by D. Yan at Shanghai Jiao Tong University, China, demonstrates the occurrence of *fusion* between two synthetic polymersomes. Polymersomes, in fact, were shown to mimic lipid-based assemblies not only in terms of physical properties and structural features, but also in a *functional-biological* context. Polymersomes hosted a plethora of biological molecules that retained their functional properties, including membrane-associated proteins, receptors, and channels.

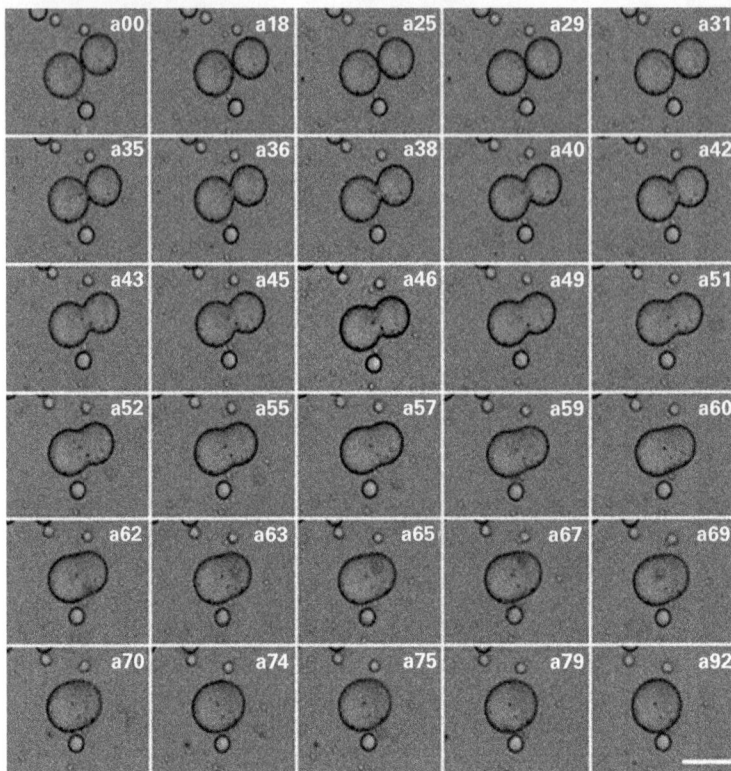

Fig. 7.8: Polymersome fusion. Time-lap experiment showing fusion of two giant polymersomes. The times elapsed are indicated in seconds. Scale bar corresponds to 50 μm. Reprinted with permission from Zhou Y. and Yan D., *J. Am. Chem. Soc.* **2005** *127*, 10468–10469. Copyright (2005) American Chemical Society.

An interesting experimental observation has been the induction of polymersome *compartmentalization*. Sub-cellular compartments, such as the nucleus, proteasome, endosomes, and mitochondria, are key features of eukaryotic cells. Intracellular compartments are necessary for maintaining separate environments for performing specific tasks and biological processes inside the cell. Creating artificial compartments in vesicles, however, is not an easy undertaking. Fig. 7.9 illustrates an experiment carried out by D. A. Weitz and colleagues at Harvard University in which a water/oil/water "double emulsion" containing diblock copolymers yielded "compartmentalized polymersomes". The sub-compartments formed within the polymersome maintained distinct phases and did not merge over time, pointing to the possible use of such compartments for carrying out orthogonal reactions.

Membranes are not "lock-tight" barriers but rather permeable to numerous molecules and ions that are essential for cell functioning. Transport through the membrane is facilitated through two mechanisms – "passive" diffusion, in which the ion or molecule penetrates into or out of the cell via diffusion through the membrane

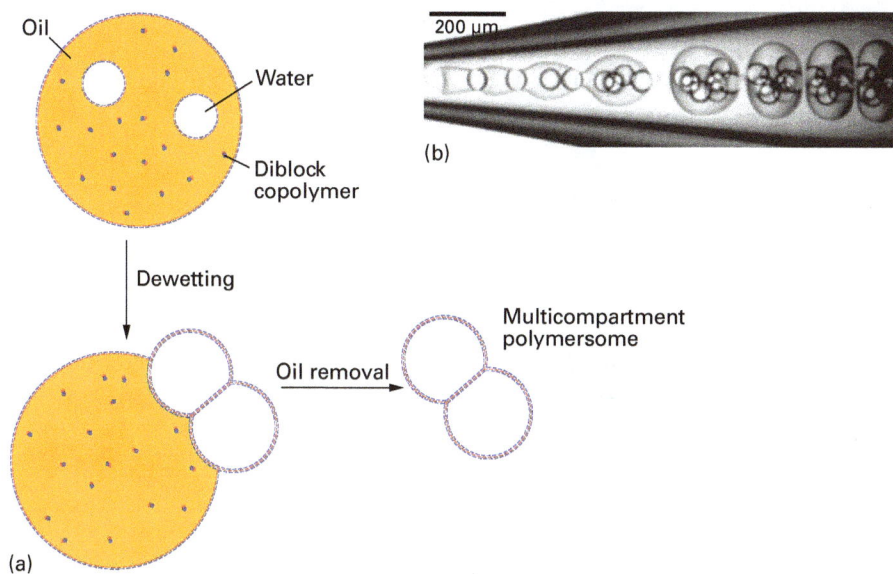

Fig. 7.9: Compartmentalized polymersomes. **(a)** Formation of separate compartments through a "double emulsion" scheme; **(b)** Visual image of polymersome compartments formed in a microfluidic device. Reprinted with permission from Zhao, Y. et al., *Angew. Chem.* **2011** *50*, 1648–1651. Copyright (2011) John Wiley and Sons.

bilayer, and "active" transport, which is mediated through specific receptors or pore-forming proteins. Pore-mediated transport processes, in particular, have attracted interest both from a scientific standpoint – understanding the relationships between transport phenomena and pore properties such as size and surface charge – as well as for practical purposes, for example identifying pharmaceutical substances that could block or open pores.

Pores play particularly important roles in regulating transport of proteins and oligonucleotides into and out of the cell nucleus, critical for maintaining the protein translation machinery and feedback loops between the cell nucleus and cytoplasm. The sophisticated macromolecular entity responsible for this task is called the "nuclear pore complex" (NPC) and its activity is believed to be finely modulated by the interplay between pore size, surface charge, and protein composition. Studying the nuclear pore complex is highly challenging, due to its multiple protein composition and the difficulty of deciphering its structural features in real time. In this context, C. Dekker and colleagues at Delft University, Netherlands, have recently created a "minimalistic" synthetic NPC (Fig. 7.10). The biomimetic NPC comprised a nanopore drilled in an inorganic sheet, whose dimensions resembled its natural counterpart, thus achieving *physical* selectivity. In addition, since molecular transport through the NPC is dependent upon interactions of the translocating molecules with the proteins lining up the pore, the inorganic pore was further coated with peptides abundant in the nuclear membrane. To complement the biomimetic design, the artificial pore provided means for *monitoring* transport – attained through measuring ionic conductance. The high sensitivity of the ion conductance technique made possible observation of opening and closing on a *single pore* level.

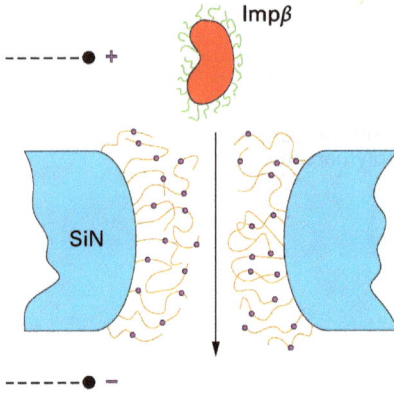

Fig. 7.10: Biomimetic nuclear pore complex (NPC). A nanopore with ~40 nm diameter, approximating the dimensions of the nuclear membrane pore, was drilled in a SiN film. The nanopore was then coated via covalent attachment of nucleoporins – abundant peptides present in the nuclear membrane. Ionic conductance was measured, providing the means for evaluating pore opening and closing. The figure shows a nuclear transporter – Impβ – successfully translocating through the artificial pore. Control proteins (such as bovine serum albumin) did not pass through the pore, confirming its biomimetic nature.

7.2 Artificial cell division

Induction of *physical division* is among the most difficult physical processes to mimic in artificial cells. Cell division, in fact, is a highly complex process from a mechanical or morphological standpoint and is particularly difficult to reproduce even in simple model vesicles. To obtain two cells from a single, parent cell, the lipid bilayer of the parent has to undergo controlled deformation and separation in order to create two distinct (and functional) bilayers. In nature, this function is generally carried out by specific membrane-associated proteins that initiate duplication of the bilayer membrane. Attempts have been made to recreate such protein-induced cascades in *vesicle* environments; however this goal is complicated by the fact that shape and structure of real cells are stabilized through coupling between the membrane and *cytoskeleton* framework inside the cell. Accordingly, including the cytoskeleton support in artificial cell design introduces an additional dimension of complexity.

Mimicking cell division in artificial systems is a formidable undertaking from other angles. Vesicle fission cannot occur *spontaneously*, otherwise the artificial cell risks premature disintegration. Indeed, cell division in nature is a highly regulated and coordinated process, involving several genetic and protein networks. In bacteria like *E. coli*, for example, cell division is orchestrated by no less than 10 genes, giving rise to almost simultaneous events including volume expansion (induced through phospholipid synthesis and cell wall expansion), membrane fission, and eventual bilayer fusion within the individual daughter cells.

Understanding and controlling the factors leading to synthetic cell (primarily vesicle) division has been an active field of research for membrane chemists and physicists. This research avenue has focused both upon identifying the pivotal molecular components – lipids, fatty acids, or membrane-associated peptides – in self-reproduction processes, as well as elucidating the "triggers" which initiate and regulate cell division. Basic physical models pertaining to vesicle division are depicted schematically in Fig. 7.11. The mechanisms summarized in Fig. 7.11, outlined by J. W. Szostak at Harvard University and others, underscore the close relationship between the *physical* parameters, such as vesicle deformation or vesicle expansion, and the *chemical* factors responsible for vesicle transformations, for example enzymatic reactions giving rise to intracellular vesicle formation.

The pathways depicted in Fig. 7.11 underscore the significance of *external cues* for inducing division of artificial cells. Interactions between *small micelles* and vesicles, in particular, were shown to constitute a powerful driving force for cell division

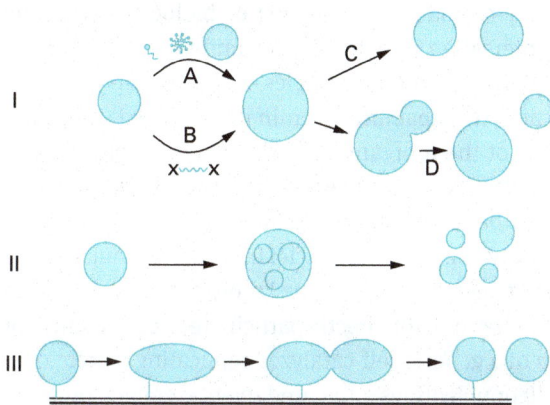

Fig. 7.11: Models for vesicle division and replication. Fundamental processes leading to vesicle division. **I.** Vesicle growth leading to replication. Growth can occur either through incorporation of membrane components such as *amphiphilic compounds*, *micelles*, or *small vesicles* from external sources (A), or through introducing precursors, such as *enzymatic substrates*, which yield bilayer components through chemical modification (B). Subsequent vesicle division can produce *symmetrical* daughter vesicles (C), or proceed via *asymmetric budding* (D). **II.** New vesicles are assembled through chemical means within a parent vesicle and subsequently released. **III.** Surface deformation leading to vesicle growth and division. Reprinted from Hanczyc, M.M. and Szostak, J.W., Replicating vesicles as models of primitive cell growth and division, *Curr. Opin. Chem. Biol.* **2004** *8*, 660–664. Copyright (2004), with permission from Elsevier.

through initiating growth of non-spherical invaginations on vesicle surfaces. Indeed, such mechanical instabilities were found to bring about formation of smaller spherical "daughter" vesicles (e.g. route "I" in Fig. 7.11). It is generally assumed that the faster growth of the vesicle surface compared to the entrapped volume is an important physical determinant for cell division.

The challenge for inducing division in artificial cells is difficult from another angle – ascertaining that the internal contents of the cell do not "escape" throughout the division process. This stipulation is, in fact, critical, otherwise consecutive divisions would result in a significant loss of vesicle content to the external solution. Several studies have addressed this issue through demonstrating techniques in which *growth* of vesicle protocells progressed *simultaneously* with vesicle *division,* rather than consecutively. The coupling between growth and division ensured that vesicle leakage was prevented.

While research into artificial cell division has naturally focused upon *biological* entities (particularly lipid-based vesicles), there have been intriguing attempts to use *non-biological* modules as scaffolds of artificial cells, which undergo both spontaneous and triggered reproduction. Examples for this concept include the pioneering work of J. W. Szostack and colleagues at Harvard University, who discovered that *clay particles* are capable of both assembling *vesicles* from fatty acid micelles, facilitate continuous vesicle growth induced by "feeding" the vesicles with more fatty acids, and that furthermore, can catalyze division of the vesicles into separate "daughter" vesicles.

The formation of spontaneously-reproducing inorganic or biological synthetic vesicles in the laboratory raises interesting possibilities for understanding "prebiotic" cells. Specifically, encapsulation of mineral particles within membrane vesicles might have harnessed their catalytic capabilities within the confined space of a membrane. The catalytic properties of the inorganic particles might be particularly useful for synthesis of *oligonucleotides* – one of the great mysteries of the prebiotic world. Individual nucleotides might be either physically attached to the clay particle within the membrane, or co-dissolved inside the vesicle. Indeed, oligonucleotides have been encapsulated within clay-generated, replicating "protocells". Even though the nucleotide guests in such assemblies did not directly participate in biochemical or chemical processes, the formation of a generic self-catalyzed replicating system containing a membrane barrier and oligonucleotides – the fundamental genetic building blocks – provides a fascinating synthetic system which could shed light on the origin of cells as independent, self-replicating entities in the prebiotic world.

Another key factor for inducing division in artificial cells concerns the *energy source* for the process. Investment of energy is an integral part in initiation of processes associated with cell division that result in significant morphological and organizational changes, such as compartment formation. Living cells generate energy through well-known biochemical processes, primarily involving adenosine triphosphate (ATP) metabolism. In principle, ATP transformations might be reproduced in a synthetic cell, however the mechanisms for ATP processing and recycling are complex and involve various enzymes that need to be successfully reconstituted in an artificial cell system. Alternative intriguing propositions are for protocols to exploit *kinetic energy* from the surrounding environment – agitation or local currents in water.

7.3 Biomolecules encapsulated in artificial cells

Constructing synthetic cells that can divide is not the only requirement for achieving a viable artificial cell. Specifically, the internal composition of a synthetic cell has to mimic the diverse biomolecule families enclosed within native cells (Fig. 7.12). Furthermore, the intricate functional and structural relationships among the molecules embedded within a cell need to be reconstructed. In practice, the enormous complexity of interconnected processes and molecules inside a cell is exceedingly difficult to reproduce synthetically; scientists have been trying, however, to determine the most fundamental processes which might, by themselves, suffice to create a functioning synthetic cell. Specifically, the requisites for a "minimal cell" can be banded together under three main characteristics: demonstration of self-maintenance, reproduction, and the ability to "evolve" over time.

Indeed, designing the *molecular composition* of an artificial cell is crucial in light of the interdependence between cell *division* and *reproduction* of the cell's internal contents. In nature, cell division goes hand in hand with reproduction (e.g. duplication) of the parent cell contents (primarily the genetic material, but also proteins and other biomolecules) so as to produce daughter cells with all the molecular substances necessary for functioning independently. Conceptually and technically the realization of both division and reproduction in artificial cells is highly challenging. Indeed,

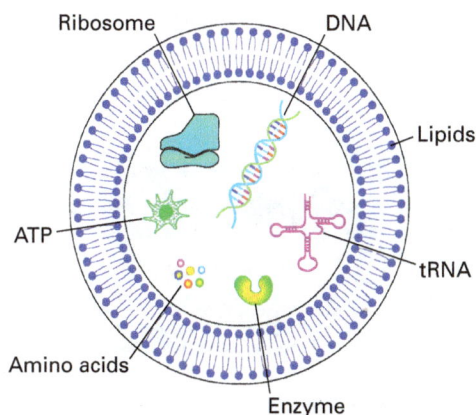

Fig. 7.12: Main molecular families in eukaryotic cells. *DNA* carries the genetic code; *enzymes* responsible for catalyzing intracellular biochemical reactions; *transfer RNA (tRNA)* and the *ribosome* are essential for biosynthesis of proteins from individual *amino acids*; *adenosine triphosphate (ATP)* represents energy source molecules.

these two processes need to be *coordinated* in order to avoid dilution and loss of the parent cell contents among the daughter cells throughout the division processes. Simply put – if an artificial cell is made to divide (through external factors and/or self-initiated) without simultaneously reproducing its composition, the molecular components embedded in successive cell generations would become highly diluted, diminishing cell functionalities.

A key "origin of life" question which is also relevant to artificial cell contents concerns the factors contributing to presence of concentrated solutions of macromolecules within the small enclosed area of a cell. One suggested scenario points to metabolic and/or enzymatic processes occurring inside the prebiotic cells over long time periods, ultimately yielding highly-concentrated enclosed solutions inside the cell. An alternative proposition raises the possibility that crowding inside a cell is a result of spontaneous encapsulation of a concentrated macromolecule solution of the "primordial soup" within a self-assembled membrane "compartment".

Even though a clear answer to the above riddle remains elusive, various experiments have lent support to both theories. Interesting experiments carried out by P. L. Luisi and colleagues at the University de Roma Tre, Italy, provide evidence for the "spontaneous crowding" model, demonstrating that random formation of vesicles in solutions containing proteins, protein-expression machinery (ribosomes), and oligonucleotides resulted in *some* vesicles exhibiting very high concentrations of biomolecules, even when vesicle formation was carried out in dilute solutions (Fig. 7.13). This result cannot be explained through simple statistical distributions of molecule concentrations inside vesicles, and suggests that self-association of macromolecules was promoted in parallel with formation of vesicle enclosures, resulting in a "crowding effect" in some vesicles. Importantly, the crowded environments within the vesicles enabled distinct *functional* behavior – for example protein expression, which appeared significantly more enhanced compared to the dilute solution in which the vesicles formed.

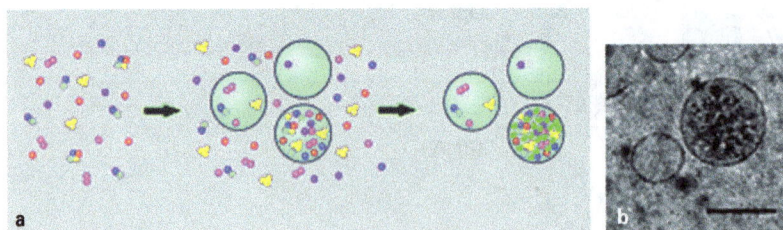

Fig. 7.13: "Molecular crowding" inside an artificial cell. Spontaneous encapsulation of molecules inside an artificial vesicle. **(a)** Schematic drawing of vesicles forming in the solution containing the molecules; some of the produced vesicles enclose high concentrations of the soluble molecules; **(b)** TEM image showing a crowded vesicle adjacent to an empty one. Scale bar is 100 nm. Reprinted with permission from de Souza, T.P. et al., *ChemBioChem* **2011** *12*, 2325–2330. Copyright (2011) John Wiley and Sons.

Mechanisms proposed for the phenomenon depicted in Fig. 7.13 invoke cooperative processes arising in the concentrated solute microenvironments; "empty" vesicles would form rapidly, while concentrated solutes might stabilize open vesicle leaflets, thereby slowing full enclosure and allowing further accumulation of biomolecules. The vesicle crowding experiments suggest that, at least in principle, primitive cells could form spontaneously, comprising in a very early stage concentrated solutions of molecules such as proteins and nucleotides that are essential for cell viability and reproduction.

An emerging field in artificial cell research concerns the effect of cell shape and morphology upon the internal organization of encapsulated biomolecules, particularly proteins. This line of study is important as it aims to link the macroscopic cell environment and microscopic parameters, such as cell morphology, to molecular features pertaining to protein structure and organization. Furthermore, while most artificial cell systems constructed thus far constitute spherical symmetrical assemblies (as discussed above), most cells in the real physiological world are, in fact, asymmetric and exhibit diverse shapes. To address this issue, C. Dekker and colleagues at Delft University of Technology, Netherlands, have introduced microfluidic platforms designed to induce shape changes of artificial cells and consequently assess the effects of shape deformation on the organization of encapsulated molecules (Fig. 7.14).

Fig. 7.14a illustrates the experimental concept. Spherical cell-like compartments (e.g. vesicles, micelles, or emulsified containers) are passed through a "squeezing device" impacting desired anisotropic shape and size. The effects of cell shape upon the organization of embedded biomolecules, particularly proteins that were fluorescently labeled for easier microscopic examination, can then be analyzed. Fig. 7.14b presents striking examples of the impact of artificial cell geometry on the organization of embedded proteins. In the experiments, the fibrous proteins tubulin and collagen were seeded within microscale water-in-oil droplets mimicking cells. In spherical microcapsules, the two proteins formed randomly oriented fibers (Fig. 7.14B, left). However, when the cell-like compartments were shaped as an elongated container, both tubulin and collagen microtubules were oriented relatively parallel to the long axis of the rod-shaped cell-like container. While the innovative experimental scheme depicted in Fig. 7.14 provides a useful tool for investigating the relationships between protein assembly and cell shape and deformation, more studies are required to identify the actual molecular mechanisms responsible for these relationships and elucidating the effects of microscopic cell features upon protein functionalities.

7.4 Artificial replication

While several studies have demonstrated encapsulation of concentrated protein mixtures inside vesicles and consequent occurrence of chemical reactions, molecular organization, and compartment formation, the incorporation of *genetic material*

A

B.

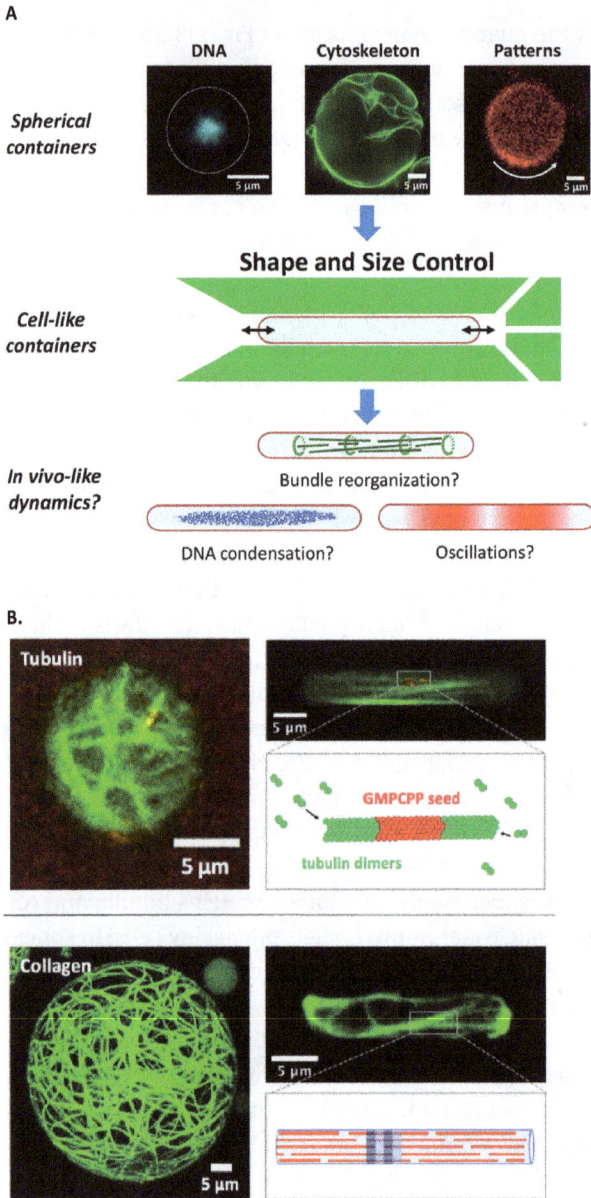

Fig. 7.14: Effect of cell shape upon the organization of embedded biomolecules. **(a)** Conceptual framework of the experiments. Spherical cell-like containers containing fluorescently labeled biomolecules are squeezed into a microfluidic device which allows modulation of container size and shape. Reorganization of the biomolecules in different modes is consequently studied. **(b)** Effects of artificial cell elongation upon organization of embedded known cellular framework proteins, tubulin, and collagen. In both cases, the assembled proteins orient along the long axis of the container. Adapted with permission from Fanalista et al., Shape and Size Control of Artificial Cells for Bottom-Up Biology, *ACS Nano* 2019, 13, 5439–5450, Copyright (2019) American Chemical Society.

and achieving translation of its information inside an artificial cell has been much more difficult. The prerequisites for realization of genetic information transfer are in a sense even more stringent than facilitating reactions of proteins; not only DNA (or RNA) molecules need to be maintained in physical contact with translation machinery (e.g. transcription factors, mRNA, ribosomes), but methods for triggering transcription and translation reactions and stopping when required have to be found.

Replication as discussed here in the context of genetic mechanisms is different to *reproduction*; the former is defined as *generating* duplicates of the original, parent molecule by chemical or biological means, while the latter refers to the reconstruction of the overall shape or morphology of a molecular assembly. The issue of "replication" poses perhaps the most formidable barrier to developing artificial cells. In its most basic, simplest form, the question one needs to ask is the following: can a *programmable* synthetic particle be assembled? The challenges here have two facets: the information-bearing molecule (DNA, RNA, or other synthetic substances) needs to both replicate itself (in a regulated manner), and also an appropriate "program" (e.g. genetic code) has to induce all chemical and biochemical processes taking place inside the artificial "cell". Furthermore, a mechanism has to exist for the artificial cell to undergo division into two "daughter cells" that will continue to carry out the same information embedded in the genetic code of the parent cell, notwithstanding the occurrence of mutations down the road (which are naturally important by themselves from an evolutionary standpoint). Adhering to all these fundamental requirements is obviously a very difficult task.

In eukaryotic organisms (including humans) DNA carries the genetic code necessary for all processes occurring in a cell. This code is an intrinsic product of base-pair complementarity in DNA structure, and the sequential nature of the DNA molecule provides the "reading frame" for the various genes. However, by itself, DNA does not convey any information or biological instructions. Indeed, the crucial components for operation of a cell, from the DNA and outward, are the *transcription* and *translation* mechanisms, carried out by specific proteins and RNA molecules. In its simplest description, DNA transcription involves RNA polymerase protein which copies double stranded DNA into a single-stranded messenger RNA (mRNA). Translation, which is carried out in the ribosome, is a more complex process, involving around 100 proteins and transfer RNA (tRNA) molecules. The translation process is also relatively extended – tRNA induces translation of one DNA codon at a time – and thus prone to errors.

Considerable research efforts into "synthetic genetics" have centered on development of "cell-free" systems in which DNA expression is carried out and regulated outside of a cell. In principle, such schemes could be subsequently reproduced in the environments of an artificial cell. Cell-free DNA *replication* is routinely carried out through the polymerase chain reaction (PCR) technique (see Chapter 6). Briefly, PCR relies on short DNA fragments – also known as "primers" – which attach to a denatured DNA double strand, resulting in replication of the DNA domain between the two primers. PCR has been implemented for replicating DNA also in lipid vesicles, however the technique is likely irrelevant to artificial cell research since it relies on high temperatures for melting, or opening, the DNA double helix to be replicated. The elevated temperature would denature proteins participating in biochemical processes within the synthetic cell.

Most studies of cell-free DNA translation have utilized combinations of natural transcription factors and translation-inducing substances isolated from various microorganisms. Specifically, cell-free *transcription* of the desired genes is carried out by addition mixture of RNA polymerase, usually from bacteriophage (bacteria-infecting virus) sources, to the DNA; the *translation* machinery added to the mixture (i.e. tRNA, ribosomes, chaperons) is usually isolated from cytoplasmic extracts of microorganisms. Mixing all the necessary components in a test tube and providing the necessary chemical energy (through addition of ATP molecules) could, in principle, initiate DNA translation and protein production.

Even though "on paper" cell-free systems should work quite efficiently, significant technical hurdles have limited both the biotechnological potential and possible implementation of these platforms in actual artificial cells. The concentration of ATP, and thus available energy, has been found to considerably affect the kinetics and yields of DNA expression reactions proceeding outside of a cell; indeed, even minute changes in ATP concentrations were found to result in dramatic changes to protein synthesis rates. Alterations of other environmental parameters such as the pH were also shown to significantly, often adversely, affect the expression efficiencies. Ultimately, one has to realize that on a basic level cell-free DNA expression systems do not truly mimic artificial cells, rather they more resemble an upgraded reaction chamber. In a sense, protein synthesis in these systems would eventually stop due to the consumption of the finite resources present. Additional drawbacks of cell-free schemes include low reactant concentrations as compared to *intra*cellular situations which reduce reaction rates and yields.

Rudiments of vesicle-encapsulated cell-free DNA expression have been demonstrated. Similar to cell-free systems, the essential elements include the genetic material and transcription/translation components; the crucial difference is that the entire translation process is carried out in a vesicle-enclosed environment. Fig. 7.15 shows an elegant construct developed by D. G. Anderson at MIT in which translation of a protein was accomplished through embedding the genetic and translation elements within a vesicular enclosure. Furthermore, a generic "regulatory component" was included; the DNA was initially protected in a "cage" configuration, and could be released and activated through irradiation of a photo-sensitive chemical moiety acting as a "lock".

While successful protein expression in an artificial cell setting has been demonstrated, as shown in Fig. 7.15, it seems that achieving truly biological replication of essential proteins and associated DNA is a much more difficult task. Several studies have suggested that a minimal number of 200–400 genes would be needed for an effective

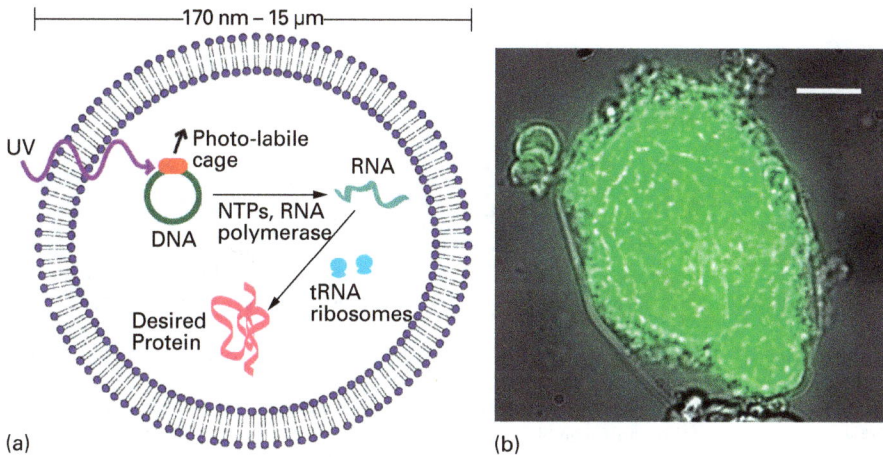

Fig. 7.15: Vesicle-encapsulated DNA translation. **(a)** Synthetic vesicles loaded with DNA and the transcription/translation cellular machinery. The DNA of the protein to be translated (green fluorescent protein, GFP) is present in a "caged" configuration, "locked" with a photo-labile compound. Irradiation of the vesicles with light activates the photolabile "lock", triggering DNA transcription. **(b)** Confocal microscopy image of an activated vesicle, showing the intense fluorescence of the expressed GFP. Reprinted with permission from Schroeder, A. et al., *Nano Lett.* **2012** *12*, 2685–2689. Copyright (2012) American Chemical Society.

and reliable replication process. Constructing such a genome and maintaining its functionality and fidelity over many generations is not easy. The occurrence of mutations is naturally expected, however one has to ascertain that any line of artificial cells would be still viable even after a large number of divisions (and thus accumulated mutations).

Combining replication of the informational substances encapsulated within an artificial cell (e.g. oligonucleotides, genes, etc.) with reproduction of the synthetic cell is considered one of the next frontiers in artificial cell research. Indeed, while replication and reproduction (i.e. cell division) have been demonstrated in synthetic systems, these two processes have been usually carried out independently. Recent elegant experiments have accomplished *replication* of DNA encapsulated within a *self-reproducing* giant vesicle (Fig. 7.16). Specifically, the experimental system developed by T. Sugawara and colleagues at the University of Tokyo facilitated both replication and amplification of vesicle-embedded DNA through conventional PCR, *and* spontaneous fission and division of the vesicles through chemical catalysis. Particularly important, it has been shown that amplification of the encapsulated DNA *accelerated* the division of the vesicles – thus linking the self-replication of the genetic material with self-reproduction of the encapsulating unit (the vesicle).

An alternative proposition for achieving *DNA-programmed* artificial cells would be the use of **RNA** to create a replicating protocell. This idea stems from the well-known catalytic (and auto-catalytic) properties of RNA which makes it attractive for inducing and regulating chemical reactions inside an artificial cell. Moreover, the

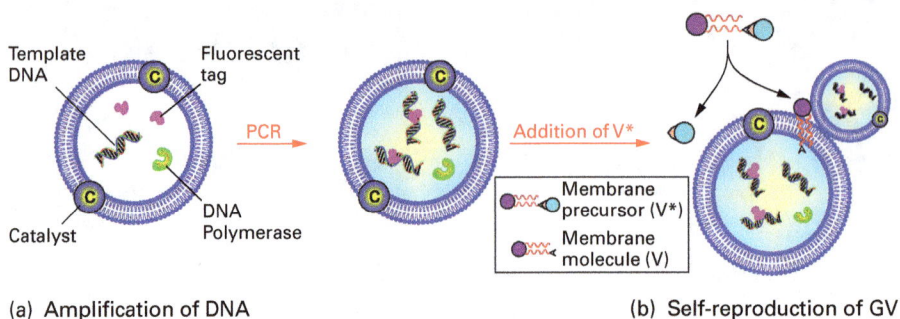

(a) Amplification of DNA   (b) Self-reproduction of GV

Fig. 7.16: Integrated self-replicating and self-reproducing vesicle system. **(a)** Amplification of DNA was accomplished through assembling the giant vesicle in a solution containing the PCR constituents (template DNA, primers, deoxynucleoside triphosphates, DNA polymerase) and a fluorescent DNA tag; **(b)** self-reproduction of the vesicles induced by hydrolysis of membrane lipid precursor V*, induced by a bilayer-embedded catalyst, yielding additional lipids. Adhesion of the amplified DNA to the inner leaflet of the vesicle accelerated vesicular growth and division.

concept of RNA-based synthetic cells is also inspired by the "RNA world" hypothesis promoted by some evolutionary scientists – the proposition that early life forms had actually started from *RNA replication* in primitive cells, rather than DNA. The main problem, however, with RNA-based artificial cells is that replication mechanisms will have to be essentially invented – i.e. transcription and translation scaffolds, regulation proteins, etc.

The pursuit of the "minimal cell" has certain philosophical dimensions, with it the question arises as to what the simplest chemical system created in the laboratory is that can be construed as "living". This issue goes to the core questions presented at the beginning of this chapter – can life itself be created from a certain combination of chemical building blocks? What actually *were* the natural processes which have led to the creation and evolution of early cell life (and which a scientist needs to mimic in the laboratory in order to create synthetic cells)? These fundamental questions take specific chemical as well as conceptual meanings: is construction of a semipermeable membrane the precondition for an artificial cell? Or is the facilitation of spontaneous chemical and biochemical processes the defining point of functional living artificial cells? How sophisticated does the information transfer between generations (e.g. replication) need to be for a cell to be considered viable? Are processes such as self-reproduction and self-replication sufficient for defining a molecular assembly as "viable"? These questions are among the main pillars of the ongoing scientific debate and driving forces for progress in this exciting field.

8 Drug delivery

The ultimate objective of any effective therapeutic treatment is to successfully deliver the "healing cargo" to its physiological target – be it a specific organ, tissue, or cells. However, transporting therapeutic molecules, either synthetic or biological, inside the human body is not an easy task. Through the eons of evolution, our body has developed sophisticated and powerful mechanisms designed to rapidly detect and destroy "non-self" molecules and invaders. This "defense umbrella" includes the physical barriers within the body (e.g. the "blood-brain barrier" (BBB)), the non-specific defense mechanisms (e.g. white blood cells, natural killer cells, non-specific proteases), and the "adaptive" immunity, functioning through recognizing and "memorizing" specific molecular units and structures of foreign substances (i.e. antigens).

Numerous drug delivery methods have been developed and this field has sprouted a multi-billion dollar industry. Considerable challenges still exist, however. Two seemingly opposing goals need to be fulfilled. For one thing the drug carrier needs to be stable in the physiological environments, particularly the bloodstream, so as not to disintegrate prior to reaching its target. However, stability should not come at the expense of effective release of the drug load at the desired target. Furthermore, the drug carrier has to elude the defense mechanisms of the body (i.e. the immune system) in order to be effective.

Due to its biomedical importance and potential health and financial benefits, the scientific field of *drug delivery* (and related technologies of *drug release)* is certainly among the most heavily invested and researched. Drug delivery technologies have reached a high degree of sophistication – in the capability to transport the biological active substances to their physiological targets, optimizing their release, and overcoming physiological barriers and enzymatic degradation. This chapter discusses, however, few examples in this vast field of basic and applied research, primarily systems in which *biomimetic* schemes constitute the core of drug delivery vehicles.

Proteins can be employed for fabricating nanoscale functional structures for drug delivery. Proteins, in fact, are particularly attractive candidates for drug delivery applications since they are intrinsically biocompatible, and possess structural elements and reactive sites that can be functionalized through biological or chemical means. Different types of protein *nanocapsules* (sometime denoted "bio-nanocapsules") have been fabricated. Fig. 8.1 presents an example of such an assembly, constructed by S. Wang and colleagues at the University of California, Irvine, from an engineered variant of a subunit of pyruvate dehydrogenase, a bacterial enzyme.

Incorporation of *guest molecules* with therapeutic functions in protein capsules can be achieved either through *chemical* methods, i.e. covalent binding to the protein surface, or through non-covalent attachment, usually inside the protein cavity. Covalent binding strategies in protein capsules are sometime hampered because of the limited chemical conjugation sites on a protein surface. Non-covalent approaches for

https://doi.org/10.1515/9783110709490-008

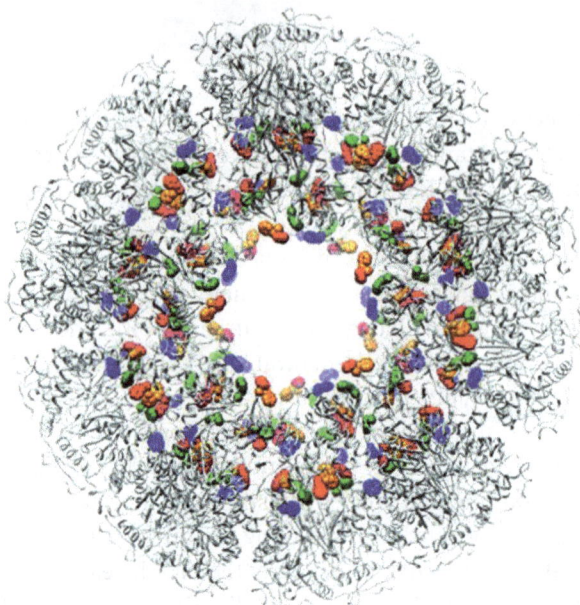

Fig. 8.1: Protein nano-capsule. Capsule assembled from the 60-subunit E2 component of pyruvate dehydrogenase. Introduced phenylalanine side-chains constituting possible binding sites for guest molecules are depicted in color. Reprinted with permission from Ren, D. et al., *Adv. Funct. Mater.* **2012** *22*, 3170–3180. Copyright (2012) John Wiley and Sons.

transporting therapeutic substances within protein capsules generally exhibit greater flexibility. For example, hydrophobic amino acids such as phenylalanine, known to anchor a broad range of structurally divergent molecules, can be introduced (through protein engineering) into specific positions within the cavities of protein nanocapsules, enhancing the transport capacity of the capsules.

Even though protein-based carriers seem ideal candidates for drug delivery due to their biocompatibility, they have certain limitations. Primarily, non-host proteins can be targeted and destroyed by the immune system. In addition, many of the proteins that are utilized in biotechnological applications are water-soluble; this property might lead to premature dissolution of complex protein-based particles, such as the nanocapsules discussed above, in physiological environments (i.e. bloodstream), and early release of the therapeutic cargo.

Different approaches have been introduced to overcome barriers to protein drug carrier applications. Inclusion of synthetic polymer "shields" within protein particles has been pursued to enhance biological inertness. Chemical functionalization of the protein molecules themselves has been carried out to increase stability inside the body. Such a chemical treatment, however, poses a conundrum, since *too high* stability might translate into insufficient release of the therapeutic cargo. Scientists have attempted to address this possible problem through further incorporation of targeting and triggered-release mechanisms. Additional approaches focus on making the protein particles' stability dependent upon the conditions of the surrounding aqueous solution, i.e. pH and ionic strength, taking advantage of the fact that variations exist in solution properties at different tissue locations or between the intracellular and extracellular space.

Lipids – the building blocks of the cell membrane – have been promoted for a long time as promising candidates for constructing drug carriers. Their appeal stems from the natural biological roles of lipid membranes – on the one hand comprising the protective layer of cells and maintaining cell stability, on the other hand facilitating transport (in and out of the lipid-enclosed space) of a variety of molecules, inorganic ions, and particles. Biomimetic lipid assemblies, i.e. *liposomes,* have long been exploited as possible vehicles for drug delivery. Similar to the cellular membrane, liposomes can display recognition elements on their external surface, including short peptides, proteins, carbohydrates, and others, which could then be employed for targeting the liposome to its biological destination and releasing its therapeutic cargo.

Despite the promise, liposomes as practical drug delivery vehicles are rare; in fact decades of research in this field has produced only two clinically-used liposomal carriers. The main problem is the extremely low stability of liposome carriers in physiological environments (which reflects a broader, often encountered divergence between bright ideas that are successful in laboratory settings, but fail in "real world" scenarios). Indeed, while in biological buffers or similar artificial solutions liposomes have been often found to be stable for extended periods, in blood serum most liposome formulations disintegrate within minutes, thus severely limiting their usefulness. Liposome degradation is traced to substances in the blood that are responsible for lipid sequestration and metabolism, primarily lipid-cleaving enzymes (i.e. *lipases*) and lipoproteins (i.e. cholesterol-extracting low-density lipoprotein [LDL] and high-density lipoproteins (HDL)).

Various chemical modification schemes have been developed towards improving liposome stability *in vivo*. Examples of such technologies include the addition of non-lipid constituents to the liposomes designed to endow "stealth" properties to the lipid assemblies. Another innovative platform has been the creation of "two-layer" vesicles, denoted "vesosomes" (vesicles-liposomes, Fig. 8.2). Vesosomes comprise small vesicles encapsulated within a larger vesicle enclosure. Vesosomes exhibit notable advantages in comparison with conventional vesicles. Specifically, the bilayer compositions of the inner layers and outer layer can be different, on the one hand enabling high stability of the external layer and display of targeting moieties while on the other hand producing smaller vesicles that can easily release their drug contents at the site of action. Moreover, the internalized vesicles can be protected from lipid degrading enzymes. Also, a single vesosome can, in principle, deliver *different* drugs simultaneously in different internal compartments.

While liposomes have performed rather poorly as drug delivery vehicles, other lipid-based biomimetic particles show some promise. Prominent among these are *lipoprotein*

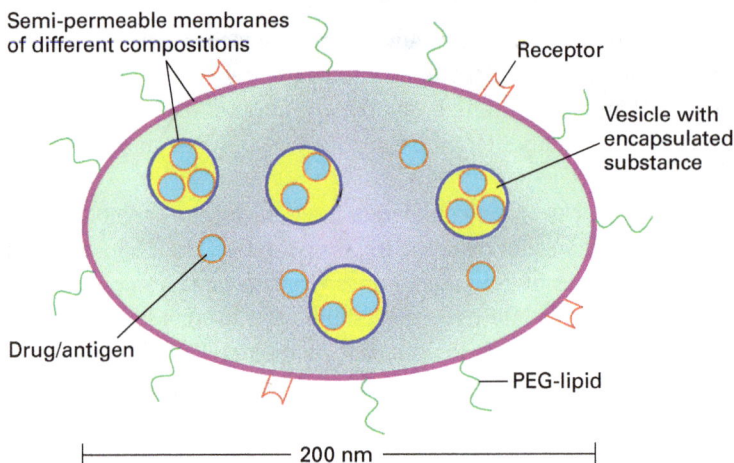

Semi-permeable membranes
of different compositions

Receptor

Vesicle with
encapsulated
substance

Drug/antigen

PEG-lipid

200 nm

Fig. 8.2: Vesosome structure.

nanoparticles (NPs). Lipoproteins – primarily the low density lipoprotein (LDL), and high density lipoprotein (HDL) – are sometime referred to as "Nature's nanoparticles", as they form compact particulate structures. These NPs could be ideal for drug- and imaging agents delivery applications (Fig. 8.3). The compact structures of lipoprotein assemblies make them inherently stable, resilient, and less prone to enzymatic degradation. Moreover, lipoproteins circulate in the bloodstream for extended time periods before removal in the liver. As shown in Fig. 8.3, different parts of the particles can be used as delivery platforms including the *hydrophobic cores* which are amenable for embedding hydrophobic and water-insoluble drug molecules, the *surfaces* can similarly serve for covalent or non-covalent placement of therapeutic substances and recognition elements.

Drug delivery vehicles comprising biological molecules and *inorganic* materials have attracted interest as possible alternatives to purely biological carriers such as liposomes. *Metal NPs,* in particular, offer certain advantages for drug delivery applications. Gold NPs can be synthetically functionalized with diverse molecular

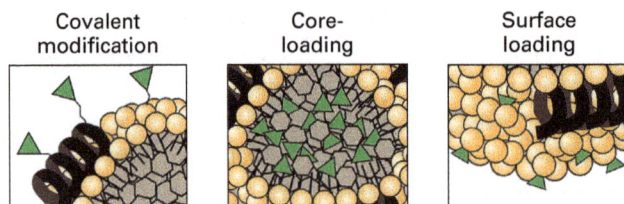

Covalent
modification

Core-
loading

Surface
loading

Fig. 8.3: Lipoprotein nanoparticles (NPs) for drug delivery. Different modes of incorporating drug molecules and bio-active agents within lipoprotein NPs. The green triangles represent pharmaceutical substances to be delivered. Reprinted with permission from Ng, K.K. et al., *Acc. Chem. Res.* **2011** *44*, 1105–1113. Copyright (2011) American Chemical Society.

units designed to embed drug cargoes or steer NP-conjugated biomolecules to their physiological destinations. Such targeting residues have included small peptides, antibodies, protein receptors, nucleic acids, and others. Fig. 8.4 depicts an example of a hybrid nanocarrier (for delivering DNA in this specific example) developed by C. S. Thaxton and colleagues at Northwestern University. The biomimetic aggregates comprised gold NPs acting as templates and stabilizing agents, and coated by phospholipids and apolipoproteins which together aim to mimic native lipoprotein structure. Cholesterol-coupled nucleic acids were subsequently also added to the biomimetic Au-lipoprotein aimed at induction of gene expression at the target cells.

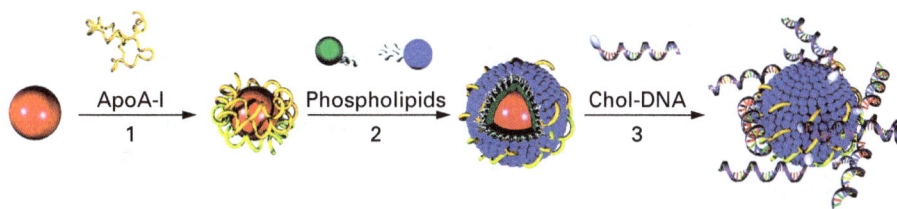

Fig. 8.4: Gold-NP templated particles for drug delivery. Gold NP is coated through non-covalent attachment of apolipoproteins (proteins comprising natural lipoprotein particles) and phospholipids. Further display of oligonucleotides is facilitated through incorporating cholesterylated-DNA conjugates. Reprinted with permission from Ng, K.K. et al., *Acc. Chem. Res.* **2011** *11*, 1208–1214. Copyright (2011) American Chemical Society.

Metal NPs themselves have great potential as diagnostic and therapeutic tools, and a thorough discussion of this subject is beyond the scope of this book. Briefly, NPs can be employed as *imaging* agents, particularly iron NPs (serving as reagents in magnetic resonance imaging experiments). Other promising avenues are the use of gold NPs for *localized tissue heating* through infrared light irradiation (promising therapeutic route for targeted obliteration of cancerous tissues). In such applications, however, the big challenge is to successfully deliver the NPs to their cell or tissue targets; this requires both the design of targeting mechanisms as well as surmounting the various physiological barriers encountered.

Biomimicry has aided the efforts to develop therapeutic uses for metal NPs, primarily through attachment of a variety of biological molecules to the metal surface. Diverse synthetic routes have been introduced, for example, for achieving *covalent* conjugation of biomolecules to NPs. Proteins and peptide fragments can be covalently bonded to gold NPs through the *thiol* (S-H) moiety of the sulfur-containing amino acid *cysteine*. Metal binding of proteins through the *amino* moieties has also been exploited. Similarly, nucleic acids can also be attached to gold NP surfaces through capping the DNA or RNA sequences with thiol groups. Lipids and lipid conjugates (such as lipoproteins) have been mainly associated with metal NPs through non-covalent interactions such as electrostatic attraction.

Amyloid peptide fibrils, discussed in more detail in Chapter 2.2, have been proposed as an intriguing drug delivery platform. Fibrillar peptide assemblies exhibit unusually stable and degradation-resistant structures and as such might not seem such favorable candidates for drug transport. However, dissociation and slow release of peptide building blocks has been observed in several amyloid fibril systems, providing a possible drug delivery route. Indeed, one could potentially perceive drug compounds embedded in peptide fibril assemblies that could provide both protection from physiological degradation as well as a slow-release mechanism achieved through the dissociation of fibrils.

Fig. 8.5 depicts a schematic picture of fibril-enabled slow release of bioactive compounds. In that work, R. Riek and colleagues at the Salk Institute assembled fibrils from analogs of the gonadotropin-releasing hormone (GnRH), a broad-activity peptide hormone involved in regulation of growth, reproductive activities, and other functions. The researchers demonstrated that fibers constructed from the hormone units constituted scaffolds for slow release of the peptide. As shown in Fig. 8.5, the release mechanism was based upon slow disintegration of the hydrogen-bonded peptide monomers at the fibrils termini. Indeed, sustained release of hormone monomers was recorded upon injection of the fibrils in animal models.

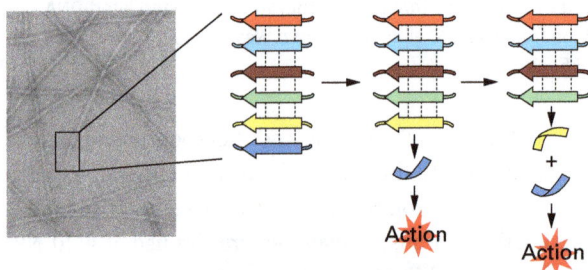

Fig. 8.5: Amyloid fibers as vehicles for delivery and slow release of biomolecules. Micrometer-long fibrils of gonadotropin-releasing hormone (GnRH) comprised of peptide monomers held together by hydrogen bonding. Monomers can be slowly released through disintegration at the fiber terminus, enabling their biological action. Reprinted from Maji, S.K. et al., *PLoS Biology* **2008** *6*, e17. Copyright (2008) Maji et al.

Other classes of biomolecules have attracted interest as potential biomimetic drug transport vehicles. *DNA molecules* can be designed to form a variety of organized structures that might be adapted for drug delivery purposes (see Chapter 9 for detailed discussion on DNA nanotechnologies). DNA-based nanotubes at various diameters and lengths have been synthesized, and some designs might be employed for drug delivery applications. In particular, DNA nanotube scaffolds, such as shown in Fig. 8.6, allow incorporation of guest molecules and exhibit high stability due to the hydrogen-bond-based complementarity between the individual strands forming the structures.

DNA nanotubes exhibit important properties that conform well with the requisites for drug delivery vehicles. First, the nanotubes can encapsulate different molecular cargoes within the enclosed space, stabilized through modulation of the electrostatic and van der Waals interactions with the tube components, or through binding to specific recognition elements embedded within the tube (such as the biotin moieties as shown for example in Fig. 8.6). Furthermore, the modular nature of the nanotubes – i.e. the DNA sequence determining the tube dimensions – provides a means for matching the nanotube to the molecular cargo.

Other advantages of DNA-based nanotubes have recently become apparent. Several studies have shown that the curved, high aspect ratio shape of the nanotubes enable their efficient penetration through the cell membrane and similar physiological barriers. DNA nanotubes do face, however, significant problems for practical therapeutic applications. One of the major issues concerns digestion by nuclease enzymes, a particularly acute problem for exposed DNA molecules. Beside DNA, the *ribonucleic acid (RNA)* world provides possibilities for drug delivery design. RNA *nanoparticles* (NPs), for example, have attracted interest as "multitasking" assemblies comprising RNA components which act as therapeutic cargo, delivery vehicle, and targeting units – all in a single NP. Such applications build upon the remarkable functional diversity of RNA; in particular, since RNA molecules play major *regulatory* roles inside

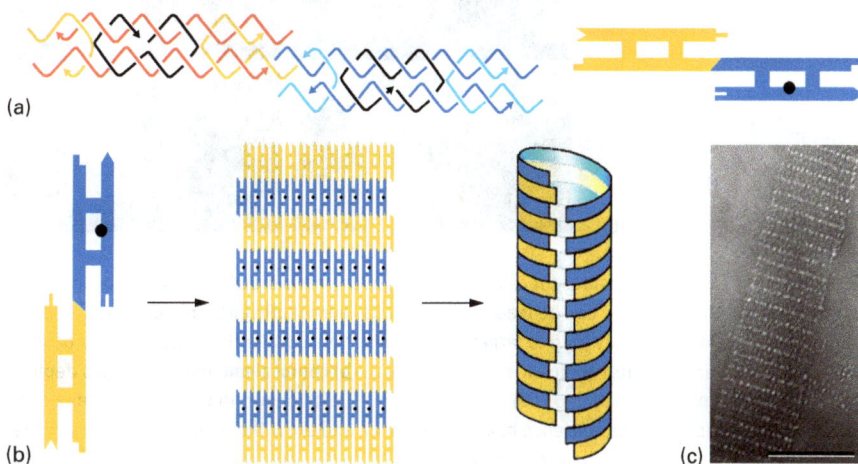

Fig. 8.6: DNA nanotubes. Self-assembled DNA nanotube. **(a)** The nanotube forms through crossover of individual DNA "tiles" comprising double stranded DNA. **(b)** The arrowheads indicate the 3' termini of each DNA strand. The black dot represents a biotin-conjugated site on the DNA strand, enabling in principle coupling of guest molecules through the biotin-avidin binding mechanism. **(c)** Transmission electron microscopy (TEM) image of a typical DNA nanotube. Scale bar is 200 nm. Reprinted with permission from Hamblin, G.D. et al., *J. Am. Chem. Soc.* **2012** *134*, 2888–2891. Copyright (2012) American Chemical Society.

the cell, the use of RNA modules in drug delivery could provide intrinsic triggering mechanisms for externally-controlled release of the therapeutic cargo.

One could conceive RNA particles comprising *small interfering RNA* (siRNA) – an RNA sequence designed to interfere and disrupt gene expression – as the bioactive constituent. Such a particle could also comprise *RNA aptamers* – short RNA fragments selected to recognize and bind to specific peptide and other biomolecular targets. Complementing the particle design, the siRNA and RNA aptamer units can be embedded within interweaved RNA structures serving as a *scaffold* stabilizing the entire particle and its components.

Fig. 8.7 presents an example of a polymeric nanocapsule carrying therapeutic substances (which could be drugs, DNA, siRNA, others) decorated with RNA aptamers as a targeting agent. The generic design, developed by O. C. Farokhzad and colleagues at the Harvard Medical School, utilized covalent coupling of the RNA units onto the capsule surface. The cell uptake experiments shown in Fig. 8.7b demonstrate the *selectivity* achieved by the aptamer units displayed on the particle surface.

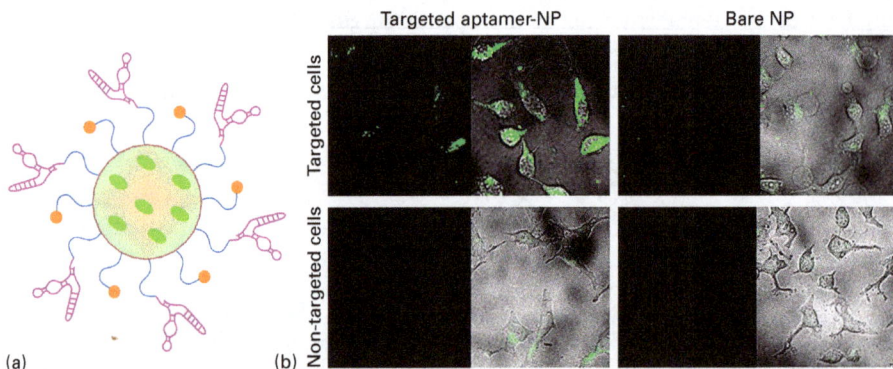

Fig. 8.7: RNA-decorated nanocapsules for cell targeting. **(a)** Schematic drawing of the nanocapsule design. The polymer nanocapsule can be loaded with various therapeutic guest molecules. RNA aptamers, designed to specifically bind to target cancer cells, were covalently attached to the polymer surface through maleimide units (depicted as yellow circles); **(b)** confocal microscopy images depicting cell uptake of the particles, labeled with a fluorescent dye (NBD, shown in green). In all panels the *left* image corresponds to the fluorescence microscopy image and the *right* image is an overlay of the fluorescent and optical images. Cell uptake occurred *only* when particles decorated with *specific* RNA aptamers were added to the *specific* targeted cells (top left panel). Reprinted with permission from Xiao Z. et al., *ACS Nano.* **2012** *6*, 696–704. Copyright (2012) American Chemical Society.

Among the most intriguing bio-inspired strategies for drug delivery are schemes involving the use of microscopic adversaries of the body – viruses and bacteria. Viruses are well-equipped to evade the body defenses (as anyone suffering from a winter common cold would readily attest) due to their small size, bare-bone protein compositions, and overall stealth properties. These properties have attracted biomedical researchers for many years with the aim of using modified viral particles as means for delivering therapeutic molecules to physiological targets.

Probably the most intensive research in virus-based drug delivery has been in the field of *gene therapy*, which has long been considered a sort of "holy grail" therapeutic approach for numerous diseases. The seemingly simple idea behind gene therapy is to treat disease through modifying the genetic content of a cell – either replacing defective (mutated) genes with functional ones, inserting new genes for producing proteins which might substitute or supplement insufficient production of the natural counterparts, or a combination of the two approaches. Despite the touted promise of gene therapy, this technology has yet to demonstrate its potential as several gene therapy clinical trials have failed in recent years.

A critical requisite for an effective gene therapy is the design of appropriate *platforms* (commonly denoted "vectors") for delivering the desired genes into the cell nucleus in which the externally-provided genes would merge and/or insert into the innate genetic material of the cell. This goal has proved quite elusive, likely due to the need of any synthetic or natural vectors to successfully pass through two membrane barriers (the plasma membrane and the nuclear membrane), avoid degradation in the cytosol, contain a targeting mechanism to the nucleus, and maintain integrity and functionality of the DNA cargo until it reaches its gene target.

Viral particles have been among the early and most widely-studied gene carriers. Indeed, "gene delivery" is the core function of a virus in nature – their life-cycle is centered upon delivering and inserting their genes into the nucleus of the host organism. Evolution has accordingly endowed viruses with sophisticated mechanisms to penetrate into the host cell, survive inside it, and eventually replicate their genetic material as part of the host genome. This last feature, in fact, is one of the main reasons viruses have been pursued as vehicles for gene therapy, essentially taking advantage of the simple and generally "promiscuous" viral genetics (which translates into high mutation rates of the viral genome, leading to more effective evasion of the immune system). Accordingly, many attempts have been reported to insert specific genes into the viral plasmid (e.g. genetic material), counting on the virus to deliver its genetic loads into defective nuclei.

The basic gene therapy approach is illustrated in Fig. 8.8. The desired gene (e.g. encoding a functioning protein which is needed for replacing a defective one) is inserted into the viral genome through conventional genetic engineering techniques, with the goal of ultimate incorporation at a specific position in the DNA of the host-human cell. Following infection by the engineered virus, the target gene becomes part of the host cell genome, producing the protein sought through the regular genetic translation machinery. It is also feasible, of course, to *over-express* the protein corresponding to the foreign gene in cases where the gene translating machinery is "hijacked" by the infecting virus. The main disadvantages of viruses as vehicles for gene therapy are the risk of toxicity, systemic pathogenecity, and unintended genetic consequences, such as mutagenesis. Moreover, many viral vectors

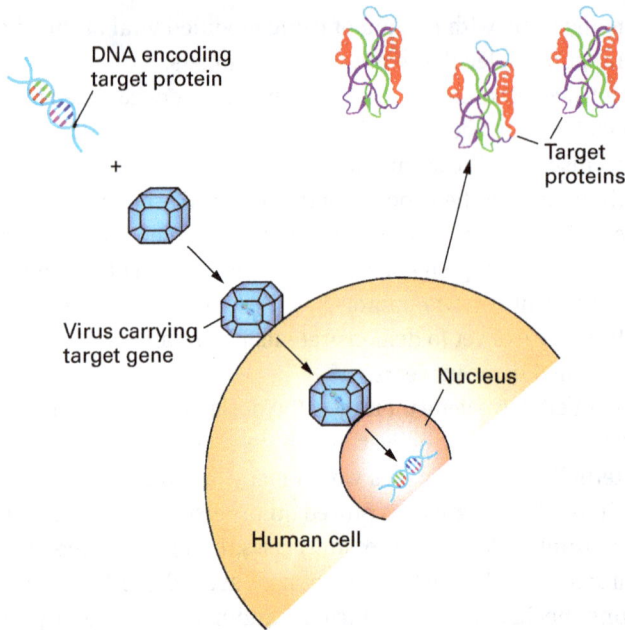

Fig. 8.8: Principle of gene therapy applied via a viral vector. The DNA encoding for the target protein is inserted into a viral DNA, transferred through infection into the genome of the human cell, which produces multiple copies of the protein.

have generated high immunogenicity, and displayed low infectivity and inefficient gene transfer. Indeed, so far the promise of gene therapy through viral vectors has not been realized.

To overcome some of the limitations outlined above, viral vectors for gene therapy applications are usually altered in order to disrupt their natural replication mechanisms and thus eliminate the infectivity hazards associated with their use. Another approach has been to engineer the virus in a way that it would replicate only in specific target cells (for example tumor cells) consequently killing only the infected cells. A related scheme aims to engineer the infecting virus so it amplifies expression of therapeutic proteins within a diseased cell or tissue, leading to targeted cell or tissue death.

Virus-like particles (VLPs) constitute a novel platform for targeted drug delivery and vaccine development (Fig. 8.9). VLP has become a generic description of modified viruses, mostly targeted for drug delivery applications. VLPs that *do not contain* any viral genes have attracted particular interest as drug transport vehicles. Such particles resemble the viral structure – essentially the "capsid" shell (the coat proteins encapsulating the genetic material), but *without* the genetic component. VLPs are assembled through expression of specific viral coat proteins (usually in bacterial expression vectors, but also in mammalian cells and plants, or in "cell-free" systems) programmed to reconstruct the viral organization – without the enclosed DNA and RNA. The *absence*

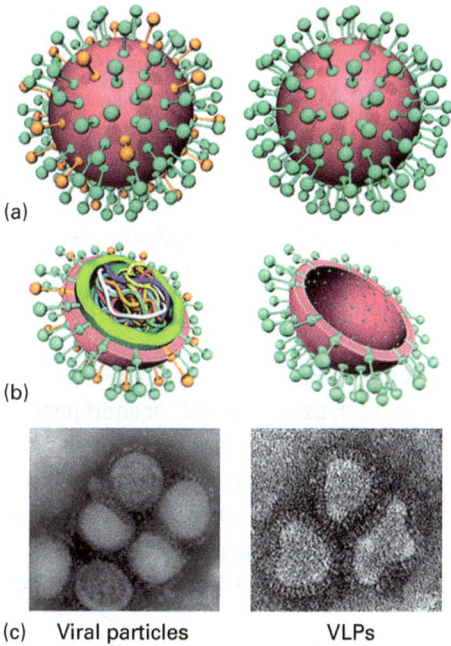

Fig. 8.9: Virus-like particles (VLPs). Influenza virus-like particles; native virus depicted in left column, VLPs in the right column. **(a)** External surface; the native influenza particle displays both the hemagglutinin protein (green) and neuraminidase (yellow), while the influenza VLP in this case displays only hemagglutinin; **(b)** internal cross section; the VLP does not contain genetic material; **(c)** TEM images of the native influenza viral particles and VLPs. Reprinted with permission from D'Aoust, M. et al., *Plant Biotech. J.* **2010** *8*, 607–619. Copyright (2010) John Wiley and Sons.

of genetic material in VLPs consequently reduces (or eliminates altogether) viral infectivity and pathogenicity.

In parallel, VLPs exhibit many of the features that make viruses a highly successful life form and are also valuable in a context of drug delivery applications. VLPs can be produced in large quantity in host cells, and this substantial "copy number" could be employed for delivering a high drug load. Also, varied *genetic manipulation* routes have been developed for expression of different peptides on the viral coat. Such recombinant peptides can be used both for attaching drug cargoes, as well as for targeting the VLPs. The compact, nanometer-scale dimensionality of many VLPs further aids their transport in the bloodstream and in other physiological environments. Another advantage concerns the intrinsic *evasion mechanisms* of viruses, evolved over eons to avoid capture and destruction by the human defenses, which have been exploited by drug designers to achieve unhindered delivery of VLP-associated drugs.

Several virus species, particularly bacteria-infecting *bacteriophages*, are among the most promising candidates for VLP-based drug delivery platforms. The MS2 bacteriophage VLP is a case in point, its solution structure is shown in Fig. 8.10

(calculated by N. A. Ranson and colleagues at the University of Leeds, UK). Like many other phages, MS2 particles can be produced in large quantities using their host bacterial cells as "viral-production factories". MS2 VLPs can self-assemble in the presence of therapeutic cargoes which are subsequently encapsulated inside the viral particle, including for example semiconductor "quantum dots" employed for cell and tissue imaging, small-molecule drugs, peptide and protein toxins, and others. A particularly attractive structural feature of MS2 VLPs is the natural occurrence of *nanosize pores* within the viral external coat, which could facilitate spontaneous release of the embedded drug cargo without the need for complex triggered-release mechanisms.

The spherical MS2 capsid can be modified via chemical derivatization or genetic alteration to display a variety of molecular units, such as amino acids or short peptide sequences, nucleic acids, antibodies, small molecules, and others. In particular, the large diameter of the viral particle makes possible encapsulation of relatively bulky guest molecules, as well as attachment of guest species to the *inner* capsid surface, as compared to the more conventional external display. This delivery scheme is potentially significant, since internally-attached molecules would be less prone to recognition and attack by the immune mechanisms.

Synthetic viruses offer an intriguing alternative for the use of modified indigenous viruses in gene therapy, since they can significantly reduce the risk of pathogenicity. Current efforts in this field are in fact less directed towards modifying the genome of human cells, but rather aim to create artificial viral entities that would infect *bacteria*. The "infected" bacterial cells (which will consequently become genet-

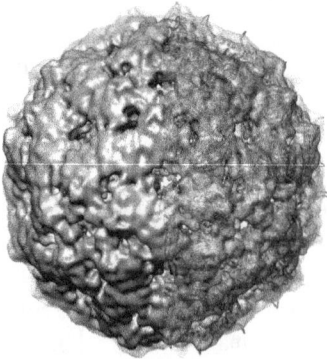

Fig. 8.10: MS2 virus-like particle. Surface view and atomic structure of the MS2 VLP determined by cryogenic-electron microscopy (cryo-EM). The left side shows the cryo-EM density envelope as a solid grey surface. On the right, the atomic coordinates are embedded within a transparent surface. The nanosize pores are apparent as dark spots on the surface. The resolution of the map was 8.9 Å. Reprinted from Toropova, K. et al., The three-dimensional structure of genomic RNA in bacteriophage MS2: Implications for assembly, *J. Mol. Biol.* **2008** *375* 824–836. Copyright (2008), with permission from Elsevier.

ically modified) could be programmed to carry out tasks such as absorbing toxic substances in contaminated environments, or produce proteins and other biomolecules having therapeutic properties. Creation of artificial, synthetic viruses does have a "sinister" aspect, as the technology could be used to assemble pathogenic viruses used in biological warfare.

Non-viral gene carriers, including polymers, biological macromolecules (proteins, glycosaccharides), and nanoparticles, have been also pursued. Such assemblies aim to accomplish the biological benefits of viruses while minimizing immunogenicity and toxicity. Among the different non-viral vectors synthesized, *polymers* have attracted particular interest since they exhibit useful features when compared to both viral vectors and biomolecule-based carriers. Specifically, the chemical properties of polymers can be tailored for optimal drug loading, cell targeting, and extracellular and intracellular "stealth" transport.

Bacteria have long attracted interest as a potential vehicle for drug delivery. Beside their evolved capabilities to evade or overcome natural defense mechanisms, many bacterial species can be turned into miniature "factories" of proteins and other therapeutic substances. Fig. 8.11 schematically illustrates the main approaches pursued for using bacterial cells as vehicles for drug delivery. *Recombinant bacteria* (Fig. 8.11a) are widely used in biotechnology. In the context of drug delivery, bacteria can be genetically engineered to eliminate their toxicity, display specific targeting elements on their surface, and also simultaneously produce therapeutic proteins that could be delivered to the targeted tissue or cells.

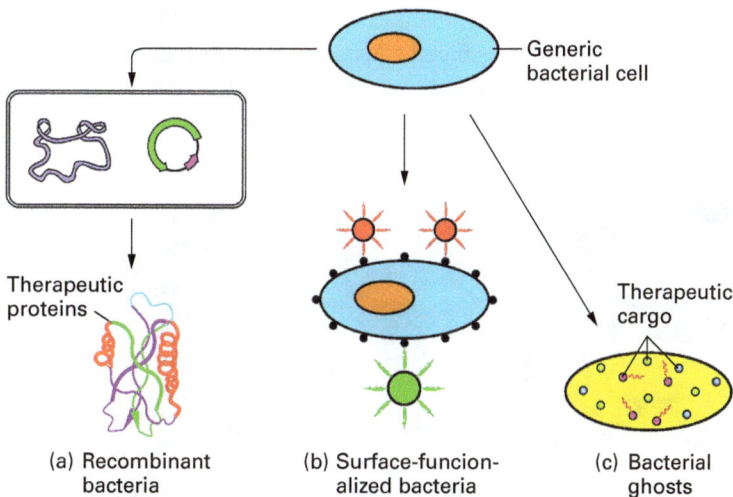

Fig. 8.11: Methods for engineering bacteria for drug delivery applications. **(a)** Using bacteria as an expression system for pr oduction of therapeutic proteins; **(b)** chemical surface functionalization for attachment of therapeutic and imaging agents; **(c)** "bacterial ghosts" produced through removal of the internal cell contents and using the bacterial shell for drug loading.

A second approach has focused on *chemical conjugation* of the bacterial surface, designed to attach to the bacteria molecules having desired therapeutic, imaging, or targeting functions (Fig. 8.11b). An upshot of such schemes has been the generation of *nanoparticle (NP)-loaded bacteria* (sometime coined "microbial robots", or "microbots"). In these systems the NPs are attached, usually through chemical means, to the cell surface (and also internally in some cases). Such bacterially-coupled NPs endow flexibility since chemical cargo can be loaded onto the *NPs*, rather than latching directly onto the bacterial surface. The NPs are subsequently internalized within the target cells or tissues following bacterial infection or capture, resulting in the release of the cargo (genes, diagnostic markers, pharmaceutical molecules). Mice injected with such "microbots", for example, were found to successfully express genes coupled to the NPs.

A third approach involves the use of bacterial *shells* for drug encapsulation and delivery (Fig. 8.11c). The removal of the internal cytoplasmic contents in such "bacterial ghosts" both creates sufficient volume for loading of molecular guests, as well as minimizes the risk of bacterial-induced toxicity. Fig. 8.12 presents electron microscopy images of *Salmonella typhi* ghosts prepared by Y. Zhou and colleagues at the Beijing Institute of Microbiology and Epidemiology, China. Bacterial ghosts, such as the ones shown in Fig. 8.12, are prepared through expression of *lysis-inducing* proteins which digest the internal contents of the bacterial cell. Following washing, the bacterial ghosts can be loaded up with biological molecules and delivered in physiological settings. The microscopy images in Fig. 8.12 demonstrate, in particular, that the overall shape and morphology of the bacterial cells were retained in the ghost cells.

Cells have also attracted interest as potential vehicles for gene therapy. Intrinsically mobile (or "motile") cells, such as blood cells, immune cells (i.e. macrophages, phagocytes), and others, have been pursued as drug delivery carriers. Such cells

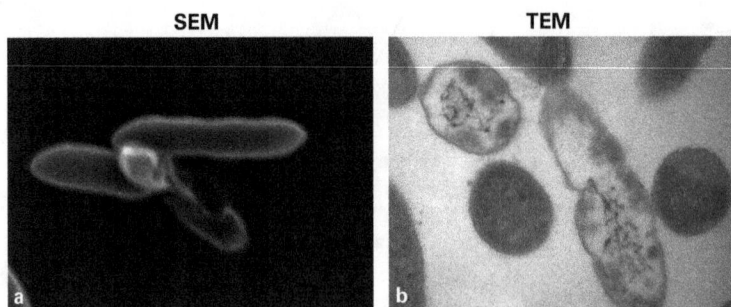

Fig. 8.12: Bacterial "ghosts". Electron microscopy images of *Salmonella typhi* ghost cells. **(a)** SEM, magnification 20K; **(b)** TEM, magnification 24K. Reprinted from Wen, J. et al., Salmonella typhi Ty21a bacterial ghost vector augments HIV-1 gp140 DNA vaccine-induced peripheral and mucosal antibody responses via TLR4 pathway, *Vaccine* **2012** *30*, 5733–5739. Copyright (2012), with permission from Elsevier.

exhibit clear benefits for drug delivery applications, beginning with their having stable membranes enclosing large volume in which therapeutic cargo can be loaded with minimal leakage to the outside. Furthermore, (endogenous) cells are naturally not recognized by the immune system as invaders and thus are inherently biocompatible. Cells can also be targeted to specific tissues or areas in the body through surface display of recognition and targeting units. Finally, well-known chemical and biochemical processes, such as enzymatic hydrolysis and pore formation, can be exploited as triggering mechanisms for releasing the drug cargo once the cell reaches its destination.

Red blood cells (RBCs, or erythrocytes) are ideal biomimetic drug-delivery candidates since it is technically easy to remove their nuclei, leaving behind "erythrocyte ghosts" which can be loaded with desired molecules. Furthermore, RBCs circulate in the bloodstream for days without disintegration, and obviously without being attacked by the immune system – even when loaded with foreign substances. The main problem with the potential use of RBCs as drug carriers is their targeting onto specific tissues or organs. This hurdle is encountered in many other potential drug delivery instruments, however cells, such as RBCs, pose particular challenges since any chemical manipulations (like surface display of recognition elements) carry the risk of significant cell damage and degeneration.

Several schemes have been proposed to address this barrier. Interesting recent contributions have focused on decorating drug-encapsulating RBCs with gold nanoparticles (Au NPs). The overall objective of gold NP conjugation in the context of drug release is the capability to induce local heating at the environment of the Au NPs environment through illuminating with light in the near-infrared (NIR) frequency regime. Once the Au NPs are attached to the surface of the engineered RBC, the controlled release of their drug cargo at a desired target can be accomplished through laser irradiation at the specific location, for example a solid tumor. The cells travel intact in the blood stream; however once they reach the irradiated region, the surface-attached Au NPs are heated and consequently rupture the cell membrane, resulting in localized release of the cell-encapsulated drug. While this approach is therapeutically promising, problems in implementation might occur. For one thing, decorating the RBCs with Au NPs might make them susceptible to attack by white blood cells which might recognize the engineered cells as foreign. In addition, the penetration of IR light is limited through solid tissues.

A similar approach aimed at devising hybrid Au NP/cell constructs involves the encapsulation of Au NPs within *macrophages* (Fig. 8.13). This "Trojan horse" scheme, developed by S. E. Clare and colleagues at Indiana University, relies on the natural propensity of macrophages to "phagocyte", i.e. swallow, foreign objects such as NPs. The NP-containing macrophages can be recruited to infected target areas, such as tumors, consequently delivering the NPs to the site of therapeutic action. Subsequent IR irradiation results in thermal damage to the surrounding cells or tissues (an approach denoted "photothermal therapy").

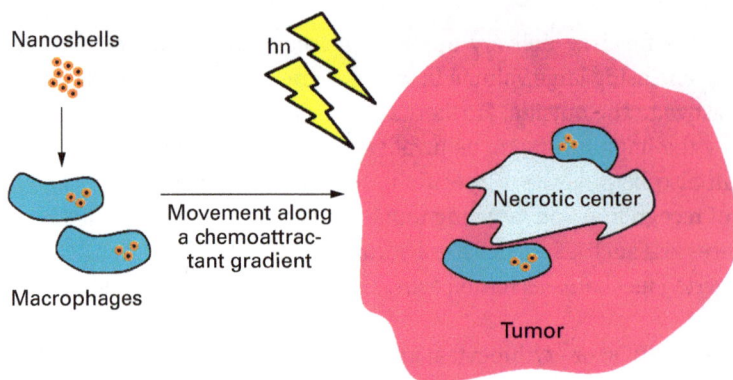

Fig. 8.13: A "Trojan horse" approach for delivering therapeutic nanoparticles. Au "nanoshells" (i.e. Au NPs having silica cores, designed to enhance light absorption in the IR range and consequent particle heating) are embedded within macrophages through phagocytosis; the engineered macrophages accumulate in sites of infection or tumors; light irradiation (IR) results in local heating and cell death. Reprinted with permission from Choi, M. et al., *Nano Lett.* **2007** *7*, 3759–3765. Copyright (2007) American Chemical Society.

Other types of human cells have been examined as drug delivery vehicles and therapeutic agents. *Lymphocytes* (primarily T cells and B cells) are prominent participants in the immune response and like macrophages are intrinsically mobile in the circulation system. Chemical techniques have been developed to conjugate various therapeutic molecular guests onto lymphocyte surfaces thereby using the engineered cells as carriers. Intriguingly, cargoes that have been grafted onto lymphocytes have included not only individual molecules but also more complex assemblies such as drug carrying liposomes, porous polymer matrixes, and others. An interesting example for such schemes is presented in Fig. 8.14. In this work, M. F. Rubner and colleagues at MIT pasted polymer patches (or "backpacks") onto lymphocytes without adversely affecting their viability or biological functions. The combination between the generic cell-surface attachment of the patches through a generic cell-adhesive layer and the minimal effect upon cell functions point to potential broad implementation of the technique for loading various cargoes, underscoring its therapeutic impact.

Stem cells have also attracted interest as potential drug delivery vehicles. It has been found that stem cells are more amenable to expression of therapeutic biomolecules through genetic engineering. Furthermore, stem cell proliferation resembles to a certain extent tumor growth; in consequence, stem cells could respond to chemical cues and signaling molecules released by cancer cells, leading to more effective honing onto tumors as compared to other cell types. Mesenchymal stem cells (MSCs), in particular, are capable of *migration* between different cell populations and tissues, and were shown specifically to be attracted to certain tumors. These features open intriguing avenues for using MSCs and other types of stem cells as "stealth" carri-

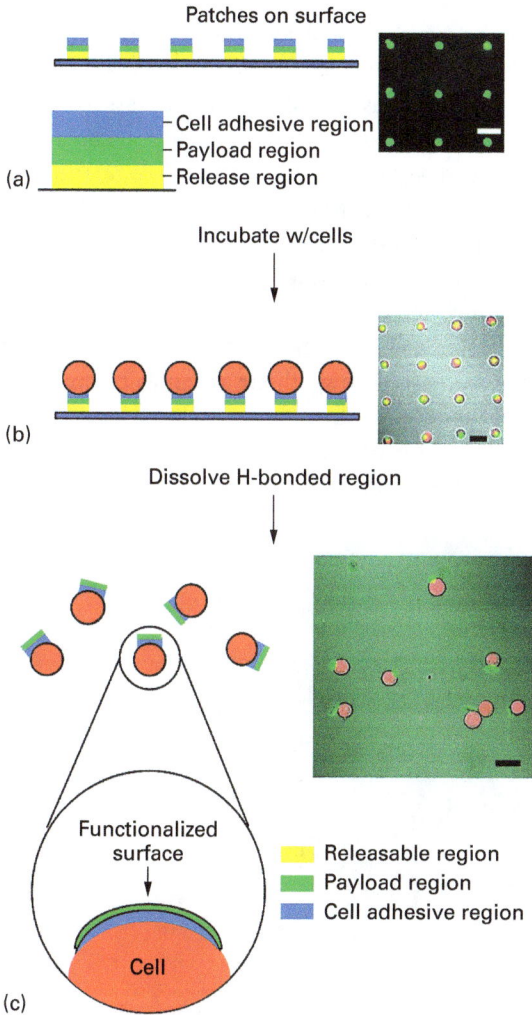

Patches on surface

Cell adhesive region
Payload region
(a) Release region

Incubate w/cells

(b)

Dissolve H-bonded region

Functionalized surface

Releasable region
Payload region
Cell adhesive region

Cell

(c)

Fig. 8.14: Cell-attached polymer patches. Procedure for attaching polymer "backpacks" to the surface of lymphocyte cells. **(a)** An array of polymer patches is deposited on a surface, each patch comprising a cell-attachment region, the payload (a fluorescence dye used in the experiment as a proof-of-concept), and a releasable layer for detaching the functionalized cells from the surface. **(b)** Following incubation of the cells with the surface they bind to the patches, and the cell-polymer constructs are subsequently released **(c)**. All scale bars are 25 µm. Reprinted with permission from Swiston, A.J. et al., *Nano Lett.* **2008** *8*, 4446–4453. Copyright (2008) American Chemical Society.

ers, delivering genes or therapeutic substances to targeted tumors or other disease-inflicted areas in the body.

A different cell-inspired drug delivery strategy is to create completely *synthetic* assemblies that would mimic cell *morphology* and *functions*, rather than employing actual cell entities. This approach has certain advantages, among them control over chemical properties, selection of only the desired molecular features and tailoring them to the actual needs, and less complexity. *Multi-compartmental* synthetic assemblies have attracted particular attention in the drug delivery community. Such systems, discussed also in Chapter 7.1, can encapsulate large quantity of therapeutic compounds. Furthermore, *different* drug compounds can be sequestered in separate compartments within a single particle. The host material can be constructed from biodegradable materials, making possible controlled, slow release of the therapeutic cargo. In many cases, multi-compartmental synthetic assemblies comprise covalently-attached polymer units (e.g. co-polymers) that form internal "pore" structures.

The "protocell" developed by J. Brinker and colleagues at the University of New Mexico embodies the above concept (Fig. 8.15). The synthetic assembly comprises a porous silicon "beehive" structure for loading the drug cargo, enclosed by a biomimetic lipid membrane. The lipid membrane further displays short *targeting peptides* designed to latch onto cell surface receptors and *fusogenic peptides* for promoting

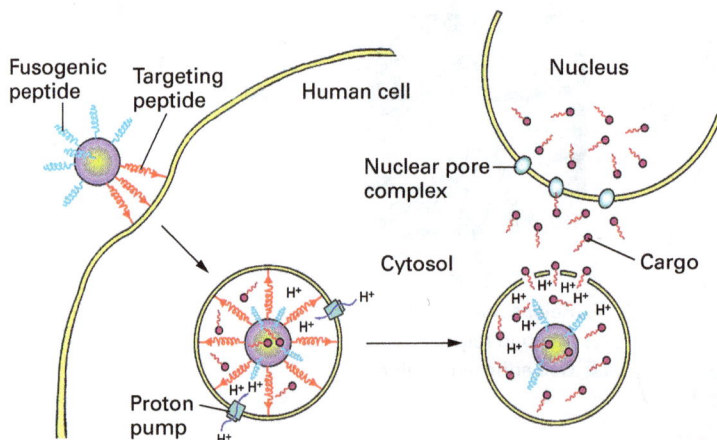

Fig. 8.15: Synthetic "protocells" for drug delivery applications. The drug-delivery construct is made of a porous silicon core within which the therapeutic molecules are loaded. The silicon nanosphere is coated with recognition peptides targeting the desired receptors on the cell surface, and fusogenic peptides for enhanced fusion with the target cell membrane. Following endosomal uptake into the cell, the protocells release their therapeutic cargo. In situations in which the ultimate target is the cell nucleus (i.e. gene therapy), the delivered molecules can further contain *nuclear localization sequences* (NLSs) for transportations through pores in the nuclear membrane.

fusion of the protocell lipid coat with the cell membrane, and subsequent uptake of the protocell cargo into the cytosol through endosome formation. The protocell-delivered payload can act on targets within the cytosol, and it could be further directed into the nucleus for gene therapy applications through inclusion of peptides promoting transport through the nuclear membrane (i.e. "nuclear localization sequences", Fig. 8.15). The protocell design in Fig. 8.15 will likely spawn more research; it is conceivable that synthetic cell-like particles could be prepared that, in addition to enabling sequestering of therapeutic substances, would integrate other cell properties such as dynamic shape, elasticity, and surface display of receptor units, to achieve highly effective and targeted drug delivery.

Biomimetic and bio-inspired design of drug-delivery systems has shown great promise, and this field is poised for further growth in the future. Indeed, the breakneck pace of R&D aimed at devising new ways to deliver bioactive molecules to their physiological targets will likely remain a major driving force for new discoveries and progress. Biomimetic drug carriers, from individual biomolecules to complex molecular assemblies and entire microorganisms and cells, will continue to be enriched from scientific research at the confluence of materials science, chemistry, and biology.

9 DNA and RNA nanotechnology

For a field that is barely 20 years old, DNA nanotechnology, and the more recent new-comer RNA nanotechnology, have achieved almost a "cult status" among nanotechnology researchers. While practical applications are still in their infancy, DNA nanotechnology highlights the core features, potential, and significant challenges of nanotechnology as a new discipline. While there has been an explosion of reported nanostructures made of genetic building blocks, practical utilization of this thriving nanoworld is still likely far off. Nevertheless, the modular nature of DNA and RNA has opened up broad new areas of research.

9.1 Basics of DNA nanostructures

DNA, through the complementary binding in the two nucleotide base-pairs, displays high propensity for self-assembly. The pairings of adenine (A) – thymine (T) and cytosine (C) – guanine (G) are the basis of the genetic code, and the determinants for the structural framework of the *double helix* organization of the DNA molecule (Fig. 9.1). Dramatic developments in recent years, however, underscore the fact that this base-pair complementarity can be exploited for creation of highly complex *non-biological* nanostructures with diverse functions.

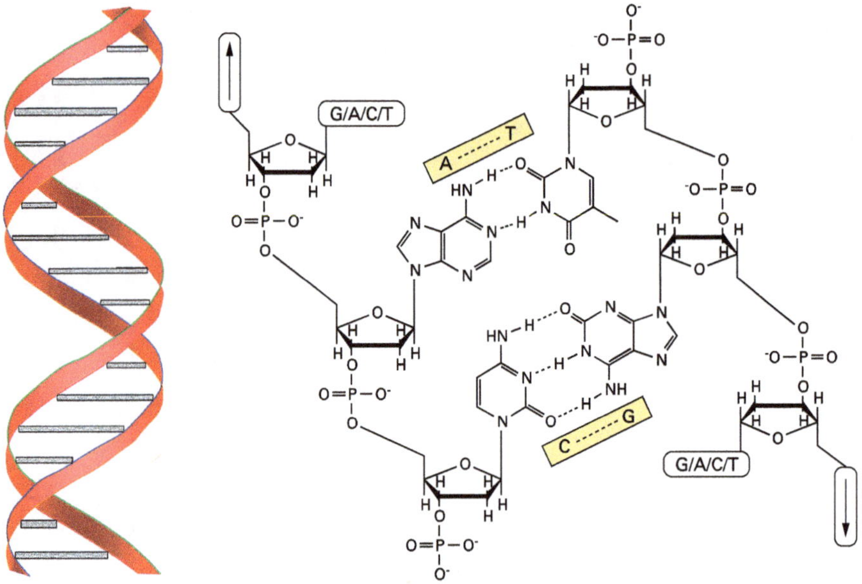

Fig. 9.1: DNA structure. Left: The DNA double helix; **right:** The T-A and G-C base-pairs assembled through hydrogen bond formation comprising the DNA backbone.

https://doi.org/10.1515/9783110709490-009

"DNA robots" combine the molecular recognition properties of DNA with mechanical motion in the nanoscale. Several variations of this concept have been reported. "DNA walkers" (Fig. 9.2) can travel on linear tracks or configured surfaces displaying single-strand DNA having predetermined sequences. The walker's "legs" comprise free (i.e. "non-hybridized") DNA strands. As illustrated in Fig. 9.2, the first "step" is induced through addition of a DNA "snippet" (A1) which exhibits two "sticky ends" – one complements the non-hybridized nucleotide sequence of one "leg" of the DNA walker, while the other end binds one of the surface-displayed strands. Subsequent addition of another sticky-end DNA snippet (A2) results in binding of the second leg to an adjacent surface-displayed strand.

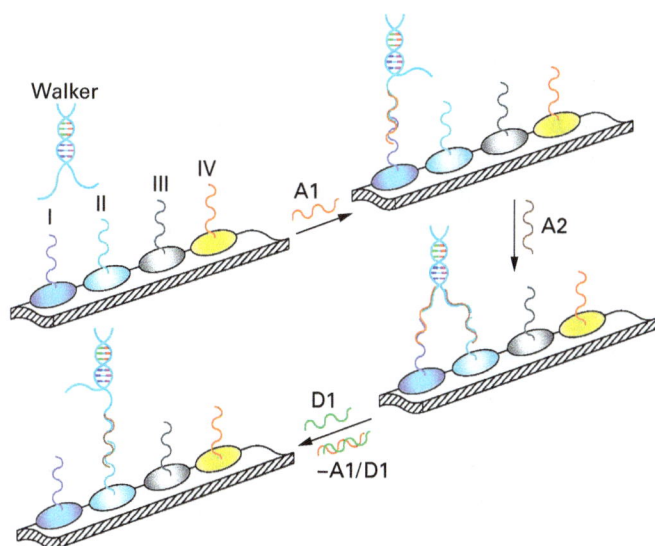

Fig. 9.2: The DNA walker. "Frayed" DNA double-strand moves along a surface track decorated by DNA strands through hybridization and dehybridization steps, see text for details.

A critical step of "lifting" one leg of the walker is subsequently accomplished through addition of another DNA snippet (D1) which exhibits higher affinity to the A1 fragment, consequently releasing the leg of the walker due to dehybridization. The liberated leg is then directed to the next surface-attached strand through supplementing a different sticky-end DNA fragment which complements an adjacent strand, and so on and so forth. Such DNA walkers have captured the imagination of researchers in recent years as means for transportation in nanospace; there have been several exciting reports in recent years of employing DNA robots for mechanically moving molecular cargoes and assembling complex structures on surfaces.

Although it took the labor of dozens of scientists, several Eureka moments, and a Nobel prize, the basic double helix structure of DNA in human (and generally eukaryotic) cells, in which two strands of complementary nucleotides form a screw-like structure, is quite straightforward from a nanotechnology point of view. A crucial observation that has led to the explosive growth in the field of DNA nanotechnology has been the fact that in certain points in the cell life-cycle, specifically during the crucial replication process, the linear double helix unzips in order to bind to complementary strands, forming separate branched strands. Indeed, by tweaking the DNA base-pair sequence, one could create varied DNA geometries. This branching is the core feature allowing the construction of, for example, squares, triangles, cubes, pyramids, and a myriad other structures.

Fig. 9.3 presents models of triangular and square DNA nanostructures assembled by K. V. Gothelf and colleagues at Aarhus University, Denmark. These DNA shapes have been created through programming of the DNA sequences, in particular the complementary strands anchoring the frayed DNA at the corners of the geometrical objects. The genesis of DNA nanotechnology has been, perhaps expectedly, difficult. It took considerable experimental work to achieve effective couplings between the *single-stranded* DNA tails (the binding elements) of complementary DNA units. A related hurdle to overcome has been the general "floppiness" of DNA, particularly the unbound strands, which makes it difficult to achieve stable structures.

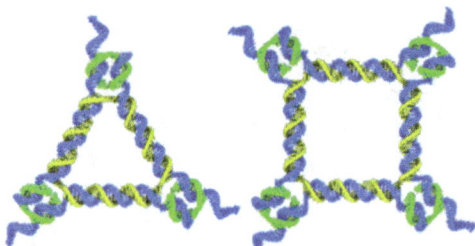

Fig. 9.3: Non-biological DNA geometries. Models of synthetic DNA structures assembled through strand complementarity. The *yellow* and *blue* strands form double helix structures which constitute the sides of the geometric shapes; the *green* strands serve as the "locks" at the square and triangle corners, respectively. Reprinted with permission from Kristiansen M. et al., *ChemPlusChem*, **2012** *77*, 636–642. Copyright (2012) John Wiley and Sons.

9.2 DNA origami

DNA origami is a recent newcomer to the field of DNA nanotechnology that could dramatically expand its scope. The innovative idea behind this technique, first introduced less than 10 years ago by P. Rothemund at CalTech (who, not coincidentally, is a computer scientist) is the use of a computer-aided design to "fold" a long DNA sequence into organized structures (Fig. 9.4). DNA origami design begins with a *single-stranded* DNA having a known sequence, usually extracted from a virus or a bacterial cell. The thrust of the technique is that addition of short DNA fragments (denoted "staple strands"), predesigned to bind at specific areas within the viral (or bacterial) genome sequence, would induce rearrangement of the long strand into desired stable structures and shapes.

Since its inception, DNA origami experiments have generated numerous complex structures, expanding quickly from two dimensions into the three-dimensional realm. Fig. 9.5 shows an example of a tetrahedron constructed by H. Yan and colleagues at Arizona State University via a DNA origami design. Essentially, the four triangular faces of the tetrahedron are assembled from a single viral "scaffold" strand held in place by numerous staple strands. The staple strands were programmed not to bind at the tetrahedron "edges", enabling folding of the four faces into a closed-face "molecular container". The transmission electron microscopy (TEM) images in Fig. 9.5 confirm that stable tetrahedron "cages" are indeed formed.

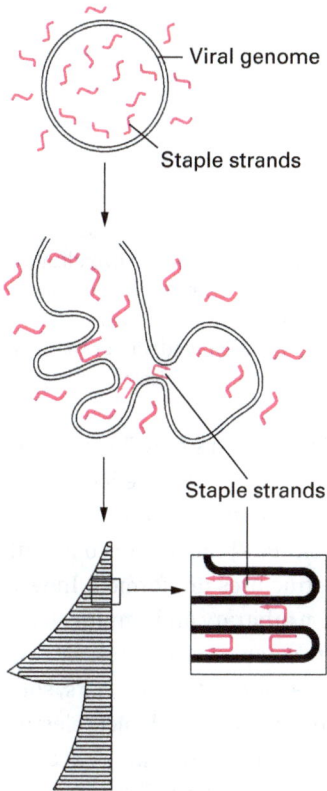

Fig. 9.4: The DNA origami concept. Single-stranded DNA sequence (extracted from a virus or bacterium) is mixed with short DNA "staple strands". The staple strands bind the long single-stranded DNA at different locations through sequence complementarity, consequently folding the long strand into organized structures.

(a)

(b)   50nm

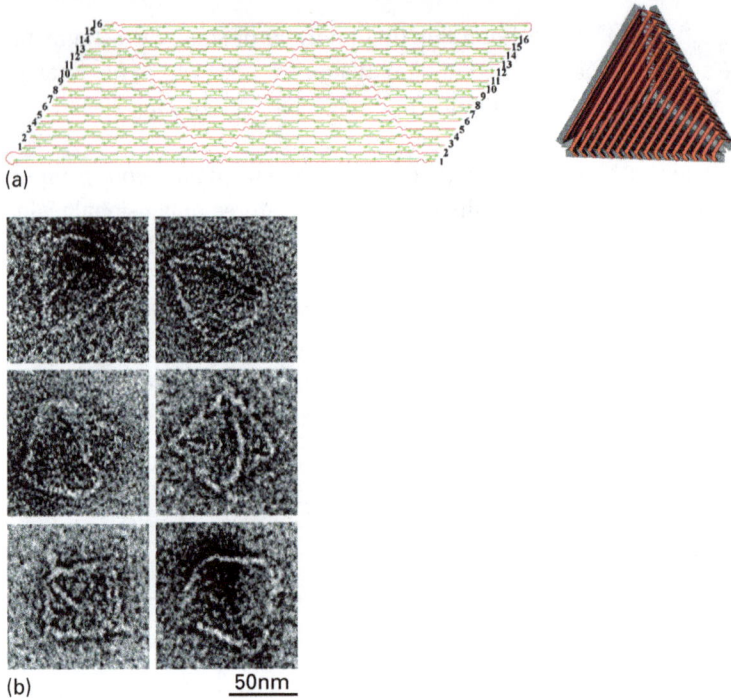

Fig. 9.5: Tetrahedron assembled through DNA origami. **(a)** Schematic drawing of the four faces comprising a long DNA single strand of the M13 virus (red), held in place through local hybridization with the staple strands (green). Note the numbering on the left and right indicating the continuous pathway of the strand. The folded tetrahedron is shown. **(b)** Transmission electron microscopy (TEM) images of tetrahedron containers formed by the DNA origami technique. Reprinted with permission from Ke, Y. et al., *Nano Lett.* **2009** *9*, 2445–2447. Copyright (2008) American Chemical Society.

Technically, DNA origami processes often rely upon a "witch's brew" approach – mixing of the entire "DNA blend" (containing also appropriate ions needed to compensate for the negative charge on the DNA molecules) and application of heating/ cooling cycles (similar to the polymerase chain reaction [PCR] procedure commonly employed for DNA amplification) until the designed structures are formed. Indeed, DNA origami experimental conditions occasionally seem arbitrary and empirical, and this has been an impediment for progress towards more complex structures.

Among the many beautiful DNA origami structures reported in recent years, some are quite unique and underscore the versatility of the concept. Fig. 9.6 depicts electron microscopy images of nanoscale *Mobius strips*, constructed by H. Yan and colleagues at the Arizona State University using DNA origami. Mobius strip is a well-known topological ribbon-like structure that has only *one side* which can be prepared through a 180° turn at the strip central axis. The DNA origami Mobius strip was created through assembling a flat surface comprising 11 DNA double strands. Through appropriate positioning of the short staple strands, one can induce the 180° turn around the

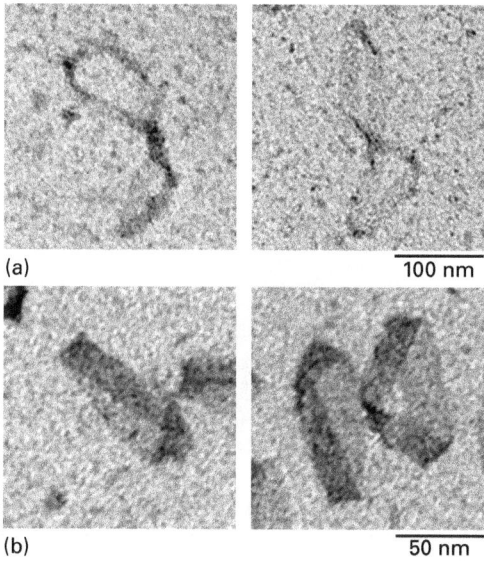

Fig. 9.6: *Mobius rings* through a DNA origami scheme. TEM images of DNA origami Mobius rings. Images courtesy of Professor Y. Han, Arizona State University.

central axis of the flat DNA plane, subsequently connecting the two open ends to form the Mobius strip.

Beside the creation of intricate structures, the DNA origami concept has inspired diverse applications. *Nanorulers*, recently demonstrated by P. Tinnefeld and colleagues at the Ludwig-Maximilians-University, Munich, provide a fine and rather conceptually simple example (Fig. 9.7). In this experiment, a DNA rectangle is assembled through the origami concept. Staple sequences containing fluorescent dyes are also included in the reaction mixture, located at predefined positions within the rectangle – with specific (and known) distance between them. The fluorescent nanoruler can subsequently be placed on surfaces, together with samples for which distances need to be determined. Subsequently, high resolution fluorescence microscopy techniques can be employed for precise distance measurements, in spatial resolution that can be even shorter than the light diffraction limit (the well-known limitation in conventional light microscopy).

Two-dimensional DNA origami can be employed as a *template* in nanolithography applications. At the core of modern lithography as applied to electronic or electro-optic device design is the capability to organize metallic structures in precisely defined patterns. In this context, DNA origami schemes have been developed in which gold (or other metallic) nanoparticles (NPs) are coupled to the staple strands (feasible through relatively straightforward derivatization of the NP surface), which enable specific positioning of the NPs on the origami surface through the complementary binding of the staple strands. Folding of the origami into the predesigned two-dimensional structure consequently yields organized metallic surface struc-

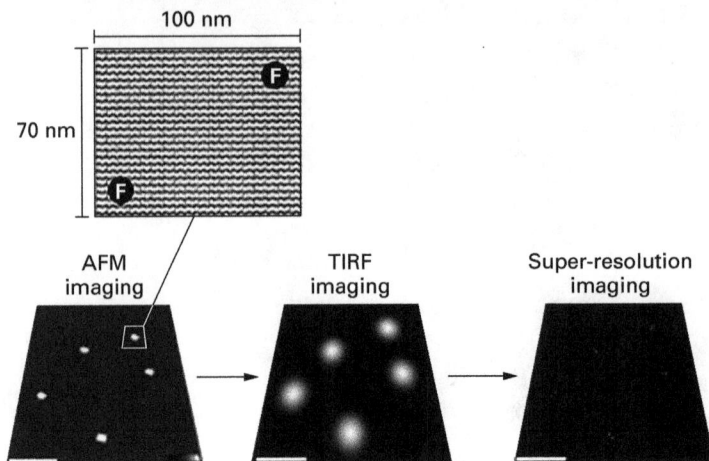

Fig. 9.7: DNA origami ruler. Two fluorophores attached to short staple DNA sequences are placed on a plane prepared through a DNA origami scheme. The origami-assembled plane is then placed on surfaces in which distances are to be measured, and the distance between the fluorophores is resolved through super resolution fluorescence microscopy. Reprinted with permission from Steinhauer, C. et al., *Angew. Chem.* **2009** *48*, 8870–8873. Copyright (2009) John Wiley and Sons.

tures. DNA origami-based nanolithography can be also accomplished via chemical means, using the DNA structures as templates for metal binding *after* the origami assembly.

An interesting system interfacing nanolithography and DNA origami is depicted in Fig. 9.8. The experiments, carried out by Y. Han and colleagues, address another notable challenge: how to integrate DNA origami, which is carried out in *solution,* with lithography – which is an intrinsically *surface* manipulation technique. In the scheme illustrated in Fig. 9.8, surface features are etched through conventional nanolithography. Within these features, DNA-binding chemical moieties (such as silanol units) are displayed through surface chemical coupling. Exposing this functionalized surface to the DNA origami solution leads to specific binding of the DNA assemblies, such as origami-produced DNA nanotubes. The combined lithography and DNA origami approach enables fabrication of organized films, using the DNA structures as templates for deposition of other materials.

Integration between the DNA origami concept and metal NPs might open new avenues for fabricating programmable *three-dimensional* metal structures. Fig. 9.9 depicts an elegant experiment illustrating this approach, carried out by B. Ding and colleagues at the National Center for Nanoscience and Technology, Beijing. First, a DNA origami rectangle is produced. The rectangle is designed to display *linear chains* of free oligonucleotide strands on one side, onto which Au NPs are captured through attachment of the complementary DNA sequences. Subsequent addition of short DNA "folding strands" folds the rectangle into a tube-like configuration, yielding helical Au NP structures exhibiting optical chirality.

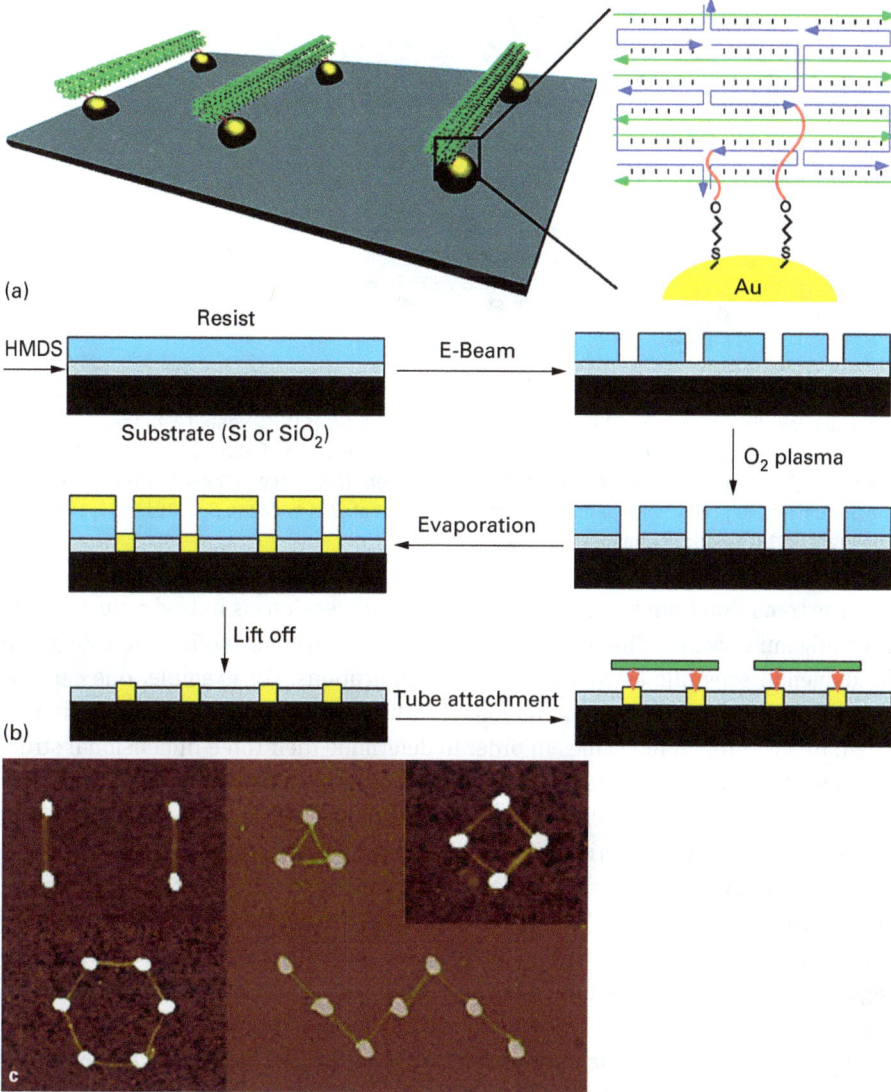

Fig. 9.8: Surface structures through an integrated DNA origami/lithography approach.
(a) Connecting the Au islands through DNA origami nanotubes. The DNA long single strand is shown in *green*, while the staple strands are depicted in *blue*. Binding the DNA tubes and the gold is achieved through thiol moieties at the ends of the DNA strands; **(b)** schematic illustration of the lithography scheme for preparation of the gold island pattern. HMDS and PMMA polymer resist layers were deposited on Si or SiO$_2$ substrate. An electron-beam created the designed pattern and O$_2$ plasma treatment was used to remove the exposed HMDS within the surface pattern. Cr/Au layer was subsequently deposited through evaporation. After lift-off with acetone, a patterned array of Au islands was produced, with the background covered by HMDS. **(c)** SEM images showing DNA nanotube patterns produced by the technique. Reprinted with permission from Ding, B. et al., *Nano Lett.* **2010** *10*, 5065–5069. Copyright (2010) American Chemical Society.

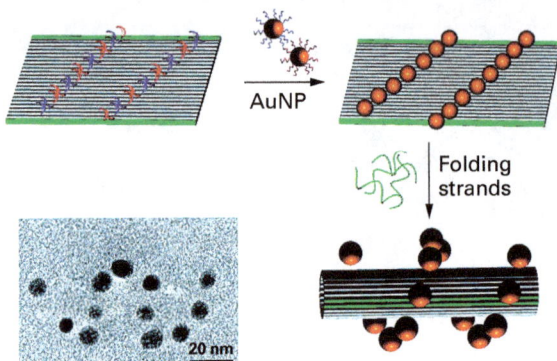

Fig. 9.9: Nanoparticle helical structures through DNA origami approach. DNA origami rectangular plane displays linear arrays of DNA binding sites. Au nanoparticles (NPs) coated with the complementary strands create linear NP tracks. Subsequent addition of DNA "folding strands" folds the rectangle into a tube, forming a helical NP configuration. TEM image of the NP helix is shown at the bottom left. Reprinted with permission from Ding, B. et al., *J. Am. Chem. Soc.* **2012** *134*, 146–149. Copyright (2012) American Chemical Society.

The transition from two dimensions to three dimensions is indeed natural in the DNA origami universe. Three-dimensional DNA nanostructures offer interesting and fundamental scientific as well as practical applications. For example, one can use three-dimensional DNA origami templates, whose sizes and shapes can be modulated, to host protein molecules in order to determine their three-dimensional structures. Such a "molecular vise" could also provide a crystallization platform, particularly useful for proteins which do not readily crystallize freely.

Applications of DNA origami structures have been proposed in drug delivery. G. M. Church and colleagues at the Harvard Medical School have recently designed a sophisticated origami-based "container" structure that can enclose molecular cargoes (Fig. 9.10). The use of a DNA origami platform for such purposes is particularly advantageous since the DNA container structure can further include short oligonucleotide recognition elements called "aptamers" which can be designed to bind to specific targets on diseased cells (for example cancerous cells). The container can be further designed to initially use the aptamers as "locks"; following recognition and binding to their molecular targets (such as antigens on the cell surface), the aptamers "open up" the container, releasing the therapeutic cargo. These "search and destroy" DNA origami drug carriers would be highly specific, and potentially employed for targeted delivery of drugs to different tissues and cells.

DNA origami containers can be also "locked" and opened with specific "keys". Fig. 9.11 presents such a "lock and key" system designed by J. Kjems and colleagues at Aarhus University, Denmark. The scientists produced a DNA origami box (containing two fluorescence dyes for evaluating intermolecular distance through energy transfer). Crucially, the "lid" of the box was kept close by two complementary oligonucleotide base-pairs, of which two strands (out of the four) presented free "sticky ends"

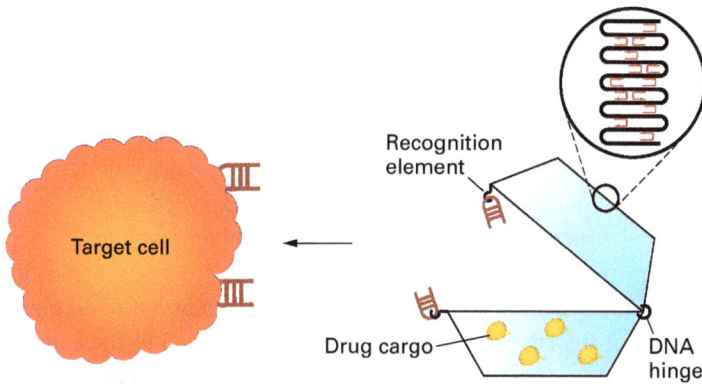

Fig. 9.10: DNA origami drug delivery. Three-dimensional DNA origami container hosts therapeutic molecules. The container is "locked" by short "latches" comprising DNA antigen-targeting strands (i.e. "aptamers"). Following binding of the aptamers to their molecular targets (for example on cell surfaces), the DNA origami capsule releases the therapeutic cargo.

for further binding of complementary sequences, denoted "displacement strands". Accordingly, when a displacement strand was added to the box, it attached to a lock strand thus "unlocking" the box and opening the lid.

The lock-and-key mechanism depicted in Fig. 9.11 can have interesting applications in the context of *logic gate* design. Note that the lid of the nano-box in Fig. 9.11 can be opened through the action of *two* keys (i.e. two displacement sequences). This means that box opening can be designed to occur *only* through interaction with *two distinct* external signals (e.g. displacement sequences) – essentially the "AND" logic operation. Similarly, other sequences can be designed to affect lid *closing* (after is has opened) – representing a "NOT" gate. And since the DNA origami box can contain *two*

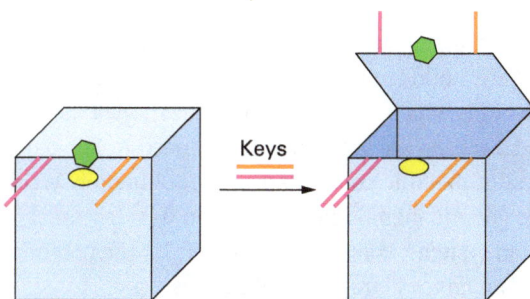

Fig. 9.11: DNA origami box with a "lock-and-key" opening mechanism. DNA origami is employed for constructing a box, further containing two complementary sequences acting as "locks" on the lid (shown in purple and orange, respectively). Following binding between the free DNA strands and their complementary displacement strands (the "keys"), the lid is being opened. The distance-dependent energy transfer between the two fluorescent dyes (yellow oval and green hexagon circle) is used for monitoring lid closing and opening.

rather than one lid, one can achieve in principle box opening through either lid – an "OR" operation.

An important issue with practical significance for future utilization of DNA origami constructs (and large-scale DNA-based nanostructures in general) in biomedical applications concerns the biocompatibility and overall physiological stability of these structures. Indeed, it has been observed somewhat unexpectedly that complex 2D and 3D DNA origami structures are stable in physiological environments. In particular, DNA nanostructures did not disintegrate even when placed for long durations in cell lysates – which contain salts, enzymes, and other biologically degrading agents. The enhanced stability of the DNA structures is generally ascribed to the extended organized structural features (primarily facets) of the origami assemblies, which confer resistance to enzymatic degradation.

9.3 DNA-based advanced materials

Natural DNA can be exploited as a "templating" agent in materials design. This line of research is based upon the fact that both DNA double strands and single strands possess remarkable means for ordering and orientation, either through physical means (e.g. base-pair complementarity), but also via chemical modification pathways. Indeed, oligonucleotides exhibit remarkable flexibility in terms of chemical derivatization, allowing attachment of varied functional units, such as *thiols* (for coupling to gold substances, for example in conducting electrodes), *silanes* (allowing conjugation with glass surfaces), *amines*, and others. Chemically speaking, one can perceive DNA molecules as programmable polymers, whose structures can be modulated in the nanoscale through the intrinsic information-bearing features of oligonucleotide complementarity.

Fig. 9.12 depicts a representative experiment for using an elongated DNA framework as a template for fabrication of a metal nanowire. The experiment, designed by Q. Gu at Pacific Nanotechnology Inc., builds upon electrostatic attraction between the negatively charged DNA and Pd^{2+} ions which results in immobilization of the ions upon the DNA template. The palladium ions are subsequently reduced to Pd^0 particles which serve as nucleation sites for metallic *cobalt* deposition, ultimately yielding a continuous Co nanowire. This generic methodology has been demonstrated in varied metal ion and metal deposition systems, comprising the basis for the potential use of DNA in *molecular electronics* and *nanoelectronics* applications.

DNA has been used in other "nominally" non-biological devices. Intriguing applications in which DNA has been utilized as a biomimetic "antenna" have been reported. Solar cells and other photonic devices would increase their efficiency if artificial units – i.e. *molecular antennae* – were available to absorb incoming light and funnel it through appropriate circuitry towards a storage device. Fig. 9.13 illustrates an original platform developed by S. O. Kelley and colleagues at the University

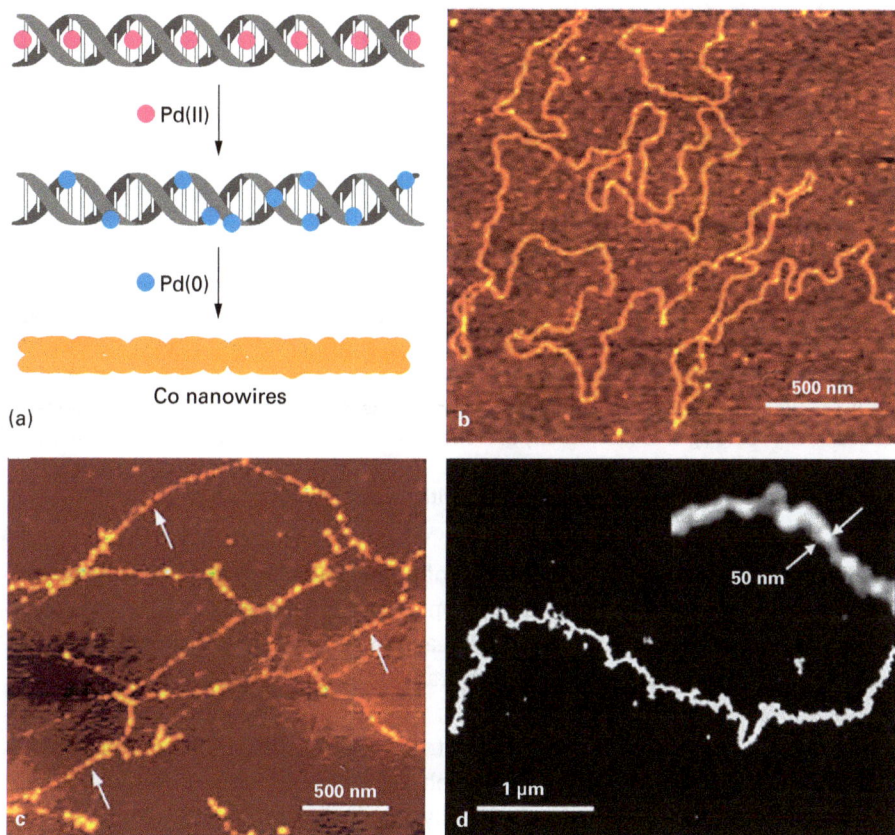

Fig. 9.12: Metal nanowires via DNA templates. **(a)** Scheme of the nanowire deposition method (see details in the text); **(b)** atomic force microscopy (AFM) image showing the starting DNA wire; **(c)** AFM image of the DNA wire after reduction of the embedded palladium ions; individual Pd nanoparticles are apparent; **(d)** scanning electron microscopy (SEM) image of the continuous cobalt nanowire, the inset shows a magnified view (width of the wire 50 nm). Scales in (b) and (c) are 500 nm, in (d) 1 μm. Reprinted from Gu, Q. and Haynie, D.T., Palladium nanoparticle-controlled growth of magnetic cobalt nanowires on DNA templates, *Mater. Lett.* **2008** *62* 3047–3050. Copyright (2008), with permission from Elsevier.

of Toronto in which semiconductor *quantum dots* were linked through complementary DNA strands, forming a programmable light absorbing and emitting system. The optical properties of quantum dots are highly dependent both upon the *sizes* of the individual dots, as well as the assembled complexes. Specifically, the researchers showed that quantum dots having different *sizes* and *valency* — from one to five DNA strands per quantum dot — constituted building blocks for a variety of geometries of quantum dot complexes exhibiting distinct optical properties.

A different type of DNA-scaffold antenna has been used for mimicking *photosynthesis*. Creating synthetic systems that would imitate photosynthesis is a formidable

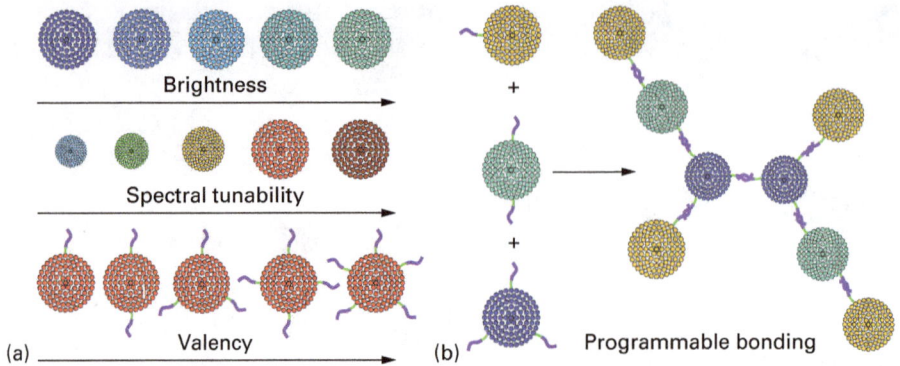

Fig. 9.13: DNA-mediated assembly of quantum dot complexes: CdTe quantum dot (QD) complexes formed through binding of complementary DNA strands covalently attached to the QD surface. The use of distinct quantum dot building blocks, having different sizes, brightness, and number of DNA strands per QD (i.e. valency) produces complexes with different optical properties.

challenge. The task involves the creation of a molecular system which allows unidirectional energy transfer from energy donors (e.g. pigments in natural photosynthetic systems) to an acceptor (e.g. the "reaction center") which converts the light energy into chemical energy. The energy transfer in photosynthetic pathways often occurs through elaborate molecular routes, with the added requirement of minimal energy losses. Laboratory-designed DNA structures provide two important benefits for artificial photosynthesis. First, chromophores acting as energy absorbers (acceptors) and emitters (donors) can be chemically attached to the oligonucleotide strands at different locations. Second, the DNA assemblies can be programmed to form three-dimensional scaffolds in which the donors and acceptors will be precisely positioned to allow efficient energy transfer towards an embedded artificial reaction center.

Fig. 9.14 shows a DNA scaffold for artificial photosynthesis developed by Y. Liu and colleagues at the Arizona State University, Tempe. The structural module employed consists of a "seven-helix bundle" (each cylinder in the figure corresponds to a distinct DNA helical structure). Energy-wise, attaching the designed molecular energy donors and acceptors in strategic locations within the DNA bundle motif resulted in efficient energy transfer throughout the assembly. Specifically, light excitation traveled from a "primary donor" positioned at the helix bundle periphery, through an "intermediate donor" attached closer to the helix core, and finally towards a single acceptor at the central helix. Importantly, the three-dimensional organization of the DNA assembly was the thrust of the "antenna" and its efficient energy harvesting.

Development of DNA and nanoparticle (NP) conjugates (Fig. 9.15) constitutes an interesting and practically applicable platform in biomimetic DNA-based materials. The concept, first introduced by C. A. Mirkin and colleagues at Northwestern University, is simple: Two *non-complementary* oligonucleotide strands are covalently bonded to the

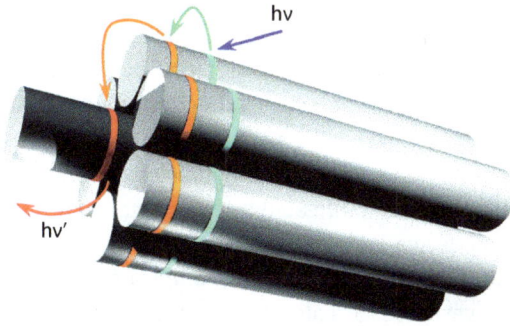

Fig. 9.14: DNA light harvesting antenna. Unidirectional energy transfer in a seven-helix DNA bundle containing three chromophore arrays at different positions. Light initially absorbed by the *primary donors* (green); energy subsequently transferred to the *intermediate donors* (orange), and on to the acceptor dye at the central DNA helix (red). Reprinted with permission from Dutta, P.K. et al., *J. Am. Chem. Soc.* **2011** *133*, 11985–11993. Copyright (2011) American Chemical Society.

surface of two different populations of gold NP through thiol-based chemical linkage. The two Au NP populations do not interact with each other. However, linking of the different DNA-derivatized Au NPs can be achieved following addition to the mixture of a DNA duplex (double strand) exhibiting free "sticky ends" at each end, corresponding to the respective *complementary* sequences of the two NP-attached oligonucleotide strands. The "double hybridization" process (i.e. binding of complementary strands on each side of the DNA duplex) is thermally reversible and allows creation of multi-scale supramolecular assemblies of Au NPs, with consequent modulation of spectroscopic and optical properties related to the aggregation state of the NPs. This elegant method has been among the first demonstrations of DNA-mediated hierarchical assemblies of metal nanoparticles.

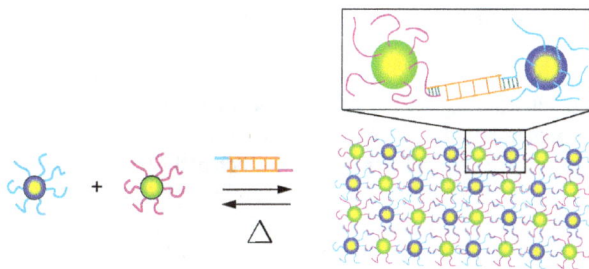

Fig. 9.15: Supramolecular nanoparticle assembly mediated by DNA complementarity. Two types of NPs are coated with different single-stranded DNA (**left**); addition of a *double-stranded* DNA having two "sticky ends" corresponding to sequences that are, respectively, complementary to the NP-displayed strands, results in binding of two NPs; optimization of the double-stranded DNA concentration produces an organized three-dimensional NP array.

Coupling between Au NPs and single-stranded DNA can be exploited for creating "programmable" geometries. In an imaginative experiment designed by J. Kim and colleagues at the University of Arkansas, specific numbers of DNA strands were covalently attached to Au NPs (Fig. 9.16). Steric hindrance stipulated distinct symmetries of the strands attached to each NP, producing DNA-NP building blocks for construction of modular structures. Specifically, NPs were fabricated with discrete "coordination numbers", i.e. number of DNA strands available for coupling to complementary strands. As depicted in Fig. 9.16b, addition of Au NPs displaying *complementary* DNA sequences to the DNA building blocks gave rise to "colloid complexes" having programmable geometries.

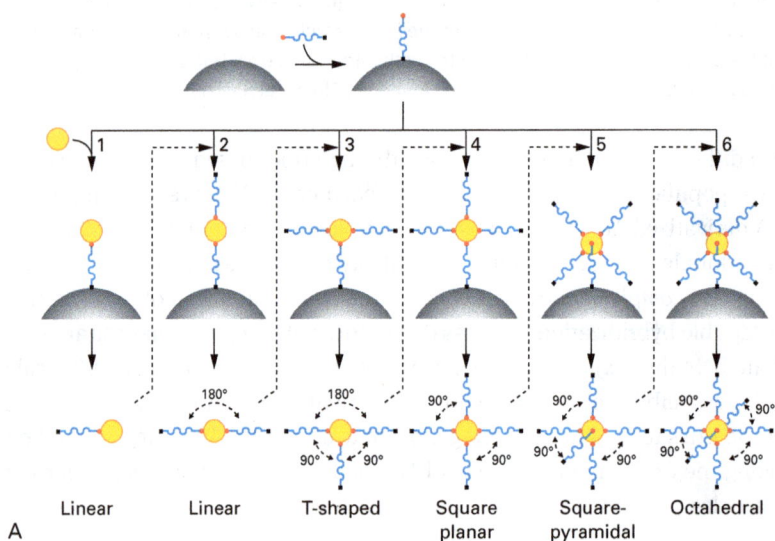

Fig. 9.16: "Nanoparticle coordination complexes" assembled through conjugated complementary DNA. **A.** Synthesis of DNA-NPs with different numbers of covalently-attached single-stranded DNA. Au NPs (yellow circles) with different symmetries were produced through successive attachment of DNA fragments having thiol ends (the thiol units covalently bind to the Au NP surface). Number of successive DNA linking steps is indicated (thus, a NP with one DNA strand is the starting material for the second DNA attachment and so on and so forth). The relative position of the DNA strand with respect to the previously attached strands for each successive step reflects the configurations which minimize the steric hindrance (accordingly, the *second* strand maintains an angle of 180° with the first strand, the *third* attached strand yields a 90° angle, etc.). Following DNA attachment the DNA-NPs were detached from the silica gel support (grey semi-circle);

Spherical DNA/NP nanostructures represent another interesting and potentially useful DNA-associated configuration (Fig. 9.17). These structures, developed in the Mirkin laboratory and denoted "spherical nucleic acids" or SNAs, comprise dense layers of chemically-functionalized nucleotides covalently attached to NP surfaces (Au NPs have been mostly used so far). Interestingly, these spherical particles exhibit

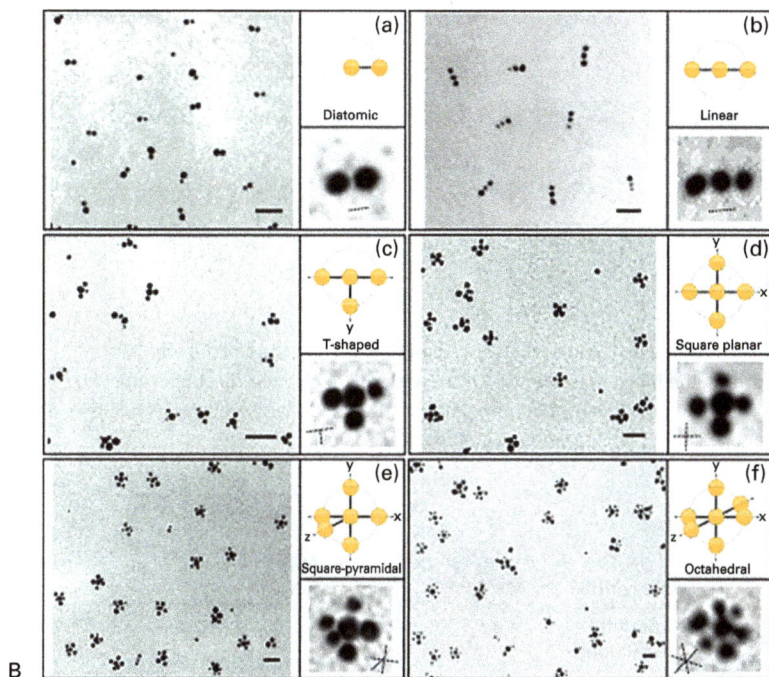

Fig. 9.16: B. TEM images and corresponding geometries of the "nanoparticle complexes" comprising the DNA-NP building blocks shown in **(A.)** and additional NPs displaying the complementary single strands. The bold lines in the schematic pictures indicate double-stranded DNA. Scale bars correspond to 20 nm. Reprinted with permission from Kim, J. et al., *Angew. Chem.* **2011** *50*, 9185–9190. Copyright (2011) John Wiley and Sons.

properties that are different from either the metal NP core or the DNA shell. Particularly important, the integration between the molecular recognition capabilities of the DNA strands and the physical characteristics of the metallic cores (i.e. plasmonic, catalytic, spectroscopic properties), provides powerful means for utilization of these new biomimetic assemblies for varied applications in sensing, biocatalysis, cell transfection, gene regulation, and others.

An interesting extension of SNA technology has been the observation that the metallic cores in SNAs can be dissolved, leaving behind the dense oligonucleotide shell which still retains its unique properties and functionalities (Fig. 9.18). These include the ability to bind complementary nucleic acids (e.g. undergo sequence hybridization), and efficiently cross the plasma membrane inducing cell transfection through delivery of target DNA sequences. Beside their practical utilization, such "ghost" DNA spherical assemblies (analogous to red blood "ghosts" produced through removal of the cell nucleus, Chapter 8) constitute a novel DNA organization (beside the naturally-occurring *double helix* and *single strand* configurations).

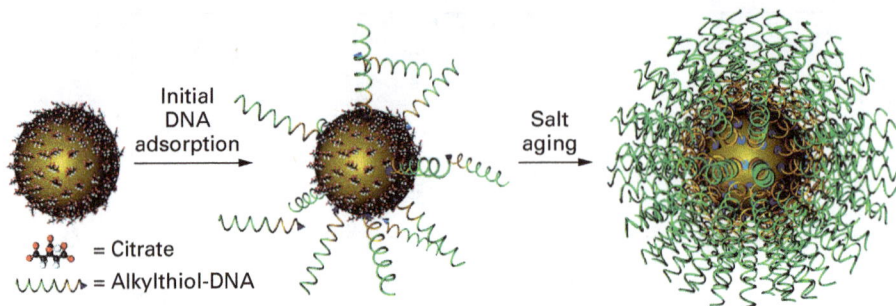

Fig. 9.17: Spherical nucleic acid (SNA) particles. Preparation and structure of the Au NP/DNA conjugate, comprising the NP core, to which DNA strands are attached via functional units for covalent binding (i.e. alkylthiol). Reprinted with permission from Cutler, J.I. et al., *J. Am. Chem. Soc.* **2012** *134*, 1376–1391. Copyright (2012) American Chemical Society.

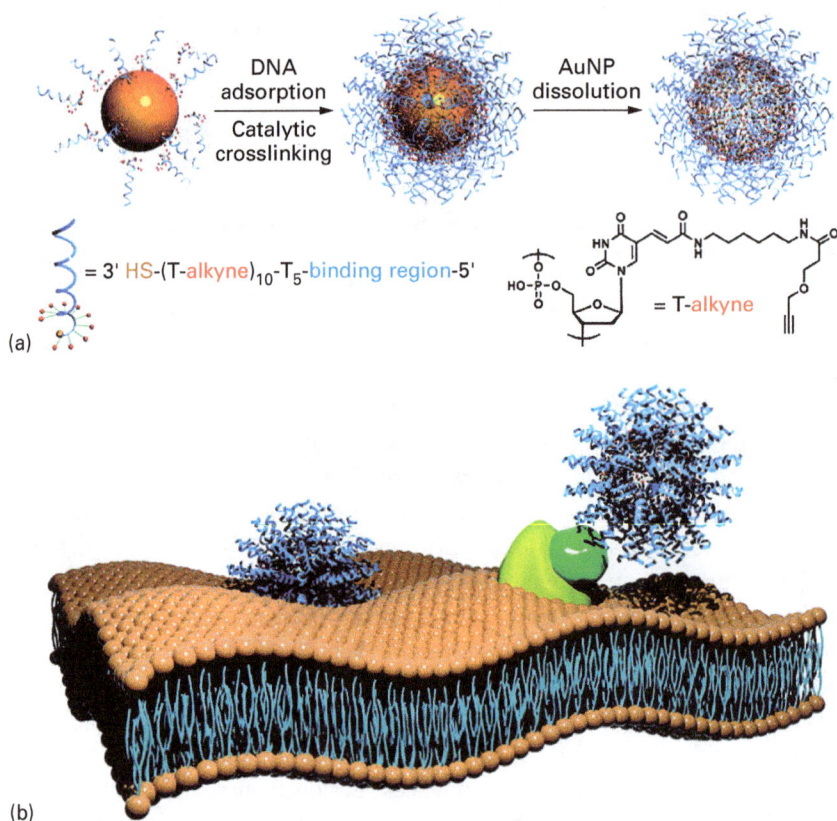

= 3' HS-(T-alkyne)10-T5-binding region-5'

= T-alkyne

(a)

(b)

Fig. 9.18: Empty spherical nucleic acid (SNA) shells. **(a)** Preparation of the "ghost" SNA through dissolution of the NP core via potassium cyanide treatment; **(b)** interactions of the ghost SNAs with cell receptors resulting in cell-internalization of the hollow DNA shells. Reprinted with permission from Cutler, J.I. et al., *J. Am. Chem. Soc.* **2012** *134*, 1376–1391. Copyright (2012) American Chemical Society.

9.4 RNA nanostructures

Similar to DNA, ribonucleic acid (*RNA)* molecules exhibit base-pair complementarity (the nucleotide *thymine* in DNA is substituted in RNA sequences with *uracil*) which enables design of programmable nano-objects similar to DNA. RNA, however, exhibits important differences compared to DNA which might render it attractive for biomimetic applications. RNA is mostly found in nature as *single strands*, rather than the double strand configurations that are abundant in case of DNA. Another notable distinction is the diverse *tertiary structures* of RNA, including helixes, loops, bulges, stems, and others. Moreover, numerous RNA sequences possess chemical reactivity, particularly *catalytic properties*. Overall, the structural and functional diversity of RNA constitutes a potentially powerful platform for constructing new materials.

Fig. 9.19 illustrates the use of RNA units in "hierarchical" design. Packaging RNA, or pRNA, schematically drawn in Fig. 9.19a, is a component of an "RNA motor" responsible for packaging the DNA plasmid of the phi29 bacterial virus. pRNA exhibits two loop regions, denoted "A" and "b'", which constitute its functional domains. In a revealing study, P. Guo and colleagues at Purdue University demonstrated that the two loop domains could serve as binding modules for construction of multimeric structures, including dimers, trimers, and "pRNA twins" comprising two monomers displaying palindromic sticky ends. These RNA configurations can be tuned through varying the experimental parameters such as monomer concentration, pH, temperature, and ionic concentrations. The RNA modules depicted in Fig. 9.19 could form a basis for higher order RNA "superstructures" and organized arrays.

Other RNA superstructure architectures are presented in Fig. 9.20, depicting elegant experiments carried out by L. Jaeger and colleagues at the University of California, Santa Barbara. The researchers employed several RNA modules that assembled into distinct square-shaped configurations. Uniform squares exhibiting different dimensions were produced, depending upon the properties of the RNA units, i.e. the positioning of the reactive loops, size of the RNA "stems" etc. Particularly interesting was the observation that the RNA squares could form *higher-order* architectures, in which individual units could be linked into more complex, tile-like structures (Fig. 9.20c).

The design strategy in "RNA nanotechnology" is conceptually similar to the approach pursued in DNA systems. First, the overall structure of the desired RNA-based assembly and functional objectives are defined. A computational scheme is then applied to determine the core RNA sequence, the smaller motifs needed to initiate folding, and the mixing steps carried out to achieve the final ternary structure. The building blocks are then generated via *in vitro* transcription of the long strand RNA or chemical synthesis of the smaller sequences, and finally the individual subunits are mixed and assembled into the desired architectures. This generic scheme can be augmented by templating the RNA modules with other materials and functional units, such as metal NPs, either in the assembly stage, or after producing the targeted structures.

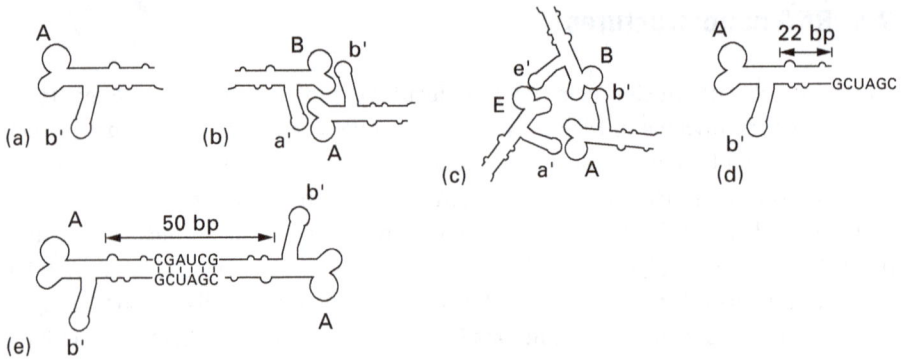

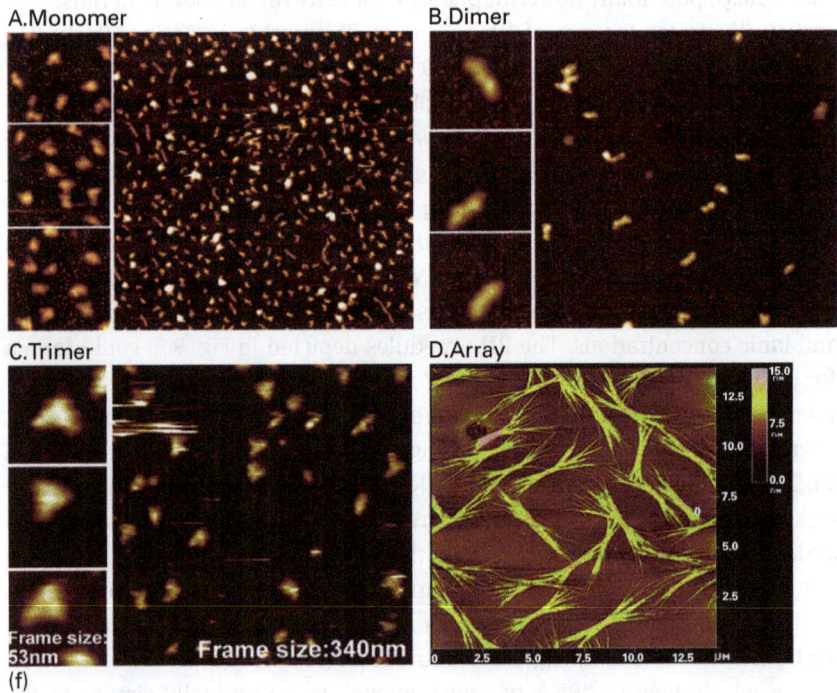

Fig. 9.19: Multi-unit RNA structures. **(a)** The pRNA motor of phi29 virus, showing the two functional loops "A" and "b'"; **(b)** pRNA dimer; **(c)** bRNA trimer; **(d)** pRNA with a 6-nucleotide palindromic sequence; **(e)** "twin" pRNA with back-to-back binding of the palindromic sequence; **(f)** AFM images of the different pRNA modules. Reprinted with permission from Shu D. et al., *Nano Lett.* **2004** 4, 1717–1723. Copyright (2004) American Chemical Society.

A potential advantage of RNA nanotechnology is the use of RNA modules exhibiting *physiological functionalities* as building blocks. Varied RNA structures have known biological roles, such as "switches" used in genetic and metabolic regulation processes, catalytic domains, DNA-packaging units, and others. Such RNA sequences

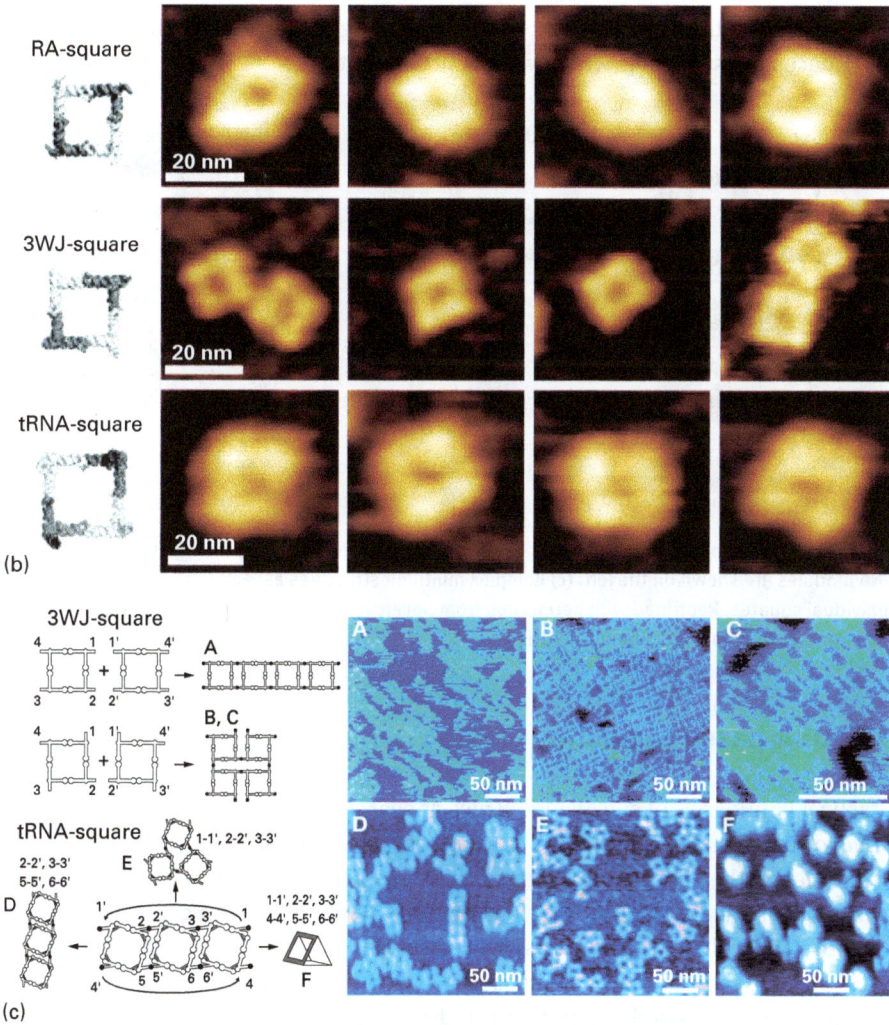

Fig. 9.20: RNA "superstructures". **(a)** Schematic drawing of three square-structures having different dimensionalities assembled from distinct RNA motifs. The L-shaped RNA modules assemble into squares through binding between the perpendicularly-oriented loops.

can be incorporated within artificial RNA-based structures, endowing such assemblies with functional capabilities. For example, insertion of ligand-recognition RNA *aptamers* can be employed for targeting drug-containing synthetic RNA assemblies into specific targets (such as viral cells). Another exciting avenue in this field concerns the design of *combined* DNA and RNA structures, taking advantage of the complementarity between RNA and DNA which is intrinsic to the genetic translation mechanisms. Hybrid DNA and RNA assemblies might further benefit from the synergy between the relative structural stability of DNA and greater chemical diversity of RNA to produce nanostructures with diverse functionalities.

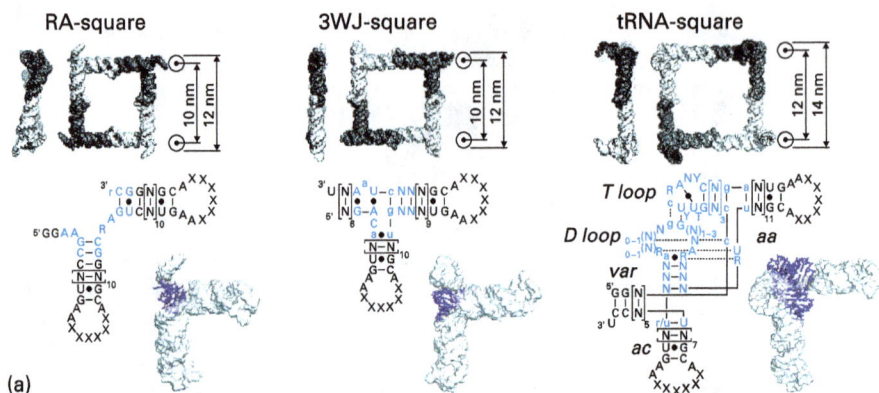

(a)

Fig. 9.20: (b) AFM images of the squares. The schematic structures corresponding to the different RNA modules are shown on the left. **(c)** Complex multiunit structures assembled through linking of individual squares. Reprinted with permission from Severcan I. et al., *Nano Lett.* **2009** 9, 1270–1277. Copyright (2009) American Chemical Society.

9.5 DNA-based computing

Practical boundaries of silicon electronics appear imminent in light of the relentless drive towards smaller and smaller feature sizes of electronic devices and circuits. In consequence, there is a growing interest in development of novel molecular platforms utilized in "nanoelectronics". Many techniques, however, rely on generation of ever smaller nanoscale features of existing devices and/or electronic materials, and, as such, face the same practical challenges of silicon-based electronics.

Biomolecule-based computing could present distinct advantages. The use of biomolecules as building blocks for electronic devices could significantly shrink their size, reduce use of wires, require less energy, and might promote new concepts for computing. Furthermore, the intrinsic *biocompatibility* of biological molecules will enable many biomedical applications, for example implantable devices that could interface with external electronics, biosensors, and inner-body diagnostics platforms. Current molecular electronics and biocomputing research generally focuses upon systems that mimic silicon-based electronics – i.e. trying to create elements such as binary gates and microcircuits that are based on biomolecules and/or biomolecular interactions.

DNA provides a particularly attractive platform for biocomputing, since the natural function of the DNA molecule is informational: the complementary nature of the double strand produces the information flow from the genetic code to protein production. In a sense, the complementary DNA base-pairs (G-C and T-A) could form a basis for information storage. Furthermore, the sophisticated, high-fidelity biological mechanisms which nature has evolved for *translating* the genetic information into biochemical molecules and processes is analogous to the conventional "microelectronic-based" computers we currently use – in which information is stored and translated into specific outputs.

In this context, the appeal of a DNA-based computer is easy to see. Beside the wide availability of synthetic DNA and the sophisticated techniques developed for preparing and modifying DNA sequences, one might consider the issue of possible "memory" size – a single gram of dried DNA contains roughly as much information as 1 trillion compact discs. In principle, the selective binding and unbinding of complementary DNA strands can even be used to process vast amounts of information in parallel. However, the initial excitement for developing DNA-based biological computers was followed by disappointments due to the reality that biological molecules such as DNA are not as robust, fast, and reliable as devices produced from semiconductor materials and other non-biological entities.

The bulk of current DNA-computing research relies upon DNA hybridization as the mechanistic driving force of "Boolean logic" operations. Boolean logic "gates" constitute the conceptual core of modern computing – essentially converting input signals into defined, discrete outputs. Specifically, gates such as AND, OR, YES, and others describe the relationship between the types of inputs and outputs generated by the gates. In DNA-computing experiments, oligonucleotides are added as *inputs* to a reaction mixture, while the *outputs* are produced following chemical transformations induced after hybridization of the input strands with their complementary sequences. Chemical events responsible for output signals include, for example, enzymatic cleavage, structural transformations, fluorescent emission, visible color changes, and other physical processes.

Many variations of the above concept have been reported in the scientific literature, and a couple of examples are outlined here. Fig. 9.21 depicts a scheme of DNA-based logic gates designed by K. S. Park and colleagues at Korea University. The output signal of the gate in this case is *DNA replication* – reported through the fluorescence of a dye which binds to *double-strand* DNA. Importantly, DNA replication is an enzymatic process carried out by DNA polymerase induced by two *primers* – i.e. short oligonucleotide strands – that attach to the respective ends of the two complementary strands of the DNA template which is being replicated.

The "trick" in the logic gates shown in Fig. 9.21 was the intentional insertion of "mutated" (i.e. substituted) bases in each of the two primers required for replication. The two substitutions utilized were adenosine-to-thymine and guanine-to-cytosine; due to the resultant base-pair mismatch (i.e. T–T or C–C) no replication could occur. However, certain metal ions can overcome these mismatches: Hg^{2+} is known to insert in-between two thymines, binding them tightly and effectively "overcoming" the mismatch. A similar feat is achieved by Ag^+ ions in the case of a C–C mismatch. Accord-

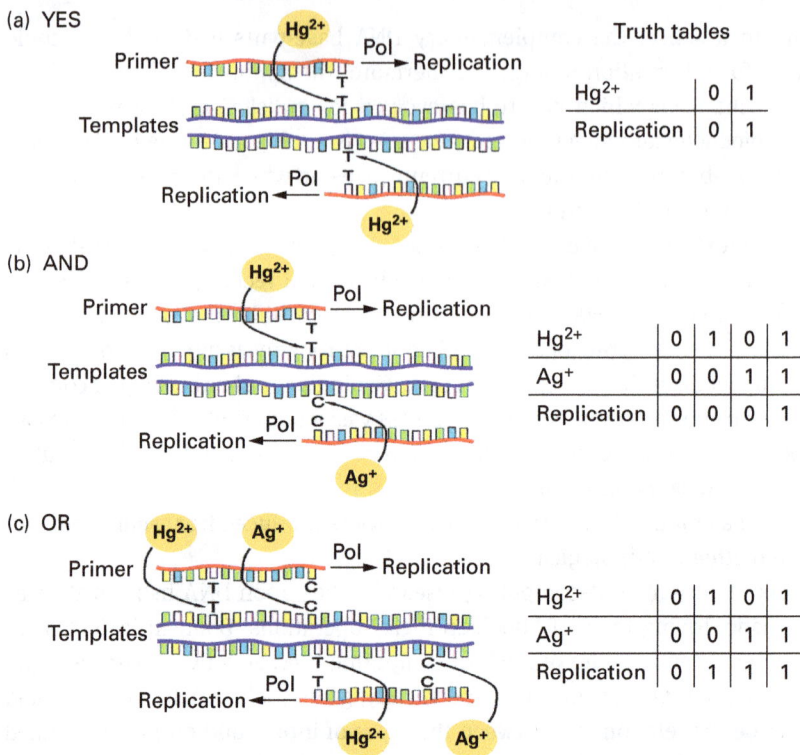

Fig. 9.21: DNA-based logic operations. Short DNA replication primers are added to a DNA double strand template, albeit each primer exhibits a single base-pair *mismatch* (either thymine–thymine, T–T, or cytosine–cytosine, C–C). These mismatches inhibit replication of the DNA template by the DNA polymerase enzyme, *unless* the metal ions Hg^{2+} or Ag^+ are added, which *override* the mismatches – Hg^{2+} for the T–T mismatch and Ag^+ for the C–C. Only in case where the mismatches are annulled does replication occur and an output signal is produced (expression of a fluorescent protein) The three different scenarios represent distinct logic gate operations, shown also in the truth tables: **(a)** *YES* gate – both primers contain T–T mismatches, accordingly DNA replication occurs only in the presence of mercury ions (Hg^{2+}); **(b)** *AND* gate – the two mismatches are present and thus DNA is replicated only in the presence of *both* Hg^{2+} and Ag^+; **(c)** *OR* gate – *Two* sets of primers are added (instead of one) – replication occurs when either Hg^{2+} or Ag^+ are present.

ingly, as shown in Fig. 9.21, these two metal ions constitute the perfect inputs for different Boolean gates.

Importantly, the *type* of logic operation (YES, OR, AND, etc.) is determined by the primers employed, more precisely – the identity of base-pair mismatch. Thus, a YES gate is achieved by using two primers exhibiting *only* the T–T mismatch and the output signal (DNA replication) is generated *only* when Hg^{2+} is provided (Fig. 9.21a). Similarly, an AND logic operation is achieved when *both* mismatches are applied, requiring addition of *both* Hg^{2+} and Ag^+ ions for producing an output signal (Fig. 9.21b). Constructing an OR gate required a clever design, depicted in Fig. 9.21c. Here, the researchers used two different primers, each with a different base mismatch. Each primer was bound to the template at a *different* location; replication was initiated when *either ion* was added to the mixture – producing an OR operation.

Applications of DNA logic devices could have diverse uses. Implementation of the logic concept for sensing of *ions* in complex solutions can be envisaged, for example identifying the presence of two specific ions simultaneously. The basic logic gate scheme could be similarly employed for reporting on other external molecules as inputs – for example pathogen fragments, molecular signaling molecules, and others. The DNA computing approach can be further extended to design systems in which the *output* of one gate (replicated DNA, for example) would be the *input* for another gate, creating sophisticated branched circuits. Despite the simplicity and generic nature of DNA logic operations as depicted in Fig. 9.21, actual use of DNA in so-called "biological computers" might be limited, however, because molecular signaling processes are inherently orders of magnitude slower than electron transport in silicon-based devices.

DNA, the "molecule of life", and its cousin RNA have inspired research in varied branches of science and technology, from nanotechnology and advanced materials, through drug design and diagnostics, towards biocomputing, and others. The combination between programmable structural features and information-bearing properties is a unique feature of DNA and RNA, exploited for diverse applications. In particular, the interface between DNA and RNA technology and related fields, both *chemical* in nature such as nanoparticle science as well as *biological* such as drug delivery and targeting, will likely continue to be a source of exciting biomimetic systems and discoveries.

10 Mimicking biological phenomena and concepts

Some of the most "awe-inspiring" advances in the broadly-defined field of "biomimetics" involve borrowing biological phenomena, and using those as *conceptual frameworks* for innovative materials, devices, and practical applications. The distinction between these R&D activities and the more general discussion of biomimetic systems in other chapters of this book is blurred. The discussion below focuses on artificial systems which have been conceived and developed in order to mimic broader conceptual contours of fundamental biological phenomena, rather than focused on specific structural or functional applications.

10.1 Catalytic antibodies

Catalytic antibodies (or "abzymes" – antibody-enzymes) first appeared on the science stage in the mid-1980s in the laboratories of P. Shultz and R. Lerner, and exemplify the evolution of a new scientific endeavor at the intersection of chemistry (more specifically organic chemistry and catalysis) and biology (enzyme science). The catalytic antibody concept can be depicted as "biomimetic" since it aims to tie *biological antibodies* to catalysis of *non-biological* entities. Catalytic antibodies, as their name implies, are molecules which are designed to recognize and bind specific molecular targets. Unlike antibodies produced in the human body which play a major role in the immune response, catalytic antibodies do not function in a biological context but rather are aimed to participate in, and *accelerate* chemical reactions. As such, catalytic antibodies essentially act as *enzymes* – accelerating biochemical reactions through binding to the reactants.

The thrust of the catalytic antibody concept is shown schematically in Fig. 10.1, which depicts typical *reaction coordinates* of a non-catalyzed and catalyzed chemical or biochemical reaction. The *rate* of the reaction is generally determined by the height of the energy barrier, i.e. the relative energy of the *transition state*, which is influenced by parameters such as the structural features of the molecular species formed at this intermediate state including bond rearrangements, charge distribution, etc. A catalyst (either an enzyme or a non-biological entity) essentially functions by *lowering* the energy of the transition state thereby increasing the reaction rate.

The innovative idea behind the catalytic antibody technology is to harness the immune system to identify the most effective molecule (antibody) for accelerating a specific chemical reaction. In the "catalysis terminology", the optimal catalytic antibody should bind the *transition state* structure, lowering its energy as much as possible (in the same way as an enzyme achieves catalytic action) and thereby increasing the reaction rate. The thrust of a catalytic antibody experiment is the design of a "transition state analog" (TSA), a molecule which is supposed to closely mimic the

https://doi.org/10.1515/9783110709490-010

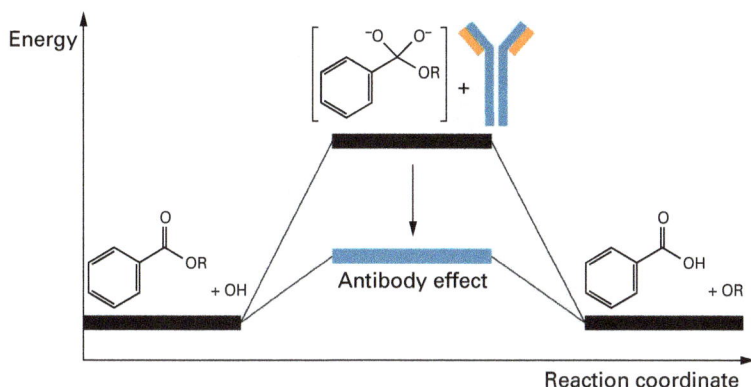

Fig. 10.1: Catalytic antibodies. Energy diagram of a representative reaction (shown here is *ester hydrolysis*). A catalytic antibody (green "T-shape" molecule) lowers the energy barrier through binding to the transition state structure, thereby accelerating the reaction.

transition state conformation of the target reaction. This molecule is subsequently used for generating the antibodies using conventional methods – injection into mice or other animals and harvesting the antibodies generated. The TSA-triggered antibodies are screened for the one that can accelerate the target reaction most efficiently. The huge intrinsic diversity of antibodies produced by the immune system against foreign substances (such as a TSA) means that, in principle, selection of molecules that could lower the transition state and thus exhibit catalytic activity should be possible.

Fig. 10.2 presents an example of an antibody catalyzing the insertion of a metal ion into a macrocyclic chemical compound (mesoporphyrin). To generate the antibody, P. G. Schultz and colleagues at the University of California, Berkeley, used *N-methylprotoporphyrin* (depicted in the inset in Fig. 10.2) as the antigen, since the distorted porphyrin ring of this molecule was believed to mimic the putative transition state in the metal insertion reaction. Indeed, an antibody generated by the researchers that was specific to *N-methylprotoporphyrin* was shown to successfully catalyze metal insertion into mesoporphyrin at a rate that was similar to natural enzymes.

Since its inception, the field of catalytic antibodies yielded some notable successes, but also its share of disappointments. Indeed, the critical step of designing the appropriate TSA is an "Achilles' heel". For one thing, knowledge of the reaction pathway and perceived structure of the actual transition state is important but in many cases difficult (or impossible) to achieve. Secondly, constructing a TSA that would closely mimic the transition state – in terms of structure, polarity, chemical bonds, etc. – can be very challenging. Moreover, even an optimal TSA does not guarantee generation of effective catalytic antibodies; in many cases the antibodies produced are found to rather favor the *end products* of the reaction under study, binding them tightly and thereby inhibiting the reaction rather than accelerating it.

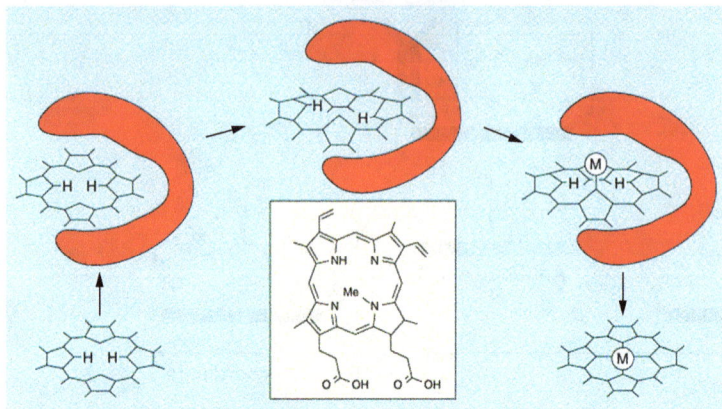

Fig. 10.2: Catalytic antibody elicited by a transition state analog (TSA). Antibody generated to N-methylprotoporphyrin (shown in inset), which was used as a TSA, was capable of catalyzing metal insertion into mesoporphyrin. Reprinted from Martin, A.B. and Schultz, P.G., Opportunities at the interface of chemistry and biology, *Trends Biochem. Sci.* **1999** *24*, M24–M28. Copyright (1999), with permission from Elsevier.

Despite the significant hurdles, the methodology of catalytic antibodies has had a big scientific impact. Catalytic antibodies have also found their way into commercial applications – as catalysts to many industrially-important reactions. This field, in fact, has further expanded, encompassing now also *synthetic* catalytic antibodies (rather than antibodies produced in animal models, as initially carried out), and further optimizing the catalytic activity of the antibodies through concepts such as *directed evolution* (see Chapter 6).

Other applications of catalytic antibodies are, in fact, not directly aimed at improving reaction yields or amplification of reaction products. The approach has been used, for example, to combat drug abuse and to develop new schemes for addiction treatment through specific binding of the antibodies to drug molecules. Specifically, antibodies can be generated that bind and decompose (through common reaction pathways such as hydrolysis) controlled substances such as cocaine; when the antibodies circulate in the bloodstream, they can recognize and break up the drug molecules, reducing their uptake in the brain and the ensued damage.

10.2 Artificial photosynthesis

Photosynthesis is among the most ubiquitous and important biological processes occurring on earth. In a nutshell, photosynthesis is carried out in plants, algae, and certain bacteria via a cascade of chemical reactions, structural transformations, and electron transfer processes occurring in specific organelles (chloroplasts in plants and algae). Light (usually in the blue region of the visible spectrum – giving rise to the *green* appearance of leaves) is absorbed by pigments (i.e. chlorophyll) embedded in specialized protein matrixes called *photosystems*. A series of structural transformations, energy transfer, and electron transfer reactions ensue, resulting in production of the well-known energy-storage molecules adenosine triphosphate (ATP) and nicotinamide adenine dinucleotide phosphate (NADPH). These compounds constitute the molecular workhorses effecting reduction of carbon dioxide into carbohydrates (a process generally referred to as "carbon fixation").

The series of chemical reactions by which plants transform light into chemical energy has inspired scientists, technologists, and lay people, both for their sophistication as well as for the possibilities for harnessing or mimicking photosynthesis as an alternative route to produce energy. Extensive efforts have been directed towards creating artificial photosynthetic platforms. In particular, many studies have focused on synthesizing molecular entities that could perform the crucial chemical transformations accomplished in the natural photosynthesis cycle. This particular broad body of work belongs more to the realm of "classic" synthetic chemistry and bioorganic chemistry, and is generally beyond the scope of this book. In the following will be examples of biomimetic systems which aim to imitate photosynthesis as a generic process, or create artificial photosynthetic platforms through the use of nonbiological building blocks.

Photosynthesis has often been employed as a conceptual "template" for solar cell applications. In this context, many studies attempted to imitate the core phenomenon of photosynthesis – light energy converted through chemical means to energy. An example of a biomimetic scheme for artificial photosynthesis (one of numerous similar systems reported in the literature) is shown in Fig. 10.3 depicting a "solar fuel cell" developed by F. A. Armstrong and colleagues at Oxford University. The components of the system include a molecule responsible for *light absorption* and *generat-*

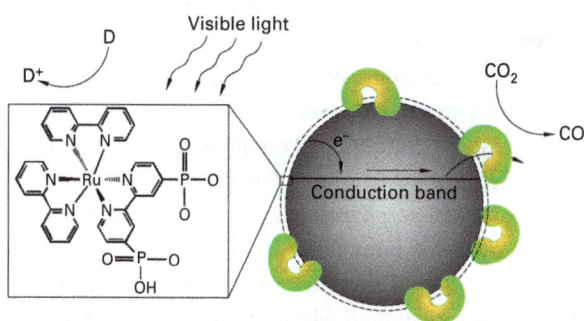

Fig. 10.3: Biomimetic photosynthesis. Chemical assembly designed to harvest light, facilitate electron transport, and utilize the electrons for carbon fixation. The ruthenium dye, which absorbs light and emits electrons, is embedded on the surface of metal-oxide NPs. The electrons are transported across the NP, consequently reducing CO_2 to CO through the catalytic action of CO dehydrogenase (green crescent-shaped units).

ing electrons (usually a metal-containing macrocyclic molecule); a chemical species which donates electrons (for regeneration of the dye molecule); titanium oxide nanoparticles (NPs) acting both as *transporters* of the donated electrons, as well as *scaffolds* for enzyme molecules (carbon monoxide dehydrogenase); the enzymes catalyze the reduction of carbon dioxide to carbon monoxide (CO).

The processes occurring in the biomimetic solar fuel cell depicted in Fig. 10.3 are the following: light is harvested by the metal-containing dye; the absorbed light facilitates electron transfer from a donor molecule onto the conduction band of the TiO_2 NPs which basically act as a "storage place" for the released electrons; the electrons are used for reducing carbon dioxide to carbon monoxide at the NP surface and this process is catalyzed by the enzyme CO-dehydrogenase. CO is the starting material for synthesis of more complex carbon-based molecules and carbohydrates. Overall, the system schematically shown in Fig. 10.3 represents an artificial system designed to mimic the natural photosynthetic "carbon fixation".

The first, crucial event in photosynthesis is the process in which light photons are "harvested", initiating the cascade of energy-, and electron-transfer reactions eventually leading to reduction of carbon dioxide, yielding the carbonaceous building blocks of life. Accordingly, designing substances capable of efficient light harvesting is a key challenge for the scientist or engineer aiming to develop an artificial photosynthesis platform. Different types of molecular light-capturing "antennae" have been developed with the aim of achieving photon uptake combined with minimal energy loss. Specifically, such antennae need to absorb light efficiently and in parallel provide reaction pathways for the photo-induced charge carriers to reach the electrodes with minimal energy losses which might occur through charge recombination or high electrical resistance.

In plants, photon uptake and electron transfer are carried out by the chlorophyll pigment within the photosystem. The sophisticated natural light harvesting and electron transport processes carried out by the chlorophyll-protein complexes are still too difficult to exactly replicate via synthesis. Instead, attempts have been reported for constructing the core features of the photosystem: synthetic light-harvesting antennae coupled to molecular species facilitating electron transfer.

Porphyrins, macrocyclic compounds that typically encapsulate metal ions, have been often used as core components of artificial pigments because of their considerable light absorption properties. *Multi-porphyrin* arrays have been among the more popular molecular units employed for mimicking light-harvesting antennae (Fig. 10.4). J. S. Lindsey at North Carolina State University and other researchers have demonstrated that the types of metals encapsulated within individual porphyrins in arrays such as shown in Fig. 10.4 intimately affect the spectroscopic and electronic properties of the complexes. Porphyrins containing zinc or magnesium ions, in particular, were shown to exhibit a tunable range of absorbed wavelengths and transferred energy, making such macrocyclic architectures attractive model light harvesting systems.

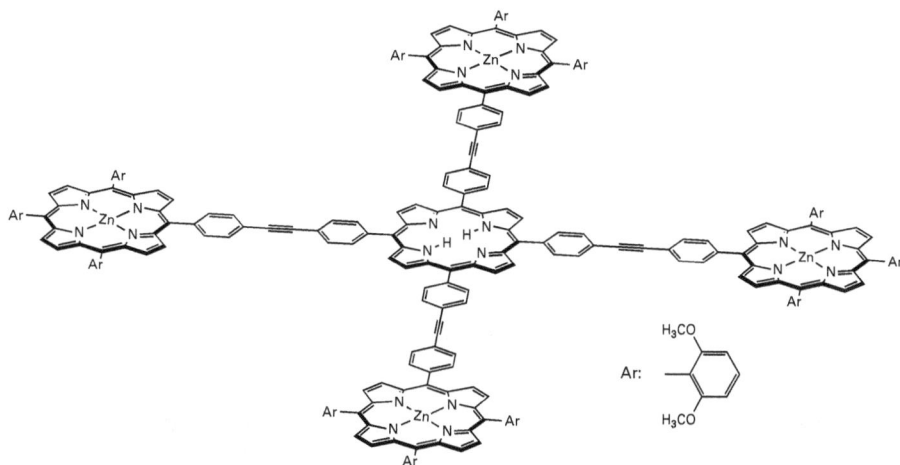

Fig. 10.4: Multi-porphyrin as a model light-harvesting antenna. Covalently-bonded zinc-containing porphyrins (shown in the periphery) and free-base porphyrin (center) enable efficient light-induced energy transfer and electron transport. Reprinted with permission from Holten D. et al., *Nano Lett.* **2002** *35*, 57–69. Copyright (2002) American Chemical Society.

While multi-porphyrins represent a relatively simple chemical antenna, they do not closely mimic the highly intricate complexes participating in light harvesting in natural photosynthesis. In particular, a fundamental property of the natural light harvesting complex, exceedingly difficult to reproduce through conventional synthetic approaches, is the existence of *hundreds* of pigment molecules surrounding the "reaction center" – the core functional unit of the photosystem – reoriented in a fashion that facilitates synergistic energy transfer. A potential route to imitate this configuration is the display of synthetic chromophores upon the surface of *dendrimers* – branched polymeric molecules that allow presentation of a large number of identical chemical units on their surface. This is depicted schematically in Fig. 10.5. Light (usually in a broad spectral range) is absorbed by the chromophores on the dendrimer periphery and energy is transferred to its core through the dendritic branches. Enhanced light is subsequently emitted from the dendrimer core.

10.3 Bio-inspired motors

Generating mechanical motion in a controlled manner is, quite justifiably, considered as one of the most significant technological advances achieved by Man – beginning from inventing the wheel. Achieving mechanical movement on a *molecular* scale has been among the most inspiring challenges of nanotechnology in general, biomimetic research in particular. Artificial molecular motors need to be able to transform input

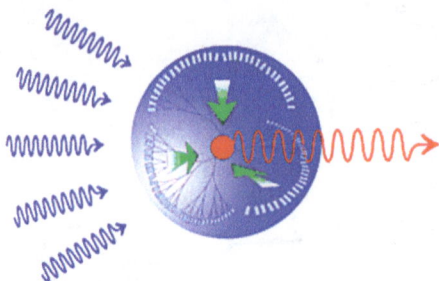

Fig. 10.5: Dendrimer as a light-harvesting antenna. Generic scheme of a dendritic light-harvesting antenna. Multitude of chromophores (pigments) is displayed upon the surface of the dendrimer absorbing light in a broad spectral range (blue arrows); the absorbed energy is transferred from the chromophores at the dendrimer periphery through the branches, and to its core (green arrows); subsequent enhanced emission is produced from the core (red arrow). Reprinted with permission from Hecht, S. and Frechet, J.M.J., *Angew. Chem.* **2001** *40*, 74–91. Copyright (2001) John Wiley and Sons.

signals (and/or input energy) into mechanical motion, and carry out this task in a controlled, regulated manner.

Mechanical molecular motion is, in fact, a fundamental physiological phenomenon. *Molecular motors* perform important tasks in biological systems including mechanical work and cargo transport. *Kinesin* and *myosin* are the most well-known biological motor systems. These proteins move along defined molecular tracks, and their mobility underlies important functions – transporting molecular cargoes in the case of kinesin, inducing muscle contraction by myosin. Fig. 10.6 depicts the basic scenario of biological mobility, in which unique nanofilaments (e.g. "microtubules" comprising the protein *tubulin*) function as "tracks" upon which cargo-carrying proteins such as kinesin, dynein, and others are moving. The mobility mechanisms of motor proteins and their cargo delivery capacity are intimately related to their struc-

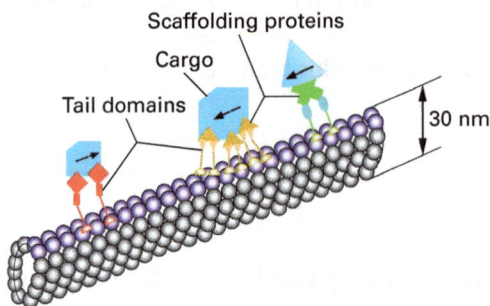

Fig. 10.6: Transport of molecular cargo on microtubule tracks. Motor proteins such as kinesin and dynein (red, yellow, blue) interact with the microtubule surface through the head regions, while their tails attach to the molecular loads through specific scaffold proteins. Figure inspired by Goel A. and Vogel V., *Nature Nanotechnology*, **2008** *3*, 465–475.

tural features. As shown in Fig. 10.6, motor proteins can carry different molecular loads either through binding directly onto the "tail" domains (the "heads" are the protein regions attached to the microtubule surface), or through specific "scaffold proteins" displayed on the protein tails.

The seemingly simple and efficient mechanical motion process depicted in Fig. 10.6 has inspired non-biological applications, for example polymer synthesis schemes in which molecular motors deliver reactants towards each other in a sequential order. Specifically, through chemical modification of the scaffold proteins and/or the tail domains of the microtubule-binding proteins, one can exploit the microtubule track as a means for transporting and eventually reacting monomeric units (the motor proteins' cargoes). This "sequential synthesis" concept has distinct advantages over conventional polymerization techniques which rely on random collisions in solution and thus impart less control over the composition and purity of the final product. Indeed, the biomimetic motor-based platform could much better influence the reaction contours through designing precise geometrical fitting between the reactants delivered towards one another on the microtubule track. Furthermore, the biological motor concept allows intimate control of the reaction dynamics through the "energetic handle" – controlled delivery of energy source, primarily adenosine triphosphate (ATP) molecules.

Biomimetic motor designs can also exploit *inverse* configurations – in which the motor proteins are immobilized, while the microtubule is the moving element (Fig. 10.7). Such designs can be accomplished by anchoring the motor proteins (kinesin for example) on a surface via their tails. The motor proteins' heads exhibit sufficient flexibility enabling "gliding" of the microtubule. The inverse configuration

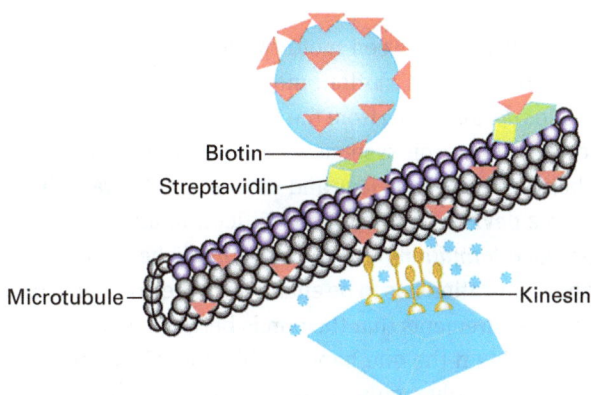

Fig. 10.7: Molecular transport through microtubule mobility. "Inverse" configuration – a motor protein (kinesin) is immobilized on a surface and the microtubule is transported through interactions with the kinesin's head domain. Molecular cargoes (represented as the blue sphere) can be loaded upon the microtubule surface through biotin-streptavidin interactions. Figure inspired by Goel A. and Vogel V., *Nature Nanotechnology*, **2008** *3*, 465–475 (2008).

has certain bioengineering advantages, mainly related to the fact that the microtubule surface is more amenable than the motor proteins to chemical derivatization and display of guest molecular species, which can be consequently transported. The scheme in Fig. 10.7 illustrates one of the various microtubule delivery schemes, in which the molecular cargo is immobilized onto the filament surface through the strong binding between avidin and biotin.

Other important parameters play roles in the design of biomimetic molecular motors. Like all mechanical motors, initiation of motion requires *energy*. In natural and artificial molecular motors *chemical energy* constitutes the input energy, rather than electrical, thermal, or mechanical stimulation. This dictates on the one hand a continuous supply of energy-rich biological compounds, such as ATP or NADH, which release energy upon molecular transformations. On the other hand, such an energy input provides a powerful means for *controlling* the motor operation: initiating motion upon supply of chemical energy, stopping the motor action upon blocking input of the energy-producing molecules.

A significant contrast between biomolecular machines and what we perceive as efficient mechanical instruments is the reliance of the former upon *Brownian motion* (i.e. natural vibrations and translation in solution induced by thermal energy). Indeed, in many instances movement of biological motors is initiated by thermal motion, and chemical energy is consumed mainly to modulate the motion (speed, direction, etc.). In this context, thermal motion combined with the viscous environment in proximity to surfaces upon which biological motors usually operate mean that *surface smoothness* is not a necessary prerequisite for motion. Furthermore, the fact that the biological motor is usually *not* covalently attached to the surface underscores the intricate *kinetic balance* that needs to be maintained between motor and substrate.

Indeed, the non-covalent interactions between motor proteins and their tracks underlie a critical mechanistic feature of their motion: mobility is *processive*; i.e. at any time during motion the proteins *do not detach* from the surface (i.e. the track). Processive motion is commonplace in macroscopic systems – for example when a person walks there is always contact between one of the legs and the ground. However, processive mobility in *molecular-scale* devices is much more difficult to attain. In such assemblies, *bond breaking* and *bond formation* need to be accomplished simultaneous, and furthermore, directional translation has to be enabled via the bond breaking or formation process. Indeed, the requirements that the bonds between a "molecular walker" and the surface will be labile on the one hand, stable enough to maintain contact on the other hand, and additionally regulated externally, have proven a formidable undertaking.

Strand complementarity in DNA molecules provides an effective mechanism for achieving "molecular walking", and "DNA walkers" are discussed in Chapter 9. Other generic mechanisms for walking on a surface have been proposed. Fig. 10.8 depicts a chemical system designed by D. A. Leigh and colleagues at the University of Edin-

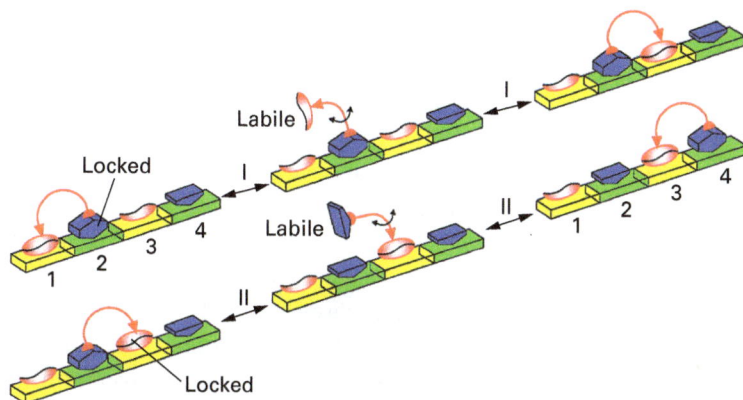

Fig. 10.8: A generic molecular walker. The molecular walker has been designed so that bond brea-king and bond formation of the two "feet" (red and blue, respectively) could be induced through changes in solution acidity. To assure that processive walking occurs, the two feet cannot be detached simultaneously from the surface (therefore, detachment occurs *either* through process I *or* process II, induced by different experimental parameters). *Directional* processive transport of the walker along the track can be accomplished if the experimental conditions are made to favor the forward direction of either chemical equilibrium.

burgh, UK, which produced a molecular walker. The key feature of the system, ena-bling processive walking, was the induction of the bond breaking and bond forming events by *different* (orthogonal) experimental conditions; for example one affected by acidity in the solution while the other by basicity. In such a scenario the two sets of experimental conditions are applied consecutively, leading to molecular motion. Importantly, in order to achieve *directionality* of movement rather than random motion back and forth the researchers had to identify experimental conditions in which the *ratio* of forward to backward transformations of the walker could be externally varied (and accordingly selecting the conditions that will cause forward motion).

10.4 Artificial muscles

Efforts to create *artificial muscles* have been pursued through different disciplines, from engineers who build actuators, e.g. motion-generating devices, and all the way to physicians, aiming to reintroduce motion capabilities to paralyzed patients. As we know, particularly the body builders among the readers, a good muscle needs to possess strength, toughness, but also flexibility. Muscles exhibit other physiologi-cally important properties; through changing their length in response to nerve stimu-lation they can exert controlled amounts of force – from the movement of a butterfly wing to lifting heavy weights. Muscles are recognized for their *scale invariance*: the operating mechanism of a muscle tissue is the same regardless of whether it works in an insect or an elephant.

Artificial muscles build upon the fundamental concepts dictating muscle physiology for creating systems that will be able to perform mechanical work. Artificial muscle technologies have generally focused both on identifying chemical substances which can achieve desired mechanical properties – motion and/or energy produced through stimulated deformations – as well as engineering the different components of a muscle (core "muscle" units, triggering mechanisms, etc.) for practical purposes. Several chemical systems employed in artificial muscle design will be discussed below.

From a bioengineering standpoint, the generic design concept for artificial muscles is quite straightforward. In fact, various devices conforming to an artificial muscle definition have been around for decades. *Electro-active polymers (EAPs)* are among the most common constituents in artificial muscle design and have been widely used in actuators and sensing devices. EAPs display the core property of a muscle – a change in size or shape when stimulated by an electric field. In particular, many EAPs can endure significant deformation while sustaining large forces. The combination of structural deformation and withstanding strain makes EAPs excellent candidates for artificial muscles that exhibit superior properties compared to their physiological analogs.

A typical EAP-based actuator consists of a "sandwich" arrangement in which elastic insulated EAPs are placed between electrical conductors (Fig. 10.9). *Dielectric elastomers* – a subgroup of EAPs which undergo substantial field-induced deformations – have been often examined in this configuration. When an electric field is

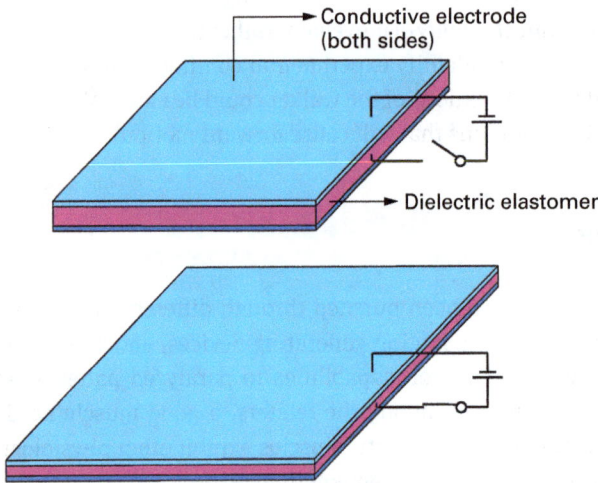

Fig. 10.9: Artificial muscle comprising a dielectric elastomer. The dielectric elastomer (a type of electro-active polymer) is placed between two electrodes. Upon application of an electric field the elastomer expands in all three dimensions.

applied between the conductors, charge builds up on the polymer surface, producing a compressive stress which results in contraction or expansion of the polymer just like a real muscle. The shape change of the polymer provides a force that can be further exploited, for example to move a prosthetic limb or a robotic arm.

In order for artificial muscles to fully mimic their physiological analogs, particularly in the body environment, it would be advantageous to create integrated and compact assemblies – in consideration of the limited space and the fact that body-implanted artificial muscles are usually not conducive to bulky external electrodes. This bio-engineering challenge can be tackled through different design efforts. Specifically, EAP-based artificial muscles have been built with associated internal sensors, triggering mechanisms, and electrical circuitry. Such artificial muscles are capable of motion control through embedded feedback loops, overall closely imitating their natural counterparts.

Taking advantage of their capability to transform mechanical energy into chemical energy and vice versa, artificial muscles have been proposed as components of energy storage devices, converting mechanical energy to electrical energy. Indeed, just like ordinary motors, artificial muscles can be used in reverse as generators, transforming movement into electricity. Fig. 10.10 illustrates the mechanism of an electricity generator based upon an artificial muscle concept, developed by T. McKay and colleagues at the University of Auckland, New Zealand. The process often utilizes dielectric elastomers which, even while structurally deformed, *do not* change their volumes and thereby could generate electrical potential through the changes of charge distribution at their surfaces.

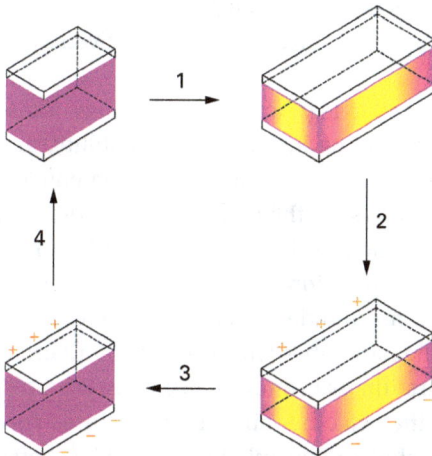

Fig. 10.10: Dielectric elastomer generator. Scheme depicting the cycle of converting *mechanical* energy to *electrical* energy by the generator. The different stages of the cycle are described in the text.

Converting mechanical energy into electrical energy in a dielectric elastomer generator (DEG) is initiated by stretching the polymer, providing elastic strain energy (state 1 in Fig. 10.10); external voltage is then applied, injecting charges to the polymer surface (2); following relaxation of the deformed polymer, the charges become denser and separation between the positive and negative charges *increases* (3) – resulting in greater potential energy. The charges are then drained from the system and used externally (i.e. the generator action). It is obvious that energy needs to be supplied to the system for the initial polymer deformation, however such energy can be harvested from varied sources such as ocean waves, cars passing on a road, a person walking, and many others – making this artificial muscle application attractive as a renewable energy technology. The DEG concept can be further expanded to achieve a "self-priming" artificial muscle – essentially transferring the mechanically-induced electrical energy *back* towards affecting the mechanical deformations of the polymer, which consequently produce electrical energy, *ad infinitum* (until, of course, the losses due to charge leakage and heat result in ceasing of the mechanical deformation).

While the systems discussed above represent imitations of the muscle *concept* for practical applications, other interesting constructs have been reported which mimic *both* function and structure of muscles. The core biomolecular components in physiological muscles are filamentous contractible proteins, such as actin, myosin, titin, and others. Accordingly, mimicking the anisotropic, fiber-like organization of the muscle using non-biological components has been a major research route. Indeed, in some instances artificial muscles comprising *inorganic* fiber-like structures exhibited superior properties compared to their physiological analogs. A recent example is the fabrication of an artificial muscle comprising *carbon nanotubes* (CNTs) developed by R. H. Baughman and colleagues at the University of Texas, Dallas (Fig. 10.11).

CNTs are among the most prominent "workhorses" of nanotechnology. These carbon allotropes exhibit remarkable tensile strength, mechanical stability, and thermal and electrical conductivity, overall making them useful components in diverse new materials and applications. CNTs by themselves are not flexible, however macroscopic assemblies of CNTs, such as the intertwined CNT sheet depicted in Fig. 10.11, exhibit significant compression and stretching properties, surpassing conventional rubber by orders of magnitude. Specifically, the researchers discovered that through intertwining the nanotubes within sheets of a porous "aerogel" matrix, the material had steel-like stiffness in one dimension (in the direction of the nanotube spines vector), while the two other dimensions featured considerable rubber-like flexibility and expandability that could be induced by charge injection. Intriguingly, the CNT-based assemblies shown in Fig. 10.11 further demonstrated *torsional rotation*, a rare feat observed in native muscles, ascribed to the tubular packing.

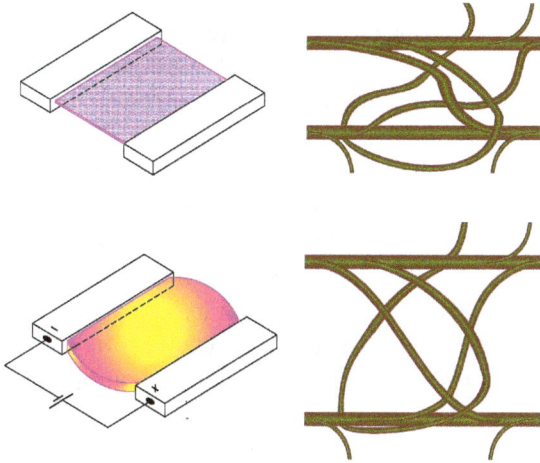

Fig. 10.11: Artificial muscle from a carbon nanotube assembly. Application of voltage to a sheet comprising intertwined carbon nanotubes (CNTs) gave rise to dramatic expansion of sheet area and increased thickness.

10.5 Biomimetic pores and channels

Pores and channels are fundamental constituents of cell physiology and have vital biological roles. Pores are present in cellular membranes, and have crucial functions in regulating ionic and molecular transport between the outer environment and the cell interior, and into or out of sub-cellular compartments. Examples of naturally-occurring channels include ion channels that regulate ion flow; pores induced by infective viruses which serve as conduits for transporting viral genetic material into the cell; the nuclear pore complex (NPC) that controls the transport of mRNA and proteins through the nuclear membrane; and pores facilitating molecular passage into cellular organelles such as mitochondria.

Designing *artificial biomimetic pores* has been an active field of research, mostly aiming to recreate the transport properties and selectivity profiles of natural pore systems for varied applications, including biosensing, molecular separation, extraction, and others. Attaining this goal, however, has been elusive although recent advances in miniaturization and nanotechnology have significantly contributed to realizing the potential of this field. Progress in nanofabrication methods has probably been the single most important technological aspect underscoring the expanding applications of biomimetic pores and channels. Nanometer-sized pores are now fabricated quite routinely, and constitute scaffolds for further functionalization with biological molecules. Indeed, pore-containing films can be assembled from non-biological materials, such as polymers, organic, and inorganic matrixes. Such thin films can be molded using common "top-down" nanolithography techniques, yielding arrays of nanopores with controlled dimensionalities and spatial distances. Other approaches employ both synthetic and biological components, creating conjugated systems that display the characteristics of biological channels in artificial settings.

Fig. 10.12, for example, depicts an experimental approach, demonstrated by D. Losic and colleagues at the University of Southern Australia, for fabricating nanopore arrays using heavy ion beams (a technique sometime poetically referred to as "ion-beam sculpting"). Specifically, a thin-film polymeric substrate is bombarded with heavy ions, consequently creating pores or channels in the film. The sizes and distribution of the pores can be controlled through the type of ions and intensity of the ion beam. Subsequent chemical (or biological) modification of the *internal surface* of the channels is also carried out to endow molecular recognition and selectivity properties to the artificial channel.

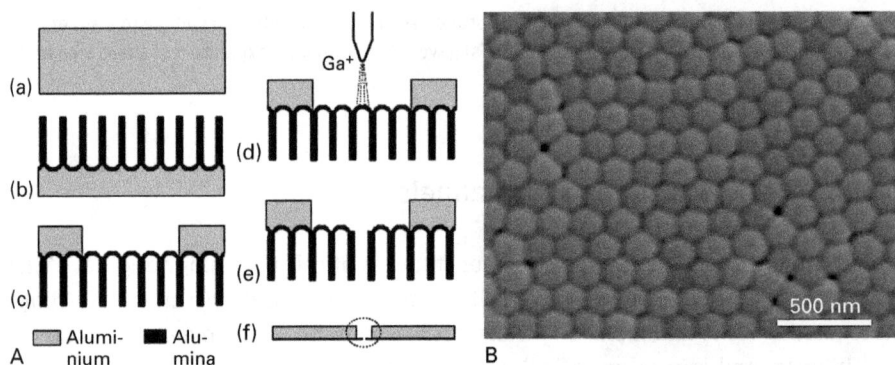

Fig. 10.12: Synthetic nanopore arrays for biological applications. **A.** The *focused ion beam* method is employed for drilling a single nanopore or nanopore arrays on anodic alumina (AAO) as substrate. **(a)–(b)** Fabrication of the AAO substrate by anodization of aluminum; **(c)** an exposed substrate prepared for drilling with the ion beam; **(d)–(e)** drilling a nanopore with Ga⁺ beam; **(f)** the resultant nanopore in the AAO substrate. **B.** Scanning electron microscopy (SEM) image of a nanopore array. Reprinted from Lillo, M. and Losic, D., Ion-beam pore opening of porous anodic alumina: The formation of single nanopore and nanopore arrays, *Mater. Lett.* **2009** *63*, 457–460. Copyright (2009), with permission from Elsevier.

A promising biological application of artificial nanopore arrays is their use for *DNA sequencing*. In this experimental approach single-strand DNA is passed through a nanopore upon application of electrical voltage (analogous to electrophoresis). Since the dimensions of the nanopore can be made to be slightly greater than the diameter of individual DNA strands, the strands move through the pore in single file, enabling determination of the oligonucleotide sequence through sensors identifying the basepairs passing through the pore. DNA sequencers operating through this scheme consisted of wholly synthetic pore structures, however other biomimetic designs have employed natural pore-forming proteins (such as α-hemolysin which displays the appropriate pore dimensions) embedded in lipid bilayers.

Functionalization of synthetic nanopores with *biological* molecules gives rise to biomimetic systems in which the pore performance and overall properties could be closely modulated. Since most physiological pores function in membrane environments, coupling of synthetic pores with *lipid molecules* has been an active branch of biomimetic pore research. Some studies have used the *coating* of the pore internal surface with lipids to promote penetration of desired biomolecules into the pore and enhance transport properties. Conjugation of synthetic nanopores with the lipid layers exhibits several notable advantages. Specifically, the use of lipids having different *chain lengths* provides means for adjusting pore diameter. Also, borrowing from physiological membrane functions, coating nanopores with organized lipid layers (i.e. lipid *bilayers*) enables incorporation of biological ligands which confer molecular specificity for the pores (concurrently reducing *non-specific* translocation through the pores). The lipid layer also prevented pores from clogging, most likely due to the fluid nature of the bilayer giving rise to constant motion of adsorbed species into the pore.

Another nice example of a non-biological pore scaffold functionalized with biological molecules is shown in Fig. 10.13. The goal of the research was to construct an artificial ion channel, exhibiting sufficient size selectivity for effective ionic transport, regulated by *zinc ions* – a ubiquitous ionic species involved in numerous regulation pathways in physiological systems. The synthetic channel designed by L. Jiang and colleagues at the Chinese Academy of Science achieves both goals through fabricating a polymeric nanochannel scaffold which was further chemically conjugated with *zinc-finger* peptides. Zinc fingers are well-known peptide motifs that generally exhibit extended, disordered conformations in the unbound states, folding into tight "finger-like" structures upon coordination with Zn^{2+} ions. Indeed, the pronounced zinc-induced structural transformation of the peptide was the key feature contribut-

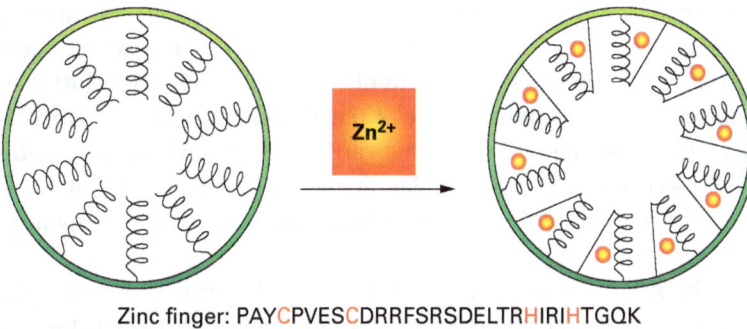

Zinc finger: PAYCPVESCDRRFSRSDELTRHIRIHTGQK

Fig. 10.13: Synthetic nanopores coated with zinc-finger peptides. Coating the nanopore surface with the zinc-finger motifs provides a selectivity mechanism. Only when Zn^{2+} is added to the pores do the zinc-finger peptides fold into compact "finger" structures thereby opening the channel. The zinc-finger sequence is depicted, highlighting the amino acids involved in binding of the metal ions.

ing to the functionality of the biomimetic channel. As illustrated in Fig. 10.13, when *no* zinc ions are present, the peptides extend into the channel interior, largely blocking the pores and restricting molecular passage. However, following addition of Zn^{2+}, the zinc fingers lining the nanochannel walls adopt compact conformations, effectively "opening" the channel.

10.6 Biomimetic signaling pathways

Signaling processes include molecular cascades, occasionally quite complex, designed by nature to provide means of regulating cellular processes and transmitting biological information to – and from cells. Manipulating signaling pathways is a common strategy in biological research, aimed at both attaining better understanding of biological processes as well as identifying therapeutic routes for combating diseases. Several studies, however, have explored utilization of *biomimetic* signaling cascades employing non-biological entities (such as nanoparticles) as possible disease therapies.

An example of such a biomimetic approach is presented in Fig. 10.14, depicting the use of the *blood-clotting signaling cascade*, a thoroughly studied physiological mechanism, for targeted cancer therapy. This novel therapeutic scheme, developed by S. N. Bhatia and colleagues at MIT, made use of synthetic species such as gold nanorods, which served as the "signal" initiators. The choice of gold *nanorods* (or other types of nanoparticles) was based on the observation that metal NPs could migrate freely in the circulatory system and accumulate at the porous angiogenic blood vessel network around tumors. Near-infrared (NIR) heating of the nanorods (generally feasible since NIR light can penetrate through human tissues) produced local disruption of the blood vessel – and consequent activation of the coagulation cascade. The induced coagulation process is the signaling mechanism harnessed for reporting the tumor location to other circulating synthetic substances carrying the therapeutic cargo. When the drug carriers encounter the coagulation products secreted at the tumor site (for example protein fibers creating the blood clot) they release their molecular cargoes. Indeed, one of the challenges in this biomimetic signaling scheme has been the design of drug carriers that would be specifically targeted to molecules involved in the coagulation process. One can synthesize, for example, peptides or other molecules that bind *fibrin,* the main protein component of blood clots.

Importantly, the *molecular amplification* inherent in the coagulation cascade, which evolved to achieve a quick and efficient plugging of blood leakage, is the key factor contributing to effective delivery of the therapeutic substance to the target tissue. This is because the "receiving" species – the drug carriers – can be maintained in high concentration in circulation without performing any biological task *until* the occurrence of the coagulation event. Therapeutic approaches based upon mimicking

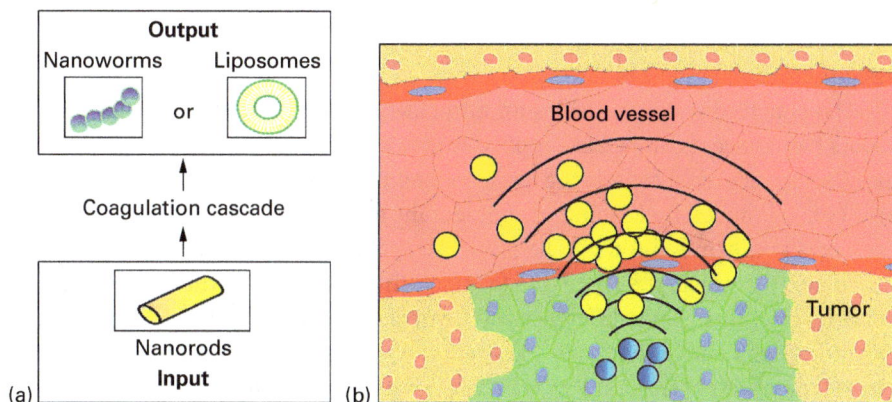

Fig. 10.14: Nanoparticle-induced signaling pathway. Both the *input* and *output* of the targeted signaling pathway (the blood coagulation cascade) are comprised of synthetic entities. **(a)** Signal pathway scheme. Gold nanorods accumulate in infected tissue sites (i.e. tumors). Localized heating of the rods through light irradiation induces bursting of small blood vessels and initiation of the coagulation cascade. Coagulation recruits synthetic substances circulating in the bloodstream, such as "nanoworms" (made of inorganic particulates) or liposomes, loaded with anti-tumor drugs. The drug-carrying vehicles consequently release their cargo at the tumor site. **(b)** Schematic drawing of the biomimetic signaling cascade showing the Au nanorods (*signaling* species, depicted as blue spheres), the coagulation cascade (expanding arcs), and recruited drug-delivery assemblies (green-yellow spheres). Figure adapted from Maltzahn et al., *Nature Mater*, **2011** *10*, 545–552.

biological signaling exhibit other strengths. In particular, signaling events (such as the clotting pathway depicted above) are generally *localized* – i.e. occurring within single cells or specific regions in a tissue – and thus provide an inherent *targeting* mechanism for drug action. Furthermore, signaling cascades possess intrinsic means for *fine tuning* of the different reactants and products formed throughout the process underscoring another lever for regulation of therapeutic processes.

10.7 Electronic noses

"Electronic noses" denote diverse chemical systems and devices designed to sense and identify volatile (vapor-phase) substances through pattern recognition and other statistical algorithms. This subject is somewhat peripheral to *biomimetics*, however as a sensing concept that is truly borrowed from human (and generally mammalian) physiology it deserves a respectable place in this book. In its most basic format, bio-logical or chemical **sensing** involves recognition between a molecule (ligand) and a specific receptor, with the binding event producing a detectable signal. Although this generic sensing concept generally provides sufficient specificity, its broad-base practicality is rather limited because the requirement of a different receptor for each ligand is problematic both in a biological context ("evolutionary expensive" to maintain numerous receptors for all possible ligands), as well as a practical platform for designing artificial sensors.

The *olfactory system* provides an attractive alternative mechanism for sensing vapors (the concept can be also applied to identifying *non-vapor* substances). When the olfactory neuronal cells on the nasal epithelium are exposed to odor, they produce a response induced by binding of the odorant molecules onto olfactory receptors. Each odorant, exhibiting different shape, size, charge, and other chemical properties, induces a unique "response code" upon interactions with the entire library of olfactory receptors. Slight differences between molecules or even a change in the concentration can change the "code". Overall, the multitude of olfaction cells generate *patterns* of signals corresponding to different odorants, which can be analyzed in the brain and assigned to a particular compound or mixture.

Electronic noses (also denoted *e-noses*) aim to reproduce the olfactory mechanism (Fig. 10.15). In e-noses, an *array* of sensor elements that are usually cross-reactive (i.e. each element can respond to the same odorant, albeit differently) generates a pattern of signals following interactions (or reaction) with the tested vapor. The pattern is then analyzed via pattern recognition algorithms designed to identify the individual components or specific mixtures through "signal fingerprints". The advantage of this concept as compared to individual ligand or receptor sensing is the fact that the generation of numerous patterns, dependent upon the number of sensor elements, makes it possible to distinguish many compounds, both individually and within complex mixtures.

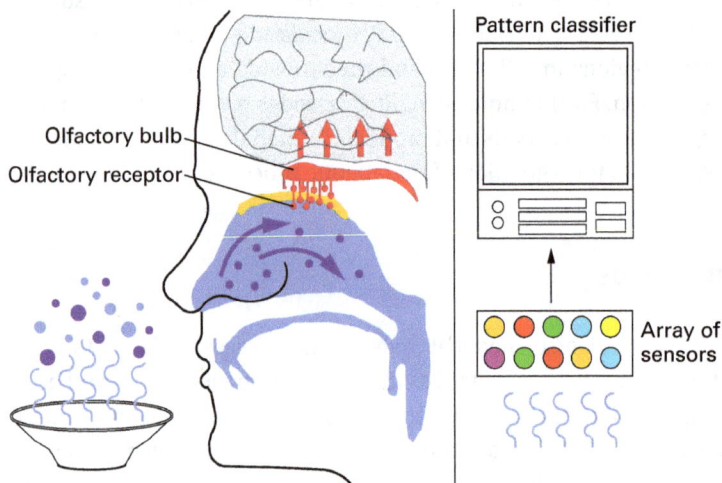

Fig. 10.15: The olfactory system and the electronic nose concept. **Left:** The physiology of smell; odors interact with receptor arrays in the olfactory bulb, the signal patterns are then analyzed in the brain which "matches" the pattern with known smells. **Right:** The e-nose concept. The tested vapor interacts with a sensor array comprising different chemical molecules; the produced patterns are analyzed using computer algorithms to identify the vapor molecules.

It should be emphasized that the thrust of artificial olfactory sensing (which has made this discipline particularly attractive as a sensing approach) is that identification of desired substances can be detached from the "chemistry" of the sensor (or more precisely the sensor elements), and significantly depends upon *data processing*, i.e. pattern analysis. Thus the sensor performance becomes more of a computational task, which in principle could be more easily addressed rather than chemical modulation of the sensing platform. However, despite the great interest in the e-nose sensing concept since the early 1980s, this field has disappointed due to problems with reproducibility, specificity, sensitivity, and "sensor training", which have confined these types of sensors to specialized applications.

Various practical applications have been promoted for e-nose sensing of volatile organic compounds (VOCs). Not coincidentally, the realities of the early 21st century have encouraged research into e-nose applications for detecting of both the sinister (e.g. explosives), as well as the more benign but equally important – disease diagnostics. The latter application is believed to hold great promise since varied diseases – from bacterial infections to cancer – are known to alter the metabolic profiles of body fluids and affect the volatile molecular components in *breath*. Both approaches will be discussed in more detail below.

The large majority of e-nose applications developed so far have mimicked the *computational* concept of the olfactory sensing mechanism, e.g. analyte identification via pattern recognition, while the actual sensing has been almost always accomplished via purely chemical means (e.g. sensor elements comprising varied functional groups designed to bind the volatile molecules screened). In a few cases, however, e-nose platforms have partly imitated the structural aspects of the physiological olfactory sensing mechanism. An example is shown in Fig. 10.16, depicting a vesicle-based assembly developed by S. Matile and colleagues at the University of Geneva. The system is designed to mimic the involvement of the *membrane* environment in sensor response and odorant identification.

This detection scheme summarized in Fig. 10.16 is based upon release of fluorescent dyes encapsulated within lipid vesicles. Dye release is activated through the action of *membrane transporters*, mimicking to a certain degree the physiological behavior of the olfactory receptors which operate within membrane bilayers. The researchers have synthesized positively-charged ionic species that reacted with hydrophobic odorants, simultaneously binding non-covalently to negatively-charged polyion membrane transporters, which consequently traverse the membrane resulting in dye release from the vesicles. The key combinatorial, i.e. pattern-generating aspect of the technique is the differential reaction rates between the odorants and the reactive cations, giving rise to distinct fluorescence patterns which can be statistically analyzed to produce fluorescence "fingerprints" of numerous odorants.

Optical detection constitutes a powerful platform for e-nose applications (occasionally denoted as "photonic noses" or "optoelectronic noses"). Examples of such

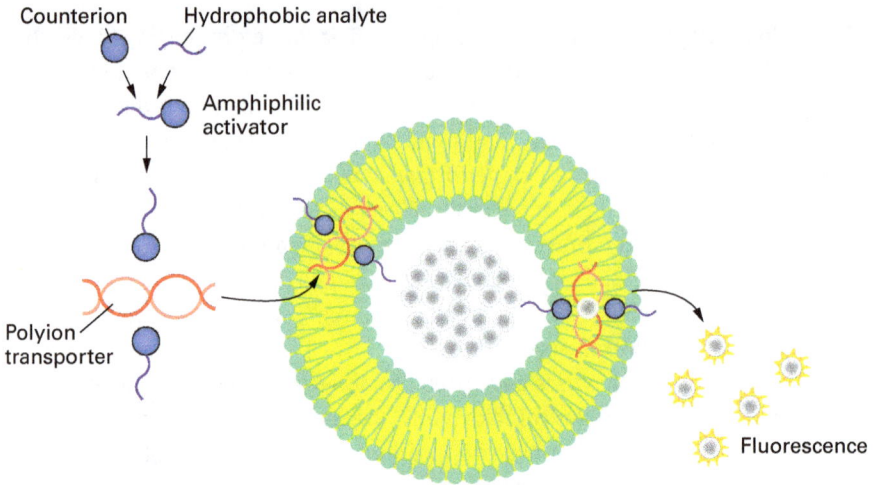

Fig. 10.16: Electronic nose sensing with biomimetic vesicles. The hydrophobic odorants (blue "strands") bind to counter-ions (blue spheres), producing amphiphilic activators. These activators in turn interact with polyion transporters that penetrate into the vesicles, inducing release of encapsulated fluorescent dyes. Pattern analysis of the induced fluorescence as a function of analyte and reactive counter-ion could distinguish among odorants.

systems include detection of color changes following odorant binding to sensor elements displayed upon the tips of optical fibers, analysis of *fluorescence* changes, *luminescent* modulation, and light refraction induced by analyte binding onto transparent beads. Optical fibers presenting microbeads comprising different compositions, surface chemistry, and attached fluorescent dyes were utilized in e-nose applications; in these systems optical transparency or fluorescence transformations could be readily recorded, fed into computer programs, and pattern analyzed.

E-noses based upon optical fiber arrays exhibit notable advantages. First, the use of *optical signals* provides an important means for high sensitivity signal generation, due to the availability of powerful lasers (i.e. *monochromatic* light) and sophisticated optical detectors. Another factor that should be emphasized is the possibility to use *different wavelengths* as output signals, thereby expanding the diversity of patterns induced by analytes and their identification. Another practical plus of optical sensing is the possibility to implement *binary* detection schemes, thus strengthening the computational aspects of the sensor in terms of pattern recognition. Optical fiber arrays present another important similarity to physiological olfactory systems. Due to their extremely small diameters, such arrays can include *more than* a hundred beads per each sensor element, providing an intrinsic signal amplification mechanism. This feature mimics the biological scenario in which many olfactory cells with the same receptor (and the same specificity) are present in the olfactory epithelium, the tissue inside the nasal passageway responsible for the sense of smell. The multiplicity of identical receptors for vapor molecules enables pronounced signal amplification, through summation of the binding events of individual receptor-ligand interactions.

An example of an innovative optical e-nose system is shown in Fig. 10.17. The thrust of the device, designed by M. Bayindir and colleagues at Bilkent University, Turkey, was an array (bundle) of optical fibers transmitting *infrared* (IR) light; volatile molecules adsorbed upon the fibers tips modulated the transmitted light following their typical IR absorption properties. The IR patterns recorded at the detectors could be processed using conventional pattern recognition techniques. Furthermore, "binary codes" for each compound could be also generated using "cutoff" intensities at specific, preselected wavelengths (Fig. 10.17b).

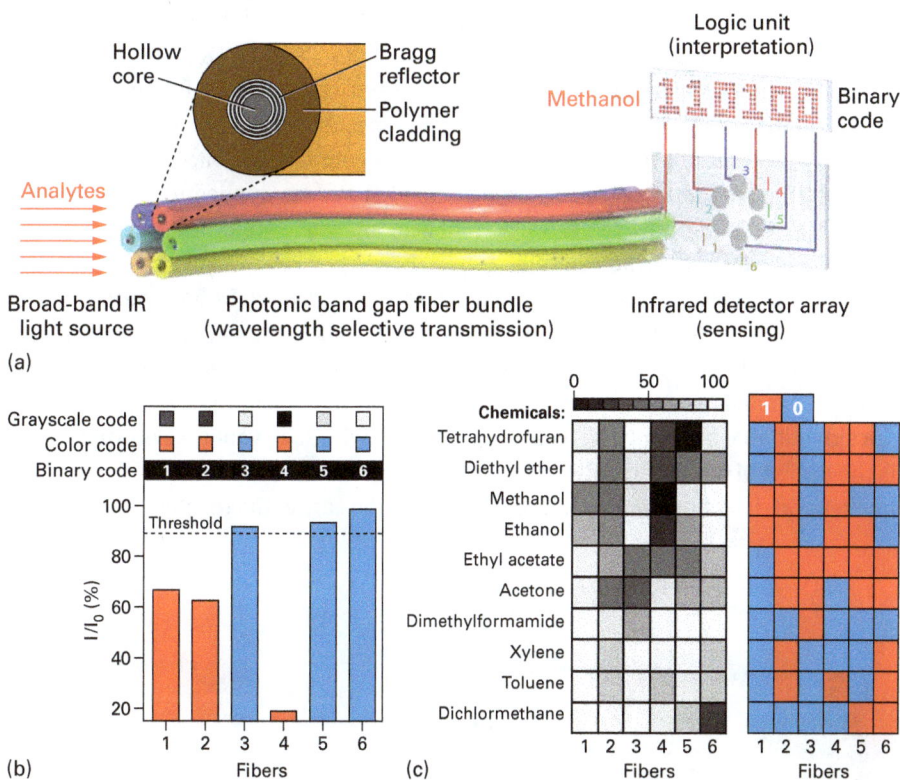

Fig. 10.17: Optical electronic nose. An array of optical fibers transmitting infrared (IR) light is used for identifying volatile organic compounds (VOCs). **(a)** Schematic description of the method. The VOCs are adsorbed at the tips of the fibers, modulating (quenching) the incoming IR light. Each fiber is tuned to record IR signals at a specific wavelength. The combination of signals induced by each molecule provides distinct "IR fingerprints" for the analytes. **(b)** The "binary code" concept: determination of an IR intensity *threshold* produces a binary response (blue or red) of the sensor element. **(c)** Signal matrixes recorded using either the "continuous" intensities or the binary scheme. Identification of analytes is made possible through pattern recognition analysis. Reprinted with permission from Yildirim, A. et al., *Adv. Mater.* **2011** *23*, 1263–1267. Copyright (2011) John Wiley and Sons.

A particularly interesting diagnostic application based upon the e-nose concept is the possibility for *disease detection* through "breath analysis". The hypothesis underlying this diagnostic scheme is that genetic and/or biochemical modification of cell metabolism, particularly associated with cancer onset, would modulate the release of VOCs from cells and tissues; this "VOC profile" might show up in pattern recognition analysis of exhaled air, thereby alerting one to the disease. This appealing idea is currently being intensively pursued with the aim of designing early-detection platforms for cancers – lung cancer, which is the natural target as it affects lung tissues, but other cancers as well.

H. Haick and colleagues at the Technion, Israel, have constructed a gold NP-based *electrochemical* e-nose for cancer diagnostics relying upon exhaled breath. The sensor comprises an array of detectors, and the surface of each detector unit contains Au NPs displaying different chemical functionalities (Fig. 10.18). The researchers found that each NP exhibited slightly different electrical resistance following adsorption of exhaled VOCs. In particular, exposure to specific VOCs associated with cancer resulted in different electrical conductivities in comparison with VOCs extracted from healthy individuals. Crucially, when exhaled air was applied to a sensor *array* comprising *several* Au NPs species, a statistical analysis could distinguish between healthy and cancer-bearing patients.

Disease diagnostics based upon pattern recognition of VOCs have been pursued not only through *breath* but also using other physiological sources – for example urine. Researchers have found, for example, that VOCs associated with tuberculosis (TB) are secreted in urine and have sufficient concentrations in vapors to produce distinct identifiable patterns. The recurring challenge with such vapor-based approaches is the design of a reproducible and easy-to-apply *detection* scheme. Unlike the electrochemical detection approach implemented in the cancer diagnostic technique dis-

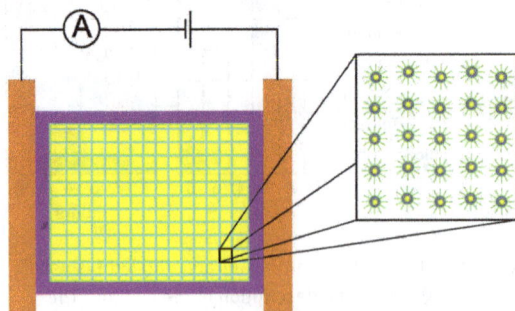

Fig. 10.18: Cancer diagnostics through an array-based electrochemical response of gold nanoparticles. Each array unit (surface "squares") comprises Au NPs exhibiting different chemical coating. Adsorption of VOCs in the breath of healthy and cancer-bearing patients, respectively, produced different patterns of electrical currents following adsorption onto the Au NPs. Pattern analysis enabled to distinguish between healthy and cancer-bearing persons.

cussed above, most current approaches employ expensive, time-consuming instrumentation, such as mass spectrometry and gas chromatography, which limit their immediate practicality. Furthermore, issues of fidelity and data reproducibility, particularly when using *different* machines (or even the same machine in different locations), have been raised.

An innovative sensing approach introduced in the laboratory of D. Walt at Tufts University focuses on *temporal encoding*, a variety of pattern recognition. Temporal encoding is a fundamental characteristic of the olfactory system, in which olfactory cells transmit to the brain not only the signal *intensities*, but also the *time dependence* of the vapor signal to identify its molecular composition. Indeed, the temporal dimension constitutes a powerful encoding parameter for vapor identification using array-based sensors. Specific elements in temporal-encoding sensors have included polymers embedding polarity-sensitive fluorescent dyes; when a vapor molecule adsorbs onto the polymer, the dye reports upon the *time modulation* of the microenvironment polarity, thereby providing another dimension to the signals pattern.

Despite the attractive features and promise of artificial nose concepts for practical sensing applications, their limitations turn out to be quite severe, relegating such sensors mostly to laboratory settings or specialized devices. A fundamental limitation of a sensing scheme based upon the mammalian olfactory system is the simple fact that our noses provide a rather *qualitative* means for odor identification. Indeed, while an analytical sensing instrument would likely be expected to provide information on *concentration* and *molecular components* of a tested (vapor) sample, the nose provides just a general identification of a substance, without any quantitative dimension. For example, we can report the scent of a rose, but we cannot determine the identity or concentrations of any of the specific molecular components comprising the odor. Another significant barrier of array-based sensing concerns the fundamental requirement that such sensors need to be *trained* repeatedly in order to achieve sufficient fidelity, specificity, and reproducibility. The repeated training cycles involving vapor application to often large arrays of chemical compounds eventually lead to early degradation and loss of the sensing properties.

10.8 Neural networks and biologically-inspired computers

The human brain is a sophisticated decision-making machine (or so we want to believe). While simple calculators can surpass the computing capabilities of the brain and starting-level computers can easily beat the human brain in terms of information processing, the trump card of our flesh-and-blood organ is the ability to *learn* and use the learned information to process future data and to solve problems that conventional computers would be totally impervious to. Physiologically speaking, *learning* in neuronal networks consists of *chemical modulation* of the synaptic connections between the neurons.

Trying to mimic neural-based decision-making and computing, in fact, has been attempted for decades, albeit with limited success. Indeed, the field of *artificial neural networks* (ANNs) has drawn significant interest both among computer scientists as well as biologists – the former community trying to advance computing capabilities and develop new algorithms, while the latter group focuses on gaining a better insight into neuronal processes and circuits in the brain. Overall, while ANN-based computers are still not commercially available, the potential of the concept and its implications for the future of computing are significant.

The core difference between neural networks and conventional computers lies in their intrinsic approach to information processing and analysis. Operation of conventional computers relies on *algorithms* – i.e. following a set of specific instructions in order to solve a problem. The instructions, e.g. stages in the problem-solving process, need to be fed to the computer in advance (via the software codes), otherwise the computer will not be able to address the specific problem it is faced with. This fundamental precondition restricts the problem-solving capabilities of conventional computers to situations which the programmer is familiar with and is capable of defining the concrete steps required from the computer to produce output from the specific inputs.

Computing based upon *neural networks* aims to mimic the information processing as carried out in the brain. A neural network is composed of a large number of interconnected processing elements (neurons) working in parallel. Crucially, in stark difference to conventional computing, neural networks cannot be programmed to perform a specific task. The network finds out how to solve the problem through a learning process based upon cumulative modulation of the signals transmitted between the processing elements. The idea borrowed from the physiological nerve system is that through the combined action of many interconnected neurons, the neural network undergoes chemical changes – which underlie the *learning* process which subsequently bestows the neural network with computing or decision-making capabilities.

The building blocks of physiological neural networks are the *neurons* – unique cells that can transmit electrical pulses. Neurons comprise several input-receiving protrusions called *dendrites*, and a single output-producing unit – the *axon* (Fig. 10.19a). At the end of the axon are regions denoted *synapses* which connect the axon to dendrites of other neurons. The synapse is where communication is established between neighboring neurons through chemical and electrical stimulation. When a neuron receives an input signal through one of the dendrites and this input is deemed sufficiently large (i.e. surpasses a certain signal threshold), the neuron sends an electrical pulse down its axon which could then excite other neurons. *Learning* is believed to occur through chemical modification taking place at the synapses which consequently affect the extent to which information (e.g. electrical stimuli) is transmitted between neurons.

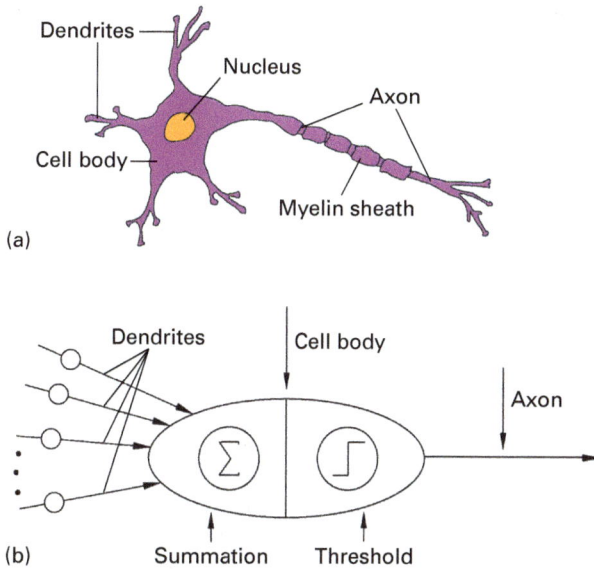

Fig. 10.19: Neuronal cells and artificial neurons. **(a)** Schematic figure showing a neuronal cell with main components highlighted. The *myelin sheath* provides insulation for the transmitted electrical signals; **(b)** an artificial neuron comprising the "dendrites" – input units, a "cell body" in which the input signals are summed up and a decision is made whether to fire an output signal, transmitted through an "axon".

An artificial "neuron" – the information-processing element in a neural network computing device – is illustrated in Fig. 10.19b. The "neuron" receives inputs through several channels (the "dendrites"), and produces output through a specific dedicated channel ("axon"). The crucial property of the artificial neuron is that decision-making – whether to "fire" (e.g. produce output signal) or not – is based upon the overall sum of inputs but it is *not* a predetermined parameter or function, but rather can be modulated in each application. Echoing natural neural networks – this "firing decision" is based upon a learning process which feeds the neuron with sets of rules determining its response to the input triggers. Importantly, *learning* means that the firing conditions can be modified, depending upon both the external conditions as well as the processes occurring within the artificial neuron.

A simple example of a neural network based upon an artificial neuron is depicted in Figures 10.20 and 10.21 (adapted from a presentation by Stergiou and Siganos, Imperial College, London). An artificial neuron has three inputs that provide signals of either "0" or "1". The firing rules for the neuron are an output of **1** when the inputs from the three dendrites are 1,1,1 or 0,1,0, or output **0** when the inputs are 0,0,0 or 0,0,1. Furthermore, the firing rule *also* dictates that when the inputs are *different* to these four combinations, the neuron will fire depending on the *total resemblance* of the inputs to the above four patterns. As an example, an input of **1,0,1** has a difference of **1 digit** compared to 0,0,1 and **2 digit** difference compared to 0,0,0 (total of

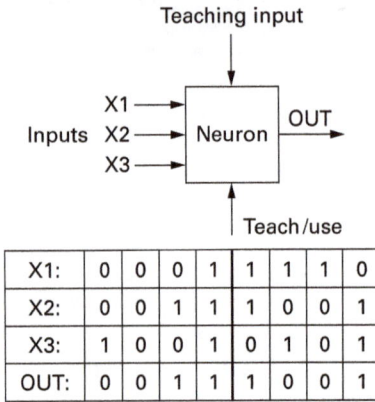

Fig. 10.20: Artificial neuron and its truth tables. The neuron has three inputs (delivering "0" or "1" signals), and one output. The output signal produced by the neuron is determined by the firing rules, which are the following (four left columns): 1. 1,1,1 or 0,1,0 give out "1" output, 0,0,0 or 0,0,1 trigger "0" output. 2. If the input signals are different to the four combinations, the output signal is determined by the overall similarity between the input signal and the four combinations (four right columns), see example in the text.

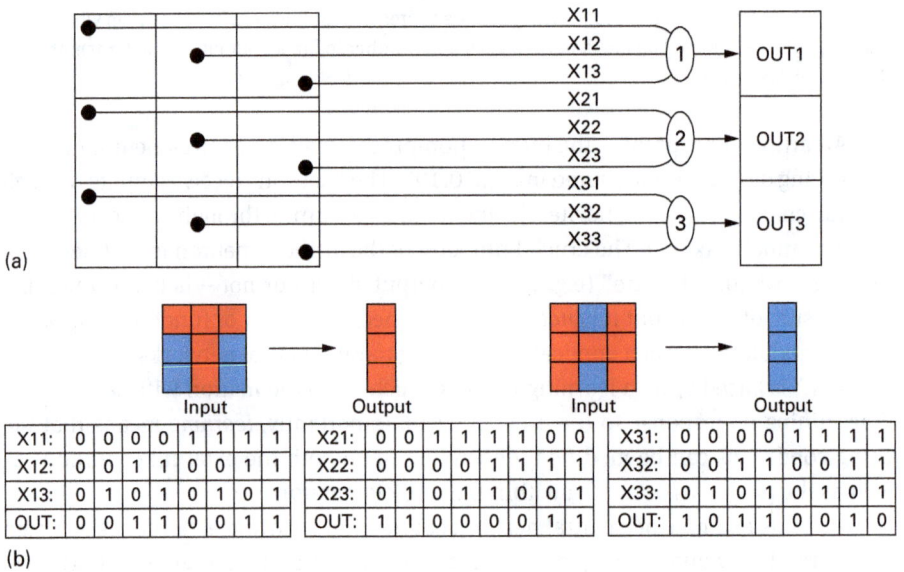

Fig. 10.21: Neural network for pattern recognition. (a) A 9-element matrix providing inputs for three neurons; (b) neural outputs of the letters "T" and "H" and the corresponding truth tables for the three neurons.

3 digit difference). However the same 1,0,1 input has **3 digit** difference compared to 0,1,0 and **1 digit** difference compared to 1,1,1 (total of **4 digit** difference). Accordingly, 1,0,1 produces a "**0**" output due to the closer similarity to the two "**0**" producing firing rules. The complete firing table (also denoted the "truth table") is summarized in Fig. 10.20.

One of the more useful applications of neural networks is in *pattern recognition*. Figures 10.21 and 10.22 demonstrate how a network of three artificial neurons, each having its own (and *different*) truth table (determined according to the learned firing rules described above) can distinguish between two simple patterns – the letters "T" and "H", respectively, written in a 9-pixel matrix. The three truth tables corresponding to the three neurons produce two distinct outputs, a "blue" column for the "T" and a "red" column for the "H".

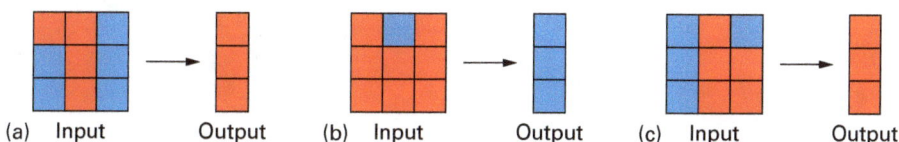

Fig. 10.22: Recognition of *undefined* patterns by the neural network. The three undefined patterns are correctly assigned by the neural network in Fig. 10.21 according to the truth tables.

The power of the neural network method is apparent when faced with *undefined* patterns (Fig. 10.22). For example, application of the truth table clearly leads to identification of the pattern in Fig. 10.22(a) as a "T", while the pattern in Fig. 10.22(b) is recognized by the network as "H". Importantly, the "decision" associated with the input pattern in Fig. 10.22(c) is the letter "T", even though this assignment appears unconvincing by brief browsing with the human eye. However, a close examination reveals that the difference between the pattern in Fig. 10.22(c) and "T" is *smaller* than the "H" – and the neural network analysis was indeed correct. This example nicely shows that the network reaches a "decision" regarding an unrecognized pattern in a similar manner as our brain "decides" (or compartmentalizes) a previously unrecognized input.

While the example above seems simple, artificial neural networks can achieve remarkable success in the analysis of complex data, detecting trends and patterns that cannot be discerned by humans or other computer techniques. The adaptive learning capabilities, parallel processing feasible in large networks, and efficient "decision-making" make this computing approach effective and promising. Overall, neural networks and conventional algorithm-based computers are not in competition but rather complement each other. Depending on the task to be performed, the two concepts can be combined to provide the most optimal solution to a problem in hand.

Further Reading

Chapter 2

Cherny I and Gazit E, Amyloids: not only pathological agents but also ordered nanomaterials, *Angew Chem Int Ed Engl.* **2008**, *47*, 4062–9.

Hyman P, Bacteriophages and nanostructured materials, *Adv Appl Microbiol.* **2012**, *78*, 55–73.

Lloyd, JR, et al., Biotechnological synthesis of functional nanomaterials, *Curr Opin Biotechnol.* **2011**, *22*, 509–15.

Lovley DR, Bug juice: harvesting electricity with microorganisms, *Nat Rev Microbiol.* **2006**, *4*, 497–508.

Luo Z and Zhang S, Designer nanomaterials using chiral self-assembling peptide systems and their emerging benefit for society, *Chem Soc Rev.* **2012**, *41*, 4736–54.

Matson JB, et al., Peptide Self-Assembly for Crafting Functional Biological Materials, *Curr Opin Solid State Mater Sci.* **2011**, *15*, 225–235.

Tao H, Kaplan DL and Omenetto FG, Silk materials – A road to sustainable high technology, *Adv Mater.* **2012**, *24*, 2824–37.

Chapter 3

Cui H, Webber M and Stupp SI, Self-assembly of peptide amphiphiles: from molecules to nanostructures to biomaterials, *Biopolymers* **2010**, *94*, 1–18.

Guo Z, Liu W and Su BL, Superhydrophobic surfaces: from natural to biomimetic to functional, *J Colloid Interface Sci.* **2011**, *353*, 335–55.

Kiessling LL, Gestwicki JE and Strong LE, Synthetic multivalent ligands in the exploration of cell-surface interactions, *Curr Opin Chem Biol.* **2000**, *4*, 696–703.

Lohmüller T, Aydin D, Schwieder M, Morhard C, Louban I, Pacholski C and Spatz JP, Nanopatterning by block copolymer micelle nanolithography and bioinspired applications, *Biointerphases* **2011**, *6*, MR1–12.

Saha K, Agasti SS, Kim C, Li X and Rotello VM, Gold nanoparticles in chemical and biological sensing, *Chem Rev.* **2012**, *112*, 2739–79.

Webb HK, Hasan J, Truong VK, Crawford RJ, Ivanova EP, Nature inspired structured surfaces for biomedical applications, *Curr Med Chem.* **2011**, *18*, 3367–75.

Chapter 4

Dvir T, Timko BP, Kohane DS and Langer R, Nanotechnological strategies for engineering complex tissues, *Nat Nanotechnol.* **2011**, *6*, 13–22.

Ferreira AM, Gentile P, Chiono V and Ciardelli G, Collagen for bone tissue regeneration, *Acta Biomater.* **2012**, *8*, 3191–200.

Garg T, Singh O, Arora S and Murthy R, Scaffold: a novel carrier for cell and drug delivery, *Crit Rev Ther Drug Carrier Syst.* **2012**, *29*, 1–63.

Hutmacher, DW, Scaffolds in tissue engineering bone and cartilage, *Biomaterials* **2000**, *21*, 2529–2543.

https://doi.org/10.1515/9783110709490-011

Matson JB and Stupp SI, Self-assembling peptide scaffolds for regenerative medicine, *Chem Commun. (Camb.)* **2012**, *48*, 26–33.

Mirensky TL and Breuer CK, The development of tissue-engineered grafts for reconstructive cardiothoracic surgical applications, *Pediatr Res.* **2008**, *63*, 559–68.

Moghimi SM, Peer D and Langer R, Reshaping the future of nanopharmaceuticals: ad iudicium, *ACS Nano* **2011**, *22*, 8454–8.

Ruvinov E, Dvir T, Leor J and Cohen S, Myocardial repair: from salvage to tissue reconstruction, *Expert Rev Cardiovasc Ther.* **2008**, *6*, 669–86.

Vinatier C, Bouffi C, Merceron C, Gordeladze J, Brondello JM, Jorgensen C, Weiss P, Guicheux J and Noël D, Cartilage tissue engineering: towards a biomaterial-assisted mesenchymal stem cell therapy, *Curr Stem Cell Res Ther.* **2009**, *4*, 318–29.

Chapter 5

Gordon R, Losic D, Tiffany MA, Nagy SS and Sterrenburg FA, The glass menagerie: diatoms for novel applications in nanotechnology, *Trends Biotechnol.* **2009**, *27*, 116–27.

Moradian-Oldak J, Protein-mediated enamel mineralization, *Front Biosci.* **2012**, *17*, 1996–2023.

Nudelman F and Sommerdijk NA, Biomineralization as an inspiration for materials chemistry, *Angew Chem Int Ed Engl.* **2012**, *51*, 6582–96.

Phoenix VR and Konhauser KO, Benefits of bacterial biomineralization, *Geobiology* **2008**, *6*, 303–8.

Soto CM and Ratna BR, Virus hybrids as nanomaterials for biotechnology, *Curr Opin Biotechnol.* **2010**, *21*, 426–38.

Yildiz I, Shukla S and Steinmetz NF, Applications of viral nanoparticles in medicine, *Curr Opin Biotechnol.* **2011**, *22*, 901–8.

Chapter 6

Brannigan JA and Wilkinson AJ, Protein engineering 20 years on, *Nat Rev Mol Cell Biol.* **2002**, *3*, 964–970.

Carter P, Site-directed mutagenesis, *Biochem. J.* **1986**, *237*, 1–7.

Chang PV and Bertozzi CR, Imaging beyond the proteome, *Chem Commun. (Camb.)* **2012**, *48*, 8864–79.

Devoy A, Bunton-Stasyshyn RK, Tybulewicz VL, Smith AJ and Fisher EM, Genomically humanized mice: technologies and promises, *Nat Rev Genet.* **2011**, *13*, 14–20.

Khalil AS and Collins JJ, Synthetic biology: applications come of age, *Nat Rev Genet.* **2010**, *11*, 367–379.

Liu CC and Schultz PG, Adding new chemistries to the genetic code, *Annu Rev Biochem.* **2010**, *79*, 413–44.

Luo D and Saltzman WM, Synthetic DNA delivery systems, *Nat Biotechnol.* **2000**, *18*, 33–37.

Lutz S, and Bornscheue UT, Protein Engineering Handbook, John Wiley & Sons 2012.

McPherson MJ, PCR: A practical approach, Oxford University Press 1991.

Nandagopal N and Elowitz MB, Synthetic biology: integrated gene circuits, *Science* **2011**, *333*, 1244–8.

Peisajovich SG and Tawfik DS, Protein engineers turned evolutionists, *Nat Methods* **2007**, *4*, 991–4.

Samish I et al., Theoretical and computational protein design, *Annu Rev Phys Chem*. **2011**, *62*, 129–49.

Tsien RY, The green fluorescent protein, *Annu Rev Biochem*. **1998**, *67*, 509–44.

Zuckermann RN and Kodadek T, Peptoids as potential therapeutics, *Curr Opin Mol Ther*. **2009**, *11*, 299–307.

Chapter 7

Hammer DA and Kamat NP, Towards an artificial cell, *FEBS Letters* **2012**, *586*, 2882–2890.

Hanczyc MM and Szostak JW, Replicating vesicles as models of primitive cell growth and division, *Curr Opin Chem Biol*. **2004**, *8*, 660–4.

Lee JS and Feijen J, Polymersomes for drug delivery: design, formation and characterization, *J Control Release* **2012**, *161*, 473–83.

Schrum JP, Zhu TF and Szostak JW, The origins of cellular life, *Cold Spring Harb. Perspect. Biol*. **2010**, *2*, a002212.

Stano P, Minimal cells: relevance and interplay of physical and biochemical factors, *Biotechnol J*. **2011**, *6*, 850–9.

Walde P, Cosentino K, Engel H and Stano P, Giant vesicles: preparations and applications, *ChemBioChem*. **2010**, *11*, 848–65.

Chapter 8

Pegoraro C, MacNeil S and Battaglia G, Transdermal drug delivery: from micro to nano, *Nanoscale* **2012**, *4*, 1881–94.

Stevenson CL, Santini JT Jr and Langer R, Reservoir-based drug delivery systems utilizing microtechnology, *Adv Drug Deliv Rev*. **2012**, *64*, 1590–602.

Yoo JW, Irvine DJ, Discher DE and Mitragotri S, Bio-inspired, bioengineered and biomimetic drug delivery carriers, *Nat Rev Drug Discov*. **2011**, *10*, 521–35.

Young LS, et al., Viral gene therapy strategies: from basic science to clinical application, *J Pathol*. **2006**, *208, 299–318*.

Chapter 9

Aldaye FA, Palmer AL and Sleiman HF, Assembling materials with DNA as the guide, *Science* **2008**, *321*, 1795–9.

Jaeger L and Chworos A, The architectonics of programmable RNA and DNA nanostructures, *Curr Opin Struct Biol*. **2006**, *16*, 531–43.

Pinheiro AV, Han D, Shih WM and Yan H, Challenges and opportunities for structural DNA nanotechnology, *Nat Nanotechnol*. **2011**, *6*, 763–72.

Saccà B and Niemeyer CM, DNA origami: the art of folding DNA, *Angew Chem Int Ed Engl*. **2012**, *51*, 58–66.

Seeman NC, Nanomaterials based on DNA, *Annu Rev Biochem*. **2010**, *79*, 65–87.

Stojanovic MN, Molecular computing with deoxyribozymes, *Prog Nucleic Acid Res Mol Biol*. **2008**, *82*, 199–217.

Chapter 10

Bar-Cohen Y, Current and future developments in artificial muscles using electroactive polymers, *Expert Rev Med Devices*. **2005**, *2*, 731–40.

Bishop CM, Neural Networks for Pattern Recognition, Oxford University Press 1995.

Kalyanasundaram K and Graetzel M, Artificial photosynthesis: biomimetic approaches to solar energy conversion and storage, *Curr Opin Biotechnol*. **2010**, *21*, 298–310.

Kowalczyk SW, Blosser TR and Dekker C, Biomimetic nanopores: learning from and about nature, *Trends Biotech*. **2011**, *29*, 607–614.

Liu SC and Delbruck T, Neuromorphic sensory systems, *Curr Opin Neurobiol*. **2010**, *20*, 288–95.

Madden JD, Mobile robots: motor challenges and materials solutions, *Science* **2007**, *318*, 1094–7.

Musto CJ and Suslick KS, Differential sensing of sugars by colorimetric arrays, *Curr Opin Chem Biol*. **2010**, *14*, 758–66.

Schultz PG, Yin J and Lerner RA, The chemistry of the antibody molecule, *Angew Chem Int Ed Engl*. **2002**, *41*, 4427–37.

Stitzel SE, Aernecke MJ and Walt DR, Artificial noses, *Annu Rev Biomed Eng*. **2011**, *13*, 1–25.

Index

https://doi.org/10.1515/9783110709490-012

www.ingramcontent.com/pod-product-compliance
Lightning Source LLC
Chambersburg PA
CBHW061350210326
41598CB00035B/5946